THE PLANT DIVERSITY OF MALESIA

Cover photograph: Primary rain forest, Sabah, Malaysia (Courtesy of Dr. Ding Hou, Leiden)

THE PLANT DIVERSITY OF MALESIA

Proceedings of the Flora Malesiana Symposium commemorating Professor Dr. C.G.G.J. van Steenis

Leiden, August 1989

edited by

Pieter Baas
Kees Kalkman
Rob Geesink

Rijksherbarium / Hortus Botanicus
Leiden University — The Netherlands

KLUWER ACADEMIC PUBLISHERS
DORDRECHT / BOSTON / LONDON

Library of Congress Cataloging in Publication Data

Flora Malesiana Symposium (1989 : Leiden, Netherlands)
The plant diversity of Malesia : proceedings of the Flora Malesiana Symposium commemorating Professor Dr. C.G.G.J. van Steenis, Leiden, August, 1989 / edited by Pieter Baas, Kees Kalkman, Rob Geesink.
p. cm.
Includes bibliographical references.
ISBN-13:978-94-010-7447-6 e-ISBN-13:978-94-009-2107-8
DOI:10.1007/978-94-009-2107-8

1. Botany--Malay Archipelago--Congresses. I. Steenis, C. G. G. J. van (Cornelis Gijsbert Gerrit Jan) II. Baas, P. III. Kalkman, Kees. IV. Geesink, R. V. Title.
QK365.5.F58 1989
581.9595--dc20 90-42122

ISBN-13:978-94-010-7447-6

Published by Kluwer Academic Publishers,
P.O. Box 17, 3300 AA Dordrecht, The Netherlands.

Kluwer Academic Publishers incorporates
the publishing programmes of
D. Reidel, Martinus Nijhoff, Dr W. Junk and MTP Press.

Sold and distributed in the U.S.A. and Canada
by Kluwer Academic Publishers,
101 Philip Drive, Norwell, MA 02061, U.S.A.

In all other countries, sold and distributed
by Kluwer Academic Publishers Group,
P.O. Box 322, 3300 AH Dordrecht, The Netherlands.

Printed on acid-free paper

Contents

List of contributors*

Bryan A. Barlow (22)
Division of Plant Industry
CSIRO
GPO Box 1600
Canberra, ACT 2601
Australia

John H. Beaman and Reed S. Beaman (14)
Department of Botany and Plant Pathology
Michigan State University
East Lansing, Michigan 48824
U. S. A.

A. A. Bidin (26)
Department of Botany
Faculty of Life Sciences
Universiti Kebangsaan Malaysia
43600 Bangi, Selangor
Malaysia

A. J. de Boer (21)
Institute of Taxonomic Zoology
Department of Entomology
Plantage Middenlaan 64
1018 DH Amsterdam
The Netherlands

Daniel R. Brooks (32)
Department of Zoology
University of Toronto
Toronto, Ontario M5S 1A1
Canada

M. J. E. Coode (6)
Herbarium
Royal Botanic Gardens
Kew, Richmond, Surrey TW9 3AB
U. K.

John Dransfield (3)
Herbarium
Royal Botanic Gardens
Kew, Richmond, Surrey TW9 3AB
U. K.

J.P. Duffels (21)
Institute for Taxonomic Zoology
Department of Entomology
Plantage Middenlaan 64
1018 DH Amsterdam
The Netherlands

I. D. Edwards (15)
Royal Botanic Garden
Edinburgh EH3 5LR
Scotland – U. K.

David G. Frodin (20)
2401 Pennsylvania Ave. 2-C-55
Philadelphia, PA 19130
U. S. A.

Rob Geesink (2)
Rijksherbarium / Hortus Botanicus
P.O. Box 9514
2300 RA Leiden
The Netherlands

Hugh C. Harries (29)
17 Alexandra Road
Lodmoor Hill
Weymouth, Dorset DT4 7QQ
U. K.

R. Hegnauer (10)
Cobetstraat 49
2313 KA Leiden
The Netherlands

* Chapter number between brackets.

E. Hennipman (11)
Institute of Systematic Botany
Heidelberglaan 2
3584 CS Utrecht
The Netherlands

R.J. Johns (13)
Herbarium
Royal Botanic Gardens
Kew, Richmond, Surrey TW9 3AB
U. K.

C. Kalkman (1)
Rijksherbarium / Hortus Botanicus
P.O. Box 9514
2300 RA Leiden
The Netherlands

Kuswata Kartawinata (12)
UNESCO/ROSTSEA
P.O. Box 273/JKT
Jakarta 10002
Indonesia

Masahiro Kato (19)
Botanical Gardens
Faculty of Science
University of Tokyo
3-7-1 Hakusan
Tokyo 112
Japan

Ruth Kiew (25)
Department of Biology
Universiti Pertanian Malaysia
43400 Serdang, Selangor
Malaysia

D.J. Kornet (30)
Institute of Theoretical Biology
Leiden University
P.O. Box 9516
2300 RA Leiden
The Netherlands

Job Kuijt (7)
Department of Biology
University of Victoria
Victoria, B.C. V8W 2Y2
Canada

Y. Laumonier (17)
BIOTROP
P.O. Box 17
Bogor
Indonesia

A.J.M. Leeuwenberg (8)
Department of Plant Taxonomy
Agricultural University
P.O. Box 8010
6700 ED Wageningen
The Netherlands

R. Löther (31)
Akademie der Wissenschaften der DDR
Zentralinstitut für Philosophie
Otto-Nuschke-Straße 10/11
Berlin 1086
Germany

L.J.G. van der Maesen (28)
Department of Plant Taxonomy
Wageningen Agricultural University
P.O. Box 8010
6700 ED Wageningen

Willem Meijer (24)
Thomas Hunt Morgan School of
Biological Sciences
University of Kentucky
Lexington, KY 40506
U. S. A.

R.W. Payton (15)
Soil Survey and Land Research Centre
Cranfield Institute of Technology
Silsoe Campus
Bedford MK45 4DT
U. K.

J. Proctor (15)
Department of Biological Science
University of Stirling
Stirling FK9 4AL
Scotland – U. K.

M. A. Rahman (9)
Department of Plant & Soil Science
University of Aberdeen
St. Machar Drive
Aberdeen AB9 2UD
Scotland – U. K.

S. Riswan (15)
Herbarium Bogoriense
P.O. Box 110
Bogor
Indonesia

S.H. Sohmer (23)
Bernice P. Bishop Museum
P.O. Box 19000-A
Honolulu, Hawaii 96817
U. S. A.

P.F. Stevens (33)
Harvard University Herbaria
22 Divinity Avenue
Cambridge, MA 02138
U. S. A.

Benjamin C. Stone (5)
Bernice P. Bishop Museum
Botany Department
P.O. Box 19000-A
Honolulu, Hawaii 96817
U. S. A.

Sukristijono Sukardjo (18 & 27)
Center for Oceanological
Research & Development
Indonesian Institute of Sciences
P.O. Box 580 JAK
Jakarta 11001
Indonesia

Elizabeth A. Widjaja (4)
Herbarium Bogoriense
Jl. Raya Juanda 22
Bogor 16122
Indonesia

Isamu Yamada (16)
Center for Southeast Asian Studies
Kyoto University
46 Shimoadachi-cho, Sakyo-ku
606 Kyoto
Japan

Professor Dr. C.G.G.J. van Steenis (1972)
to whose memory this volume is dedicated

Preface

Although the only publication with a realistic claim to the title "The plant diversity of Malesia" is Flora Malesiana itself, we have hesitatingly chosen this title for the present proceedings volume. Past, present and future work on the Flora Malesiana project was the subject of a successful symposium held in August 1989. This book contains only a selection of the papers presented at that meeting, yet it covers a much greater diversity of themes than just the inventory of botanical diversity. It even goes beyond the boundaries of the vast Flora Malesiana region in several of its chapters.

The role of the founder of the Flora Malesiana Project, Professor C.G.G.J. van Steenis, repeatedly recurs in several chapters; not only as director of and contributor to the project, but also as a pioneer in the fields of Malesian vegetation, conservation and biogeography, and as an enlightened systematist whose ideas and practical recommendations for taxonomic delimitation still largely apply.

Botanical information made available in regional and local floras is of vital importance for applications such as the exploitation of natural forests on a sustainable yield basis, for establishing gene banks for the benefit of agriculture, forestry and horticulture, and not in the least for nature conservation. Several chapters are devoted to these themes. Floristic studies are also at the basis of the biogeographical essays and vegetation studies included in this book.

Flora writers, although primarily concerned with taking stock of botanical diversity, also have to grapple with conceptual problems of plant taxonomy; therefore the inclusion of several chapters on recently changed concepts of taxa and the underlying theoretical considerations belonged to the symposium themes.

The practical problems in completing a large Flora project like Flora Malesiana within a reasonable time-span were the subject of a workshop held in conjunction with the symposium. The proceedings and recommendations from that workshop will be published separately (ed. R. Geesink, Rijksherbarium / Hortus Botanicus, Leiden, 1990).

Much work remains to be done before the first inventory of the plant diversity of Malesia will be complete. We hope that this present proceedings volume will be a stimulus, however modest, for the study, protection, and wise utilisation of the Malesian flora.

We wish to record our great appreciation to Ms. Emma E. van Nieuwkoop, who meticulously took care of the technical editing and prepared the camera-ready copy for the publishers. Thanks are also due to Prof. Dr. R.J. Johns (Kew, U.K.), Dr. P.H. Hovenkamp and Ms. G.A. van Uffelen (Leiden) for editorial assistance.

Leiden, May 1990

Pieter Baas
Kees Kalkman
Rob Geesink

1 Van Steenis remembered

C. KALKMAN

Summary

The botanical career of C. G. G. J.van Steenis (1901–1986) in Utrecht, Buitenzorg (Bogor), and Leiden is briefly reviewed. His concept of Flora Malesiana is discussed and placed in the context of developments in systematic botany. His two ideals for Flora Malesiana, to be practical and scientific, are still valid.

Introduction

It is now three years ago that van Steenis died at the age of 84 and although of course we have become used to his absence, in the research group and in the Rijksherbarium as a whole we still feel the empty place. His knowledge of plants, places and persons, his stimulating activity and efficiency, his friendship, even his outspoken criticism and impatience are still missed. This symposium, held in his honour, is an opportunity to remember him as a person, but of course also as a botanist and especially as the initiator of Flora Malesiana.

Between his first publication in Malesian botany and his last lie more than 60 years and no one has had a more profound impression and influence on the development of Malesian botany than van Steenis – with all respect due to people like C.L. Blume, F.A.W. Miquel, E.D. Merrill, E.J.H. Corner, R.E. Holttum and others.

On this occasion I will almost entirely confine myself to a consideration of his activities in Malesian taxonomy and floristics, leaving aside his many valuable contributions to vegetation science, evolutionary biology, and plant geography.

Van Steenis's very first steps as a naturalist were done among the Dutch flora, and several small reports on field observations preceded his publications on the Malesian flora. Already in 1924 he expressed his interest in the making of a herbarium as a means of documentation and to promote further study. These two aspects: field study and herbarium study, have dominated his entire life.

He became involved in Malesian herbarium botany already during his student years in Utrecht, when for his master's degree he was charged with the identi-

P. Baas et al. (eds.), The Plant Diversity of Malesia, 1–9.

fication of New Guinean Bignoniaceae, Elaeocarpaceae, Gonystylaceae, and Stackhousiaceae. The results were published in Nova Guinea. After that experience he wrote a doctor's thesis on the Malesian Bignoniaceae, acquiring the degree in 1927.

For students in Utrecht it was quite normal to think of a biological career in the tropics, in one of the Dutch colonies. The prominent professors were then F.A.F.C.Went and A.A.Pulle and they both promoted such a choice among their students. Moreover, they did what they could to find places for good students and good students for available jobs. Kees van Steenis was one of Pulle's students who wanted to see more than Europe and so he went to Java in 1927, with his young wife Rietje.

The Buitenzorg period

Although financially times were very bad for the public service, in Holland as well as in the colonies, van Steenis enjoyed himself thoroughly. At first he hardly knew any of the plants he saw around him, but soon this was remedied and he started accumulating his fabulous knowledge of the flora.

He was a gifted observer, one of those people who always see the things in their surroundings and who make other people feel ashamed that they have failed to observe what the Master was able to point out to them. He used this faculty in the herbarium and so his prompt recognition of Malesian Seed Plants and Ferns became legendary. But he also used it in the field on excursions and expeditions, as is obvious from his papers and reports.

Officially van Steenis was assigned to the Museum of Economic Botany built by K. Heyne, who left the Indies in 1927. The reality was that van Steenis went to work in the Herbarium, another part of the same Botanical Garden at Bogor, then called Buitenzorg, with a main task in identification. The first taxonomic revision he wrote at Bogor covered the Styracaceae and was published in 1932.

From the list of his publications in the period following 1927 it is obvious that he enjoyed the field experience. In the period 1927–1945 about 60% of his larger papers can be termed field studies (observations, ecology, vegetation) against only 20% taxonomical studies based on herbarium. Later (see table 1.1) in Holland this was naturally reversed. But although he then spent most of his hours between dried specimens, he always went into the field when there was a chance and many foreign visitors were taken to the dunes on Sundays.

Developments in systematic botany

Before we touch Flora Malesiana, let me try to sketch a general image of botany and especially systematic botany during the period 1925–1940, that

Table 1.1 — Categorisation of publications. Percentages of papers larger than c. 5 printed pages. See Bibliography in Blumea 32, 1987: 5–37.

subject	1928–45	1946–65	1966–87
field observations, floristics, vegetation, ecology	60%	22%	10%
taxonomy	20%	45%	38%
phylogeny, evolution, general systematics	0%	4.5%	6%
history, biography	0%	11%	16%
biogeography	5%	8%	18%

is the period in which van Steenis received his university training, started his professional work in Java, and developed the concept for Flora Malesiana.

Experimental and laboratory-based biology were of course firmly established at the time, next to the older, descriptive subdisciplines of biology as systematics and morphology. In the period mentioned all branches of botany and biology in general were bristling with activities and new discoveries. Modern biology was taking shape. It is, however, obvious that living nature as it can be observed with unaided eyes or at most a hand lens, and whole-plant botany fascinated van Steenis much more than physiological, chemical, genetical experiments. He stayed a naturalist as much as a taxonomist.

In the period mentioned, the most important development in plant systematics was undoubtedly the rise and absorption of so-called experimental taxonomy. In the early 1920s this branch started to develop in the United States and in Sweden. In the U.S. the pioneers were Hall and Clements, followed by the trio Clausen, Keck and Hiesey; in Sweden it was Turesson who was leading the way. All were attempting to introduce genetical, cytological, ecological and, consequently, evolutionary notions into systematics (see Hagen 1983).

The advent of experimental taxonomy, soon called biosystematics, was a natural contra-movement after the static, herbarium-based systematics that prevailed during the period before, dominated as it was by the influential Engler school. It also was made possible by the developments in genetics and ecology, systematics always having been rapid in absorbing new techniques and new developments in other disciplines. Of course experimental taxonomy triggered sometimes heated and fanatical discussions about the value of the old categories, especially the species, in contrast with newly proposed units, mostly on a genetical basis.

If we turn to Dutch plant systematics in the period concerned (1925–1940), we must conclude that this was very much dominated by the school of Pulle, professor at Utrecht University from 1914 to 1948. As already said, Pulle was a botanist with a distinct preference for work in and about the tropics; he himself went to Surinam a number of times and he also participated in an expedition to New Guinea. Utrecht was the Dutch centre for plant taxonomy for a long time and in view of Pulle's own preferences and achievements (editor of Flora of Surinam, editor of the botanical volumes of Nova Guinea) that implied a floristic approach.

Two other professors in the Netherlands were B.H. Danser and H.J. Lam, respectively in Groningen and in Leiden. Danser was one of the pioneers in the field of biosystematics (Danser 1929) and he came to Groningen after a stay in Buitenzorg from 1925 to 1929, beginning his professorship in 1932. He died too soon (1943) to have much influence on Dutch systematics and so in Holland a second chance was lost to participate in the development of experimental taxonomy. The first time was when J.P. Lotsy disappointedly left the Rijksherbarium in 1909, after his attempts to put systematics on a more experimental genetical basis had been frustrated (see Kalkman 1979: 15).

Lam was appointed director of the Rijksherbarium in 1932. He was one of Pulle's pupils and he had been in the Dutch East Indies from 1919 until he came to Leiden. Lam himself became more and more fascinated by phylogeny but he also continued to produce monographic and floristic papers, mostly in the families he had revised when in Buitenzorg (Burseraceae, Sapotaceae). Under his direction Leiden already before the end of the war had become specialised in Malesian taxonomy, a logical development in view of the institute's history and collections. All this makes the choice of Leiden as the later centre of Flora Malesiana self-evident.

Although van Steenis himself was not by a long chalk an experimental taxonomist, he clearly saw the uses and consequences of that approach, as is obvious from his well-known and balanced essay on specific delimitation (1957). Maybe under the influence of his friend Danser, with whom he had many discussions during the time they both were at Bogor and continued later in correspondence (van Steenis 1954), he recognised the value of the new approach, but he also sharply saw the limitations of the method for the progress in the taxonomy of tropical plants. He was quite willing to accept Danser's concept of the commiscuum (miscibility community) as the best genetically testable approximation of the so-called Linnean species but he also knew that this test could never be made for the great majority of tropical species.

One statement by the biosystematist W.H. Camp, written in 1943, after most of the steam was already off the kettle in the experimental vs. descriptive controversy, has been attacked severely by van Steenis. First I cite Camp's statement: "The day of the taxonomist who putters alone in his herbarium with an other-worldly stare is done. He must shed off his robe of academic classicism and seclusion, brush off the accumulated dust of the centuries, and come face to face with the dynamics of living populations." Van Steenis's reaction (1957: clxix) was: "A most charming appreciation of and tribute to the work of our illustrious predecessors, the Makers of Botany, and a rosy prospect for a fresh stimulant to naturalists to study taxonomy by camping." In his acid reaction to this in itself not very important and moreover caricatural statement van Steenis defended several positions at one time. It is cited here because it is a typical illustration of his standpoints and reveals his general view on taxonomy as it should be. In the first place he defended the morphological species and the herbarium as a tool of systematics, but at the same time he gave honour to our predecessors instead of putting them in the wrong and so he marked his feeling of continuity in science and his historical sense. He also defended plant taxonomy and plant taxonomists in general against attempts to put the discipline in the dusty corners of outdated activities.

That he felt offended, is not strange. Van Steenis can certainly not be accused of being a dusty herbarium worker without knowledge of the living nature. He probably knew more of the versatility and dynamics of tropical plant life than most of the biosystematists of the first period.

The concept of Flora Malesiana

We return to Buitenzorg and to Kees van Steenis, working there and gradually developing the concepts for a Flora Malesiana. Already before he went to the Indies, van Steenis was invited by Went and Pulle to a private talk (see van Steenis 1958) about "a very gradual planning of a final, regional Flora." In 1972 he cited this talk as having been a vague suggestion from Went and Pulle, that already during his journey to the Indies imperceptibly "germinated as a little seed." From the beginning of his tropical career he thus lived with this ideal, letting it grow and take shape.

The concept was and remained based on morphological taxonomy. As he mentioned several times (see, amongst others, van Steenis 1965) his authorities for the ways of monographical, taxonomical research were Bentham (1874) and Hooker (1853; Hooker & Thomson 1855), not Engler's 'Pflanzenreich' that he considered too superficial and non-professional. Research of an experimental nature into the origins of variation, into evolutionary history, into small-scale variation and its relation to habitat, into chromosome behaviour etc., he recognised as scientifically useful. The classification of tropical higher plants, the part of plant taxonomy that he valued most highly, not only emotionally, but also from a general scientific point of view, could in his opinion only make progress when the taxonomist would limit him- or herself to outer morphology and to the level of the Linnean species, recognising a wide morphological species concept.

His broad vision of taxa extended also to the other taxonomic levels of genera and families. Recognisability was important for him and in several papers he fiercely protested against the inflation of ranks. A few citations show his point: "Nothing is gained for sound classification by raising all sections of large genera ... to generic rank. Indeed, such practice would involve the loss of the structural plan embodied by the generic concept. ... Only in cases where there is reasonable evidence that species or genera are arranged in a distinctly artificial or insufficiently clear classification it is clear that there is necessity for improvement of the system and we should not shun it." (van Steenis 1965). "There is not the slightest scientific gain and sense in splitting *Nypa* from the palms as a separate family, ... *Nypa* and the other palms remain one 'lump of phylogeny', and even if raised to family rank, they remain close together, ..." (van Steenis 1978). On other occasions and on other cases (e.g. 1985) he expressed the same views.

He was not a great lover of phylogenetical research and one of his recommendations for plant taxonomic work (1957: ccxxiv) says that the latter "should confine itself to factual observation and not be hybridised with theoretical considerations on origin of species and genera or supposed phylogenetical relations. Such hypothetical deductions are not discouraged but should be kept clearly separate from the facts observed and not be introduced in their evaluation." In his opinion phylogeny certainly was a valid science, but only as long and in so far as it was based on stated fact and not on intuition and fantasy. Modern cladistics came late in his life and I am afraid that he did not like it very much. This dislike cannot be attributed to its speculative character (the argumentations and facts being clearly stated in this method, which is certainly not intuitive), but probably because in practical execution it reduces plants to rows of numbers in a matrix, the same reason why in my opinion he was not in favour of numerical taxonomy.

Intuition was a concept that he could not use in taxonomy. It could never be a guiding principle for taxonomists. On the contrary: "Zeal, time, opportunity and intelligent work are the main factors determining the quality of this kind of work. ... Brilliant ideas are welcome, but the facts should be verified." (van Steenis 1957: clxxiv).

So van Steenis emerges as a taxonomist who in the course of time – and I think that most of his wide visions came alive after he had come to Leiden – acquired a broad view of his discipline, but who at the same time advocated a deliberate and well-considered restriction in the execution of what he considered to be one of the very best goals of taxonomy: the production of a Flora Malesiana.

He saw the gain brought by evolutionary biology and biosystematics in the form of better knowledge of the processes of evolution and speciation. Later in his life he even contributed some controversial ideas about plant evolution as different from animal evolution, leaving much more room for non-adaptive play with character changes in what he called the *patio ludens*, the playground of plant evolution. He also recognised the profits a taxonomist could have from the non-morphological ways of character retrieval by means of microscopy and chemical analysis and, consequently, he enriched Flora Malesiana with paragraphs or chapters on anatomy, pollen morphology and phytochemistry, written by specialists. He considered these ancillary disciplines to be very important, mainly on the levels of genus and family, not on the species level except for the identification of incomplete material.

However, all this did not change his love for monographic but concise taxonomy that does not overflow from its self-chosen limitations of morphology, common sense, restriction and conciseness in descriptive matter, nomenclature, administration etc. And so Flora Malesiana was envisaged to be and to remain. The taxonomist had to work out his classification on the basis of relatively few basic characters that were recognised as important. In this way rapid progress would be possible.

It is too easy to say that van Steenis underestimated the difficulties and the dimensions of the taxonomic problems and that his concept was not realistic from the beginning. He was, by the way, not the only one: Pulle (1938) estimated that a Flora Malesiana could be produced in c. 25 years, with 6 full-time taxonomists, and he even made a financial picture: 35,000 Dutch guilders annually! What he and van Steenis underestimated and could not have foreseen, in my opinion, was the dramatic post-war increase of the number of specimens a modern monographer has to cope with, and on the other hand they overestimated the state of existing knowledge about the plants of the region. A touch of colonial self-overrating can be detected perhaps, at least van Steenis himself tells us so. I cite (van Steenis 1954: [9]): "In the relatively isolated scientific position of one in the tropics – and this holds for other countries far remote from the centres of learning in the Northern hemisphere – most people inevitably overestimate what can be attained; due to the absence of comparable situations, the absence of enough constructive criticism, the idea that the man on the spot is the centre of the universe and is certain of the most sound judgement – which, it is true he often has in certain respects –, and to a certain self-sufficiency. It is no use denying these very human reactions and I am quite prepared to admit that I myself did not fail to react more or less along the lines indicated during my prolonged stay in Java."

He also could not foretell the developments in taxonomy after the war. The workload and the time involved are much larger than could be foreseen between the 30's and the 50's. The large number of specimens are not only an administrative burden but they have also consequences for the knowledge and understanding of natural variation. Very often more morphological data do not make it easier to find a solution as to how to translate variation into taxa, especially species. Taxonomists, also tropical ones, have become convinced by (mainly temperate) biosystematics that the pattern of variation is intricate and the result of interplay of many ecological and genetical, directed and random factors. That pattern is, therefore, not so easily translated into a taxonomy as appeared to be the case in the times of 'Das Pflanzenreich'. This is in my opinion a more serious delaying factor than the number of undescribed species that became known by better exploration.

When in the early fifties van Steenis made his plans public, it was after careful consideration but with an optimistic view on the number of man-years involved. In 1961 he published an estimate of 300 man-years for the Flowering Plants, adding that this was only possible under the most favourable conditions. It seemed to him an attainable ideal to produce within a reasonable amount of time a Flora for the region, at the same time scientific and practical.

These two goals were repeatedly stressed by van Steenis. In 1959 he expressed it in the following words: "This principle: that a flora shall be of the highest scientific quality, but no less generally useful at that, is the basis on which the whole work rests." In his directives to prospective authors he always tried to instill in them this principle as a guideline. Scientific standard must not be sacrificed to rapid completion, but the demand of practical usefulness implies that not all scientific problems can and need be solved.

The present

Our problem is how in this time van Steenis's Flora Malesiana goals, which are still valid, can be reconciled. Our methodological resources have multiplied, materially by the advent of computers and better microscopes, immaterially by the advances in statistical methods as a left-over from numerical taxonomy, and by the advances in phylogenetical methods as a gift from cladistics. Our methodological bag of tricks has become better filled, also our problems have multiplied. I mention again the higher piles of herbarium specimens, but there is also the flood of literature, most of it without consequence and quality but still to be read or at least seen. Certainly not in the last place we have to cope with modern university life in general that does not favour ivory towers for secluded research taxonomists without other duties. This could not be foreseen in the years before 1950.

Scientific and *practical*. Scientific is not synonymous with 'complete to the last detail', it is synonymous with making choices about what is relevant and important in a given setting and what is not. These choices are not easily made but more and more it becomes clear that also in the Flora Malesiana context for today and tomorrow they have to be made.

Practical is not the same as 'comprising only the useful species.' It means that the user of the Flora must be able to find his way in the book when identifying specimens, when reading descriptions, when searching for relevant literature references and that he must not be allowed to drown in a multitude of data, most of them not essential for his purposes. Again, choices have to be made by author and editor, who have to creep into the skin of the user, who is almost never a specialist, usually not a taxonomist, and often not even a botanist.

What I said in the last two paragraphs agrees with van Steenis's opinion as repeatedly given in word and in writing. The best way to remember van Steenis is to proceed with Flora Malesiana in accordance with his views. In the workshop we will discuss how to put that into practice today. It is not impossible that we must accept a deviation of the set ways as we developed them in Leiden. We must, however, try to keep alive the ideals of van Steenis as conceived sixty years ago and shaped forty years ago, the ideals of a Flora based on sound, disinterested science and executed with consideration of the needs of its users.

References

Bentham, G. 1874. On the recent progress and present state of systematic botany. Rep. Brit. Ass. Adv. Sc. 1874: 27–54.

Camp, W.H. 1943. The herbarium in modern systematics. Amer. Naturalist 7: 322–344.

Danser, B.H. 1929. Über die Begriffe Komparium, Kommiskuum und Konvivium und über die Entstehungsweise der Konvivien. Genetica 11: 399–450.

Hagen, J.B. 1983. The development of experimental methods in plant taxonomy, 1920-1950. Taxon 32: 406–416.

Hooker, J.D. 1853. Introductory essay. Flora Novae-Zelandiae I, i–xxxix.

Hooker, J.D. & T. Thomson. 1855. Introductory essay, chapters I–III. Fl. Indica 1: 1–44.

Kalkman, C. 1979. The Rijksherbarium, in past and present. Blumea 25: 13–26.

Pulle, A.A. 1938. De inventarisatie van het erfdeel der vaderen. Meded. Bot. Mus. en Herb. Utrecht 50: 38 pp.

Steenis, C.G.G.J. van. 1924. Herbaria, plantenuitrukkerij, natuurbescherming en nog wat. Amoeba 3: 32–35.

Steenis, C.G.G.J. van. 1954. Dedication [to B.H. Danser]. In: Flora Malesiana I, 4: [6]–[13].

Steenis, C.G.G.J. van. 1957. Specific and infraspecific delimitation. In: Flora Malesiana I, 5: clxvii–ccxxxiv.

Steenis, C.G.G.J. van. 1958. Dedication [to A.A. Pulle]. In: Flora Malesiana I, 5: [6]–[7].

Steenis, C.G.G.J. van. 1959. The value of the Flora Malesiana for all who are interested in tropical botany. Mal. Nat. J. 14: 12–14.

Steenis, C.G.G.J. van. 1961. Flora Malesiana, verleden, heden en toekomst. Jaarboek 1960 van de Ned. Org. voor Zuiver-Wetenschapp. Onderz. (Z.W.O.): 143–146.

Steenis, C.G.G.J. van. 1965. Nature and purpose of botanical classification from the standpoint of the producer, particularly with respect to tropical plants. Bull. Bot. Surv. India 7: 8–14.

Steenis, C.G.G.J. van. 1972. Enige historische aantekeningen over Flora Malesiana en de betekenis van dit werk voor de bijzondere plantkunde, speciaal met betrekking tot het Rijksherbarium. In: Overdenkingen: 47–56. Univ. Pers, Leiden.

Steenis, C.G.G.J. van. 1978. The doubtful virtue of splitting families. Bothalia 12: 425–427.

Steenis, C.G.G.J. van. 1985. The undesirability of splitting Eucalyptus. Austr. Syst. Bot. Soc. Newsl. 42: 13–16.

2 The general progress of Flora Malesiana

ROB GEESINK

Summary

Growth of Flora Malesiana, expressed in number of species treated per period of five years remained surprisingly constant between 1950 and now. Also the mean number of active collaborators remained reasonably constant. In each period about as many Rijksherbarium staff contributed as there were contributions from outside the Netherlands (total around 10–15 per time slice). Despite these felicitously constant figures the estimated total execution time of the Flora Malesiana Project (Phanerogams only) is unacceptably long. Flora work of this kind can only be speeded up by decoupling the future floristic treatments from monographic inclinations of the contributors: more contributions should be an aim in itself instead of a by-product of larger monographic studies. In order to reach this goal aims and means of present larger Flora projects need pragmatic redefinitions in many cases.

Summarised history of Flora Malesiana

Flora Malesiana Bulletin volume 1, part 1, appeared in July 1947 and it may be regarded as the first material product of the Flora Malesiana Project. It contains a long essay, from the hand of Professor van Steenis, which served as an explanation of goals, aims, and methods to be applied by the collaborators in order to reach the implicitly indicated goal: a series of books in which it is possible to find the scientific and the major common names with some concise knowledge of all species of seed plants (flowering plants and gymnosperms) and 'ferns' of the Malesian region, comprising Thailand South of the Isthmus of Kra, Malaya, Indonesia, the Philippines, New Guinea and later also the Solomon Islands. Van Steenis' essay ends with a list of promised contributions by colleagues from outside Indonesia and the Netherlands, and it is noteworthy that most of these families mentioned have indeed appeared in Flora Malesiana, but in most cases 20–30 years later, and by other authors. That is lesson number one: *Things always go differently.*

We are forty-three years later now, and our shelves are bearing 9 volumes, starting with the 'green bible', the Cyclopedia of Collectors, containing knowledge of plant collections (accessible via the names of the collectors) in the Malesian region. This impressive encyclopedia was brought together mainly by Mrs. van Steenis in the difficult years during Japanese occupation and the years

P. Baas et al. (eds.), The Plant Diversity of Malesia, 11–16.

immediately after World War II. From series II ('Ferns') one volume has appeared, but I have excluded series II from this analysis for various reasons. The other seven volumes contain taxonomic treatments of classical families of seed plants.

How much is covered?

The first two volumes start with 560 pages (when taken together) general introductions and considerations, most from the hand of Professor van Steenis. In the remaining taxonomic pages nearly 5,000 species of seed plants are treated. With the appearance of part 4 of volume 10, the total number of treated species reached 4,892. In fact, over 1,000 more species are accessible in precursory papers mainly on genera of families currently under revision.

When I started this analysis with the first graphs of number of treated species over the years, I expected a drastic decrease in productivity during the last 15 years compared with the 'good old days', the fifties and the early sixties. My expectations appeared to be utterly wrong. I will not give the detailed graphs here; these will be published in the introduction of the special issue of Flora Malesiana Bulletin containing the reports of the workshop 'The Future of Flora Malesiana'.

In fact a surprisingly constant growth of the number of treated species is the case, when they are averaged over periods of 5 years or more. There is even a small but significant increase during the last 12 years. The first volumes wisely contained mainly small families with a few species only. But there were also some treatments of larger families: Convolvulaceae, Burseraceae, and Ericaceae.

The following volumes mainly contain medium-sized families, like Linaceae s.l., Ochnaceae, Bignoniaceae, Olacaceae, and Magnoliaceae, with a tendency to larger families like Anacardiaceae, Fagaceae, Cyperaceae, and Dipterocarpaceae.

The percentage of treatments by Leiden staff per volume shows a significant decrease, but with a revival in the last volume. There are some bad reasons for this decrease, but a good reason is that the Leiden staff has now mainly large families under revision, for instance Gramineae, Sapindaceae, Orchidaceae, and Legumes.

The 'semi-'monographic approach

Other families, like Winteraceae, are under a very thorough anatomical/ morphological study, which slows down production when purely measured in published species per year. The interesting results of such detailed studies, however, justify such an approach, at least in this and similar cases. Here we have arrived at the obvious lesson number two: *Detailed work disproportionally slows down flora production.*

This does not only hold for morphologically detailed work, also nomenclaturally, phylogenetically, or bibliographically detailed treatments inhibit the desired speed. The opinions of Professor van Steenis on this problem were clearly ambiguous. On the one hand he loved thorough work, and if the results were exciting he simply ignored the time-factor. In other moods, however, he was worried or clearly frustrated about the overall slow progress.

Flora Malesiana: project management

In the last forty years nearly 5,000 species found their pages in Flora Malesiana. The estimated total number of species of seed plants in the Malesian area is 25,000, and from a simple calculation it follows that the completion of the project in the present speed will need another 160 years from now. With the present format and working method this unrealistically long total execution period can only be reduced by a team of collaborators roughly increased in size to a threefold. The number of actual collaborators has always been between 10–15 in each time-slice, including advanced students and retired colleagues. It is not very realistic to expect a spectacular growth of e.g. our Rijksherbarium staff in these years of consecutive budget cuts. Our hope for an increased team of collaborators is directed to increased international cooperation with other Herbaria and Botanical Institutes and not in the least to the increased interest from the countries in the Flora Malesiana area itself. The training courses in Plant Taxonomy in Southeast Asia, efficiently organised and coordinated by UNESCO/Jakarta, and by Dr. Kuswata Kartawinata in particular, during the last few years *have to* result in a more extensive collaboration to our Flora Malesiana project.

Another equally important factor is that in the present approach a well-defined weighting of quality versus quantity per time is missing. From the point of view from project-management, the aim of Flora Malesiana is sufficiently defined, but the necessary conditions and minimally necessary qualities are not clear, and have been unclear from the beginning.

As far as long-term planning was concerned, probably from an overly optimistic expectation, Professor van Steenis estimated (in 1960) the total execution time of the project as 300 man year equivalents. It follows from the growth between 1950 and 1960 that this number is actually between 1,500 and 2,000; the time factor was underestimated no less than 5 times! In the same survey he estimated Flora Malesiana to cover about 17 volumes when completed. If it is true that the present 8 published volumes on seed plants represent roughly 20% of the total, 40 volumes will be closer to the truth. On the one hand the collaborators were expected to be highly selective in data that had to enter the published results, on the other hand an undefined 'completeness' was expected. These contradictory demands are impossible to fulfill.

I think that any Flora project must have a limited set of aims and a clear beginning and end in time; the project must be well-phased, and the several phases must be manageable. Even in very long term projects the final aim, completion of the project, must always be kept in mind.

Other large Flora projects

Really large Flora projects, like Flora Neotropica (over 50,000 species), Flora of East Tropical Africa (10,500 species), Flora Malesiana (25,000 species), Flora Brasiliensis (23,000 species), Flora Europaea (11,500 species) have (and had) total execution periods of several to many decades. They generally started as broad international cooperational projects. Both Flora Europaea and Flora Brasiliensis were finally completed by small teams under relatively unrewarding circumstances; Flora Europaea was completed by the 'Cambridge team', and Flora Brasiliensis by a small group around Ignatius Urban. The recently completed Flora of Turkey (c. 8,000 species) was produced from beginning to end by a small group of devoted collaborators, which concentrated on this work only. Professor Davis told me a few years ago, that in his opinion this is the best way to get large Floras completed. Flora Australiensis (8,125 species) was produced in 20 years from the start to the end by one person, George Bentham, and for its time it was a very good product. Its successor, the relatively recently started Flora of Australia will be over twice the size (in numbers of species covered) of its predecessor, and it is an example of a well-defined project with clear limitations in time and expected quality/quantity 'ratio'. I also want to mention the new Flora of Hawai'i, to appear in 1990, though it is considerably smaller in size. With a limited staff, but with good planning and management the 1,800 species were made accessible in a 6 years project.

Flora of Australia, Flora Europaea, and Flora Brasiliensis are strongly limited in their format: Family diagnoses, key to genera, generic diagnoses, key to species, and species diagnoses with very compact additional information. Too many modern large Flora projects try to be 'regional monographs'. In some Floras, like Flora Malesiana, it is expected that 'everything worthwhile' should be included. This is dangerous for the process of completion, as this attitude invites perfectionism in its various shapes and expressions. From the point of view of project management, such an attitude makes the project unmanageable; the time factor will become a monstrous problem.

A solution

What about a stronger separation of monographic work and Flora work? The different activities have different aims. Monographic studies aim at hypotheses of natural affinities of species and natural groupings thereof. To put my own position clearly, I am, as far as monographs are concerned, a staunch phylogenetic cladist, adhering both to the individual-concept of natural species and natural higher taxa and to the theory of non-equilibrium thermodynamics as the major cause of evolutionary change, but I realise that phylogenetic analyses with the associated theoretical considerations are very time consuming: If not distracted by administrative or educational duties (or by plain quarrels), a full-time monographer with cladistic and historical biogeographical inclinations will probably need around four years for the analysis of a group of 40 species, just to give a rough indication. A devoted flora producer, capable to resist this phylogenetic challenge, can provide floristic access to say 200 species in the same amount of time. Of course, I had the work of Dr. Sleumer in mind, when estimating these figures. For a Flora or a field guide unnatural higher taxa "do not harm" because genera and families function merely as convenience classes rather than as natural individuals to be discovered.

Both monographers and flora producers strive for naturalness of their basic taxa, the species. This part of the work, the recognition of hopefully natural species, will remain the same in both approaches. But monographers and flora producers must necessarily apply different approaches when it comes to the *discovery of natural groupings* of these hopefully natural species or the *recognition of convenient groupings* thereof.

Conclusions

To conclude: in my view, Flora Malesiana should be completed in the following 50 years. If not, the project will die, either naturally or otherwise. This time-schedule can only be realised if the present semi-monographic approach is abandoned. I do not imply to say that Flora Malesiana collaborators should not do any monographic studies any more, to the contrary even. If they meet a nice particular group which, so to say, *demands* a monographic study, they should be stimulated to do such a thorough study, of course according to the quality standards of the time in which it is planned. But only in selected cases in the form of a well-planned project, like a well-phased PhD study.

The majority of the remaining families should enter Flora Malesiana via ... let me call it the 'Sleumer-approach'. Some (but not much) time can be saved by restriction of the general sections under the family diagnoses, and by restricting descriptions to diagnoses only necessary for checking the outcome of the keys for identification. Nomenclatural activities could be restricted to the names accepted in the most important sources only. It is of little use for a regional Flora to try to solve nomenclatural problems of widespread species with natural affinities outside the Flora Malesiana region. If the collaborators could stop searching the nomenclatural pathways back to or even before Linneaus, a considerable amount of time is gained.

It is sometimes asked what the *use* of a Flora or a Field Guide is. Let me just reply that nothing can be done with unknown objects. If the species composition of a particular area is simply unknown, inventorial ecological studies are next to impossible. And so is nature conservation. And the use of natural products. It has sometimes been argued that local names and trade names are good enough, and in some cases it is true. But local names do only locally apply and certainly not in the whole area of a species. Trade names often refer to several, but sometimes quite different species. Internationally accepted Latin binomials hopefully refer to species and scientific knowledge of them. The usefulness of medical plants is obvious. If a new disease pops up in cultivated species, it may be very useful to investigate whether related species are resistant. If so, their resistant genes can possibly be borrowed, so to say. But first we must know where to find these possible resistant relatives or resistant races. In short, without basic knowledge, like names, characteristics, geographical and ecological distribution, we cannot do anything with plants; we cannot use them and we cannot protect them. Taxonomy is the beginning of knowledge of the living world. Too many species have already gone extinct before they were discovered. The destruction of so much living nature is just as stupid as a parasite species that exterminates its host species; the result is the death of both. It is now time for the third and final lesson. Some words of Professor van Steenis: *"A flora is only useful after it is completed."*

Of course partial treatments and treatments of separate families are useful, but only after completion of Flora Malesiana it will be possible to identify random collected specimens with flowers and/or fruits from the Malesian area. Even when it will appear that the actual access is mainly to the common species, the ideal of Professor van Steenis will then, but only then, be fulfilled.

3 Outstanding problems in Malesian palms

JOHN DRANSFIELD

Summary

The palm flora of the Malesian region is estimated to consist of 52 genera and over 900 species. Recent revisions and synopses have been published for 24 genera, but the larger genera have yet to be tackled. In particular *Pinanga, Licuala, Daemonorops* and *Calamus* present serious problems to the monographer and thus the preparation of an account of Palmae for Flora Malesiana. Particularly acute are the problems in the rattan genus *Calamus*, known to be very diverse in New Guinea and in the Moluccas, yet very poorly represented in herbaria. Palms seem to display a high rate of endemism in the islands of the Malesian region and as they (especially the rattans) are still neglected by the general collector, specialists are confronted with the need for detailed field work throughout the région, for which time and funding is scarce.

Introduction

The Palm family is usually regarded as being difficult, poorly known, badly represented in herbaria and so on. Some of this is true and I shall have to expand on these aspects. In drawing up a summary of what we know of the palms of Malesia the remarkable fact emerges that over a third of all palm species in the world occur in the Malesian region, and just over a quarter of the palm genera.

This great diversity of palms, economically of great importance at the local and international levels, has attracted attention throughout the history of Malesian phytography. Rumphius' account of palms is of great significance being the first detailed discussion of Malesian palms and their uses. Linnaeus was well acquainted with Herbarium Amboinense and his binomial *Areca catechu* has accordingly been lectotypified by Rumphius' description (Moore & Dransfield 1979). Griffith, Blume, Martius, Miquel and Scheffer all made large contributions to the development of Malesian palm botany. Using their work as a base together with his extensive field knowledge, Beccari, the greatest student of Malesian palms, monographed genus after genus in splendid detail. During the twentieth century few botanists have made Malesian palms their speciality, the most important being Furtado.

P. Baas et al. (eds.), The Plant Diversity of Malesia, 17–25.

Almost 20 years ago Marius Jacobs suggested to me that the preparation of the account of the palms for Flora Malesiana would be a difficult but 'glorious task'. Since then Malesian palm botany has progressed slowly. A major development in recent years has been the rise of local interest in Malesian palms; there are several researchers in the region active in collecting and doing research on the family. Nevertheless, we remain seriously hampered by one major problem – the inadequacy of herbarium collections. Whereas a botanist working on a timber tree taxon may have at hand a set of specimens, often each comprising many branches, leaves and flowers, the palm botanist has to deal with fragments – pieces of a single leaf, which itself may not be representative, because juvenile or sucker shoot leaves are often smaller and easier to collect, pieces of inflorescence and often no stem at all. The interpretation of such fragmentary material requires a degree of imagination and prowess in detection which increases with greater field experience. General collectors, faced with the desire or need to produce large sets of numbers, are reluctant to spend the time necessary to collect palms. I make a strong plea on behalf of the few palm taxonomists working on Malesian palms for general collectors to spend time collecting palms and for the collections to be distributed to herbaria with resident palm botanists. Guidelines to the preparation of good herbarium material of palms have been recently published (Dransfield 1986).

How much do we know?

Problems of generic delimitation

Generic delimitation in the palms is now quite well understood though problems remain. Worldwide, 200 genera are currently recognised (Uhl & Dransfield 1987); a further two have recently been discovered in Madagascar. In the Malesian region the group with least clearly resolved generic limits is in the Arecinae. The genera *Gronophyllum, Siphokentia, Gulubia* and *Hydriastele* as currently recognised are separated by combinations of characters known to be rather unreliable elsewhere in the family. *Gronophyllum,* in particular, includes taxa of contrasting habit, some of which appear more closely related to *Gulubia.* This complex of four genera requires reassessment.

Generic monographs

Table 3.1 indicates those palm genera present in Malesia which have been revised in the past 20 years, while Table 3.2 lists those genera for which no recent revision exists. Some of the genera in Table 3.1 clearly require reappraisal because of recent collecting activity.

Table 3.1 — List of Malesian palm genera monographed in the last 20 years.

Arenga (Mogea in press)	c. 20	(15)
Borassodendron (Dransfield 1972a)	2	(2)
**Brassiophoenix* (Essig 1975)	2	(2)
**Calospatha* (Dransfield 1979a)	1	(1)
Ceratolobus (Dransfield 1979b)	6	(6)
Cocos (Harries 1978)	1	(1)
Eugeissona (Dransfield 1970)	6	(6)
Gronophyllum (Essig & Young 1985)	c. 25	(24)
Gulubia (Essig 1982)	9	(5)
Iguanura (Kiew 1976, 1979)	18	(18)
Johannesteijsmannia (Dransfield 1972b)	4	(4)
Korthalsia (Dransfield 1981)	27	(25)
Maxburretia (Dransfield 1978)	3	(2)
Metroxylon (Rauwerdink 1986)	5	(1)
Myrialepis (Dransfield 1982a)	1	(1)
Nenga (Fernando 1983)	5	(4)
Nypa (Uhl & Dransfield 1987)	1	(1)
Orania (Essig 1980)	18	(17)
Physokentia (Moore 1969)	7	(1)
Plectocomia (Madulid 1981)	16	(8)
Plectocomiopsis (Dransfield 1982a)	5	(5)
**Pogonotium* (Dransfield 1980, 1982b)	3	(3)
Ptychosperma (Essig 1978)	28	(23)
**Retispatha* (Dransfield 1979c)	1	(1)
Rhopaloblaste (Moore 1970)	7	(5)
Salacca (Mogea in prep.)	c. 20	(17)

* = endemic genus, number in brackets indicates number of species recorded for the Flora Malesiana region.

Quite apart from taxonomy, very little is known of the natural history of most Malesian palms; their ethnobotany, however, is probably better known than for many groups – this being an indication of the great range of uses to which palms are put.

Areas least known and likely to be most productive

As for most plant groups our knowledge of the palms varies in completeness from area to area. In the following notes comments are provided on how much we know of the palms of given areas.

Table 3.2 — List of genera with no recent monograph.

Actinorhytis	2	(2)
Areca	60	(c. 50+) (partial accounts for Borneo and Philippines)
Borassus	c. 4	(2)
Calamus	c. 400	(c. 280)
**Calyptrocalyx*	38	(38)
Caryota	c. 12	(c. 8)
Clinostigma	c. 13	(1)
Corypha	6	(2) (partial revision for India)
Cyrtostachys	8	(6)
Daemonorops	115+	(100)
Drymophloeus	15	(12)
Eleiodoxa	1	(1)
Heterospathe	32	(30) (partial account for Philippines)
Hydriastele	8	(6)
Licuala	108	(c. 80)
Livistona	28	(12)
Oncosperma	5	(4)
Phoenix	17	(2)
Pholidocarpus	6	(6)
**Pigafetta*	1	(1) (living collections suggest possibility of further taxa)
Pinanga	120	(c. 100)
Ptychococcus	7	(6)
Rhapis	12	(1)
**Siphokentia*	2	(2)
**Sommieria*	3	(3)
Veitchia	18	(1)
		c. 756 in Malesia

* = endemic genus, number in brackets indicates number of species recorded for the Flora Malesiana region.

Malay Peninsula — One of the best known areas, but novelties continue to be collected.

Sumatra — Relatively well collected for palms but yet to be worked up.

Java — Now well known, though until recently the distinctness of the palm flora of East Java from that of West Java was not appreciated.

Borneo — There are still many geographical gaps in palm collections. West Kalimantan and the headwaters of all major rivers remain poorly collected for palms. For Brunei only 19 species were recorded until recent field work brought the total to over 130 species (Dransfield & Wong, unpublished).

Philippines — Recent collecting by Fernando and Madulid has put Philippine palms on a firm footing.

Sulawesi and neighbouring islands — Still very imperfectly known.

Moluccas — Very imperfectly known. Collecting by de Vogel and Mogea shows just how rich individual islands may be.

Lesser Sunda Islands — Although this is not a palm rich area, there are nevertheless a few records showing the presence of rattans and other palms in the montane forests. Most of these collections are sterile. Fieldwork would be expensive and give few returns. Anyone with the opportunity to do general field work in this area is encouraged to collect palms.

Irian Jaya — Dishearteningly poorly known.

Papua New Guinea — Palms have received quite a lot of attention in recent years in Papua New Guinea and there is currently a taxonomic survey of the rattans being undertaken by R. J. Johns and co-workers.

Rattans

The rattans provide a nice reflection of our state of knowledge of Malesian palms as a whole. I estimate there to be about 430 species belonging to 10 genera in the region; thus almost a little less than half of all Malesian palms are rattans, and most of these belong to just two genera, *Calamus* and *Daemonorops*. When these two large and protean genera are monographed then the back of Flora Malesiana palms will be broken. Recent enthusiasm for the development of rattan as a plantation crop has stimulated the collection of herbarium material of these fiercely spiny plants. Although our knowledge of the delimitation of species and their distribution has increased greatly in the last 15 years, the increased collecting has indicated the large number of undescribed taxa. Of the 105 rattans in well-botanised Peninsular Malaysia recognised by me (Dransfield 1979) 10 of these were new species which were described for the preparation of the manual. In Sabah (Dransfield 1982c) of 82 taxa, 21 were described as new and in Sarawak (Dransfield, in press) 16 out of 106. The problem is most acute in the genus *Calamus*, particularly in Sulawesi and the Moluccas, where recent collecting has suggested that up to a third of the rattan taxa there are without names. As many of such recent collections are sterile, these apparently new taxa will have to remain undescribed until fertile collections can be made. The large percentage of undescribed taxa suggests that there may be a similarly large number of new taxa in other groups of the family in areas such as the Moluccas and Irian Jaya. Carefully targeted field work in the eastern part of Indonesia may result in a dramatic increase in our knowledge of Malesian palms. It is fortunate that Johanis Mogea of Bogor has been collecting palms in this area, often making many new records.

Perhaps the most outstanding problem in Malesian palms is the largest genus *Calamus*. It is so large and complex a genus that some division seems desirable. Geographical division – working out the species of discrete geographical areas such as Peninsular Malaysia or Sabah and Sarawak – has been carried out as a result of pragmatism. Accounts of the rattans of these areas were demanded and could be tackled relatively easily, because these areas are generally quite well collected. The concentrated field work in a confined geographical area provided experience of variation of significance far beyond the flora of the one area.

Division based on systematics has a firmer philosophical base but presents several serious problems. To begin with, the genus is too large to be worked on without breaking it down into groups. So far there is no satisfactory subgeneric classification of *Calamus*. The most useful division of the genus is that of Beccari (1908 and 1913), but this is an informal grouping of species and many taxa have been described since then. Furtado's division into sections is useful, but is nomenclaturally suspect and furthermore, although it embraces the whole of the genus, it is based primarily on a knowledge of the Malayan taxa (Furtado 1956). The most recent subdivision is that of Wei (1986), published in Chinese and based purely on the species of Mainland China. From what can be made out by a non-Chinese reader, the division seems to be seriously flawed, not taking into account relationships with other rattan genera of the Calaminae and based on preconceptions of plesiomorphic states in rattans. Nevertheless subgeneric and sectional names are validly published and will have to be taken into account when the whole genus is monographed. In preparing our account of subtribe Calaminae for Genera Palmarum (Uhl & Dransfield 1987), I began to suspect that *Calamus* may represent an artificial residue, after the separation of its satellite genera *Daemonorops, Ceratolobus, Calospatha, Retispatha* and *Pogonotium*. Recent attempts at a cladistic analysis of the whole of the Calamoideae (Dransfield & Uhl, in preparation) show how very closely related these genera, as presently delimited, are.

Work on three groupings of *Calamus* is being carried out at the moment by Edwino Fernando (Los Baños) (Furtado's section *Podocephalus,* which includes several important commercial canes) and Padmi Kramadibrata (Universitas Indonesia and Reading) (Furtado's section *Macropodus* and a group of commercially important canes related to *Calamus inops* Becc. ex Heyne). There is scope for many more workers!

In contrast to *Calamus, Daemonorops* is a genus with greatest diversity in the western part of Malesia and with a sharp decline in number of species encountered east of Sulawesi. There is only one species, *Daemonorops calapparia* recorded from New Guinea. One consequence of this natural distribution is that *Daemonorops* is probably more completely represented in the herbarium than is *Calamus*. The genus is neatly divisible into two sections – section *'Cymbospatha'* (correctly section *Daemonorops*) and section *Piptospatha*. The whole of section *Daemonorops* and one species complex in section *Piptospatha* present serious problems in species delimitation – problems which seem to suggest that we may be dealing with some sort of apomixis. Populations are remarkably uniform within themselves, but neighbouring populations may be subtly and consistently different. A broad species concept would provide the most practicable way of accounting for this variation, but the need for detailed field studies is clearly indicated.

Licuala

Among the non-rattan palms the least well known genus seems to be *Licuala*. Primarily a genus of the forest undergrowth (though with a few subcanopy species) *Licuala* displays great morphological variation, and in some regions is astonishingly diverse. In particular there are large numbers of species in Peninsular Malaysia and Borneo. Vegetative differences, clearly manifest in the field, are less easily appreciated from herbarium specimens, and much of the taxonomy is based on inflorescence and flower details which are not always present in herbarium specimens. Several taxa in Peninsular Malaysia are without names and in Borneo there are numerous novelties, particularly among the massive acaulescent species much used for mat-making. The genus is ripe for monographic study.

Pinanga

Like *Licuala, Pinanga* is a genus which has speciated greatly in the undergrowth of the humid forests of Malesia. The most intriguing problem within the genus is one of sectional delimitation. Despite great diversity in habit and leaf form, inflorescence structure is rather constant. In contrast pollen is perhaps more variable than in any other palm genus. Based on gross morphology there is as yet no obvious way of dividing the genus into sections. At the species level, the greatest problem is the interpretation of species limits in the taxa occurring in the Philippines, where taxa on one island differ slightly but consistently from those of the next island, and in this way paralleling the situation in some groups of *Calamus* and *Daemonorops*. Perhaps the only short-term solution to this problem is to recognise variable species complexes.

ALICE palm database

As an important preliminary to collating data on palms for Flora Malesiana, we have developed a taxon-based database of palms at Kew using the ALICE program developed by Allkin and Winfield. The purpose of this database is to act as a source of information on the synonymy, distribution, uses, and habitat preferences of all palms. In beginning the database we have concentrated on the group for which there is greatest outside demand for information, i.e. the rattans. So far, over 500 names have been included in the database, and the preparation of data has been useful in turning up forgotten and neglected names.

The future of Flora Malesiana palms

Clearly the writing of an account of the palms for Flora Malesiana using a monographic base will require a great amount of time and resources. A more rapid synoptic approach, as has been used in the preparation of local rattan flora accounts, is in many ways a more attractive and pragmatic solution. Such an account would have immediate use and although not as taxonomically sure as a monographically-based account, would indicate where further work was required. However, I strongly believe that even this pragmatic solution is premature. With carefully targetted fieldwork over the next five years herbarium representation of this traditionally neglected family could be brought to a state where a synoptic account would have an immeasurably firmer basis and correspondingly greater validity.

References

Beccari, O. 1908. Asiatic Palms – Lepidocaryeae. Part 1. The species of Calamus. Ann. Roy. Bot. Gdn Calcutta 11: 1–518.

Beccari, O. 1913. Asiatic Palms – Lepidocaryeae. Supplement to part 1. The species of Calamus. Ann. Roy. Bot. Gdn Calcutta 11 (Appendix): 1–142.

Dransfield, J. 1970. Studies in the Malayan palms Eugeissona and Johannesteijsmannia. PhD Thesis. University of Cambridge.

Dransfield, J. 1972a. The genus Borassodendron (Palmae) in Malesia. Reinwardtia 8: 351–363.

Dransfield, J. 1972b. The genus Johannesteijsmannia H.E. Moore Jr. Gdns' Bull., Singapore 26: 63–83.

Dransfield, J. 1978. The genus Maxburretia (Palmae). Gentes Herb. 11: 187–199.

Dransfield, J. 1979a. A manual of the rattans of the Malay Peninsula. Malay. Forest Rec. No. 29. Forest Department, Kuala Lumpur.

Dransfield, J. 1979b. A monograph of the genus Ceratolobus (Palmae). Kew Bull. 34: 1–33.

Dransfield, J. 1979c. Retispatha, a new Bornean rattan genus (Palmae: Lepidocaryoideae). Kew Bull. 34: 529–536.

Dransfield, J. 1980. Pogonotium (Palmae: Lepidocaryoideae), a new genus related to Daemonorops. Kew Bull. 34: 761–768.

Dransfield, J. 1981. A synopsis of Korthalsia (Palmae: Lepidocaryoideae). Kew Bull. 36: 163–194.

Dransfield, J. 1982a. A reassessment of the genera Plectocomiopsis, Myrialepis and Bejaudia (Palmae: Lepidocaryoideae). Kew Bull. 37: 237–254.

Dransfield, J. 1982b. Pogonotium moorei, a new species from Sarawak. Principes 26: 174–177.

Dransfield, J. 1982c. The rattans of Sabah. Sabah Forest Rec. No 13. Forest Department, Sabah.
Dransfield, J. 1986. A guide to collecting palms. Ann. Missouri Bot. Gdn 73: 166–176.
Dransfield, J. In press. The rattans of Sarawak. Royal Botanic Gardens Kew and Forest Department, Kuching, Sarawak.
Essig, F.B. 1975. Brassiophoenix schumannii (Palmae). Principes 19: 100–103.
Essig, F.B. 1978. A revision of the genus Ptychosperma. Allertonia 1: 415–478.
Essig, F.B. 1980. The genus Orania in New Guinea. Lyonia 1: 211–233.
Essig, F.B. 1982. A synopsis of the genus Gulubia. Principes 26: 159–173.
Essig, F.B., & B.E. Young. 1985. A reconsideration of Gronophyllum and Nengella (Arecoideae). Principes 29: 129–137.
Fernando, E.S. 1983. A revision of the genus Nenga. Principes 27: 55–70.
Furtado, C.X. 1956. Palmae Malesicae XIX: The genus Calamus in the Malayan Peninsula. Gdns' Bull., Singapore 15: 32–265.
Harries, H.C. 1978. The evolution, dissemination and classification of Cocos nucifera L. Bot. Rev. 44: 265–319.
Kiew, R. 1976. The genus Iguanura Bl. (Palmae). Gdns' Bull., Singapore 28: 191–226.
Kiew, R. 1979. New species and records of Iguanura (Palmae) from Sarawak and Thailand. Kew Bull. 34: 143–145.
Madulid, D.A. 1981. A monograph of Plectocomia (Palmae: Lepidocaryoideae). Kalikasan 10: 1–94.
Moore, H.E. 1969. A synopsis of the genus Physokentia (Palmae-Arecoideae). Principes 13: 120–136.
Moore, H.E. 1970. The genus Rhopaloblaste (Palmae). Principes 14: 75–62.
Moore, H.E., & J. Dransfield. 1979. The typification of Linnaean palms. Taxon 28: 59–70.
Rauwerdink, J.B. 1986. An essay on Metroxylon, the sago palm. Principes 30: 165–180.
Uhl, N.W., & J. Dransfield. 1987. Genera Palmarum, a classification of palms based on the work of H.E. Moore Jr. L.H. Bailey Hortorium and the International Palm Society, Allen Press, Kansas.
Wei, F.N. 1986. A study on the genus Calamus of China. Guihaia 6: 17–40.

4 Progress in the study of Malesian Bambusoideae

E.A. WIDJAJA

Summary

Although the utilisation of bamboos has been well established in the Malesian region, intensive taxonomic studies are still urgently needed because the available treatises are inadequate to clear up the generic and specific delimitation in the light of newly accumulated evidence. Experience derived from revising one of the major Malesian genera indicated that more field observations of bamboos in Malesia are needed. A review of present knowledge on some aspects of bamboo biology (such as systematic analyses of anatomical and chemical properties, ecology and biogeography) indicates the magnitude of the research to be performed to complete a modern taxonomic account of Malesian bamboos, which would also meet current demands from industrial sectors. Some specific problems related to taxonomic delimitation of Malesian Bambusoideae are discussed.

Introduction

Bamboos and man have been in close association from time immemorial, and in Japan and China the numerous uses of bamboos have been well documented since 3000 BC. Unfortunately there is no comparable information for the Malesian region but one may safely assume that the traditional uses of bamboos in this area have started in prehistoric times. At present their uses for handicrafts, food, household utensils and construction are still frequently met with. In recent years these utilisations tend to increase in line with the introduction of newly developed technology. In Indonesia, for example, the establishment of shoot canning industries, the export of factory processed bamboo satay skewers, toothpicks and chopsticks make it necessary to plant bamboos on a larger scale to meet the increasing demands. Moreover there is now a tendency for the use of more and more bamboos in horticulture and landscape architecture as ornamental plants or for fancy fences.

In recent years more and more studies are being conducted to provide a scientific back-up for the rational utilisation of bamboos by local people. Numerous scientists, many private companies, and international agencies have shown more attention to the development of bamboos in order to know better their biological properties. Consequently a number of professional bamboo societies have sprung up to cater for these interests.

P. Baas et al. (eds.), The Plant Diversity of Malesia, 27–32.

Although bamboos represent a dominant feature in the Malesian scenery, their conspicuous presence is not a guarantee that we have an adequate scientific knowledge about them. As a matter of fact the bamboos have suffered from neglect by plant collectors and taxonomists, so that it is still possible to discover new species in a botanically well explored island such as Java. In the following an attempt will be made to assess how much work has been done and how much more will be needed before we can say that we know enough about the Malesian Bambusoideae.

State of the art of Malesian bamboo biology

The Proceedings of the Asian Bamboo Research Workshop convened by IDRC in Singapore (Lessard & Chouinard 1980) partially summarises research performed in Asia up to 1980. Since then special attention has been given to the supply and production of bamboos to meet ever increasing demands. Therefore renewed interest in studies of bamboo silviculture (incl. propagation, fertilisation techniques, and pest and disease control) and management have arisen especially in the Philippines and in Indonesia (Eusobio 1989; Dali 1980). Studies have also been undertaken on the differences in physical properties of bamboo cultivated on hill slopes and in valleys (Soeprayitno *et al.* 1989). These have been followed by further research on the physical and mechanical properties of different species of bamboos, i.e., their moisture content, specific gravity, modulus of rupture, modulus of elasticity and their modulus of strength. Some correlations have been established between these physico-mechanical properties of Malesian bamboos and their anatomical features (Widjaja & Risyad 1987).

Recently bamboo anatomy has received much attention, especially to provide useful data for bamboo taxonomy. Karelstschicoff (1868) was one of the earliest anatomists who studied the leaf anatomy of bambusoid grasses. There are a number of studies on the usefulness of the bamboo leaf anatomy and culm epidermis for identification (Tabrani *et al.* 1989; Musa *et al.* 1989). The most important uses of leaf anatomy of bamboo have been in elucidating the lines of evolution in this subfamily of the grasses (Soderstrom 1985). The study of the structure of bamboo fruits, including the details of their ovary and embryos was initiated by Stapf in 1904 and has recently been taken up again by Rudall and Dransfield (1989) on the fruits of Malesian *Dinochloa.*

The architecture of some Malesian bamboo rhizomes and culm branches which affect growth habit and ability to propagate vegetatively have been analysed by Wong (1988). It has been suggested that the woody bamboos which occur in a wider range of latitudes exhibit more ecological diversity than those occurring in restricted areas. This group (which mainly consists of *Dinochloa, Racemobambos, Nastus, Schizostachyum, Bambusa, Dendrocalamus,* and *Gigantochloa*) maintained the secondary basic number $x = 12$. According to Soderstrom (1981) the last three genera are hexaploid, and arose from the diploid herbaceous bamboos.

Phenolic, protein and isozyme characters of bamboos have received only little attention, even though they are very useful in measuring relationships between taxa of Bambusoideae. Chou *et al.* (1984) and Kumari *et al.* (1985) have shown the usefulness of isozyme patterns to measure the relationships between species of *Bambusa* in Taiwan and India. This should also be studied to elucidate the relationships of members of *Bambusa* in the Malesian region. Protein and phenolic analyses have proved to be useful to measure the similarity of species of Malesian *Gigantochloa* (Widjaja & Lester 1987).

In Indonesia and the Philippines certain insects visit the flowers of some species of *Gigantochloa, Bambusa* and *Dendrocalamus*. However, their role in the bamboo life history has not been investigated properly.

It is unfortunate that most of the applied studies performed by technologically oriented research establishments have not been backed up by proper identification of the material used, so that their implication for taxonomic purposes should be handled with the utmost care.

Malesian bamboo taxonomy: unending confusion...

The Malesian bamboos have received taxonomic attention since Rumphius described the economic plants of the Moluccas in 1750. However, despite this very early start, and although the number of species occurring in the area probably is not more than 125, Malesian bamboos are still poorly known taxonomically. New species are still to be described from Sumatra, Borneo and other areas, especially if a local bamboo flora was never attempted in the past. Most of the available standard works have been based on a limited region or island (Brown & Fisher 1913; Merrill 1925; Holttum 1958, 1967; Monod de Froideville 1968; Santos 1986; Dransfield 1980, 1981, 1983a, 1983b) and revisions for the whole Malesian region have not been attempted, except perhaps for the preliminary study in the genus *Gigantochloa* (Widjaja 1987).

The major obstacle faced in studying Malesian bamboos is the scarcity of representative collections, and if the specimens are available in abundance they are incomplete by modern standards. Intensive field work and observation often leads to a different conclusion regarding generic or species delimitation, as in the case of the scrambling Malayan bamboo *Racemobambos tesselata,* which has been transferred to *Yushania* by Dransfield (1983a). Another problem not yet solved surrounds the scrambling Malayan *Bambusa pauciflora, B. montana* and *B. klosii* which are closely related, all having slender culms, a prominent leathery sheath girdle and reduced numbers of florets in the spikelets. These features are absent from other species of *Bambusa* and hence the three species are only doubtfully maintained in the genus by Holttum (1958). Similarly

B. magica with its densely and subequal clustered branchings does not fit in the delimitation of *Bambusa* as currently understood (Wong, pers. comm.). In the genus *Schizostachyum*, Gamble (1896) confounded two different species in his description of *S. chilianthum* and accordingly Holttum separated it into two distinct species, i.e. *S. gracile* and *S. zollingeri*. The caryopsis of *S. chilianthum sensu* Kurz, originally drawn from the specimen seen by Kurz from Sumatra, did not have the same structure as that described by Holttum for either *S. gracile, S. tenue* or *S. zollingeri*. Therefore further study on the *S. gracile–tenue–chilianthum–zollingeri* complex is needed. The same problem will also be encountered in *S. jaculans*, as well as in *S. lima, S. blumei, S. iraten,* and *S. biflorum* from the adjacent Indonesian islands.

In the case of the Philippine bamboos, there are a number of taxonomic problems which can be solved only by prolonged and extensive observation of field characters. The uncertain identity of 'bayog' which was previously included in *Dendrocalamus merrillianus* needs to be studied further by comparing it with *Bambusa blumeana* and its varieties as well as with the Chinese spiny bamboos. Due to its spiny character and long rachilla on its spikelets, 'bayog' certainly cannot be included in *Dendrocalamus*. The identity of 'laak', the most useful bamboo in Mindanao for banana props, needs to be clarified further, because it has been generally referred to as *Guadua philippinensis* which has recently been transferred to *Sphaerobambos* by Dransfield (1989). Since *Sphaerobambos* is a scrambling genus, it seems that the stout and erect-culmed 'laak' has been wrongly identified. The records of *Arundinaria niitakayamensis* and *Pseudostachyum polymorphum* from the Philippines require confirmation because the two genera are not yet known to occur in Malesia. The taxonomic status of *Schizostachyum palawanense, S. textorium, S. toppingii, S. curranii, S. diffusum, S. dielsianum,* and *S. fenixii* which have climbing habits need reinvestigation because in many respects the morphological characters (climbing or zigzag habit of the culms, presence of a girdle which can be seen clearly when young, and the peculiar fruit structure) suggest affinity with other genera than *Schizostachyum*. Some of the Philippine bamboos included in *Dinochloa* have two florets, whereas the current generic concept of *Dinochloa* only allows the presence of one floret. Since the fruit structure of some of the Philippine *Dinochloa* species is also distinctive, a complete morphological and anatomical comparative study involving many species of other closely related genera is necessary. If more field work could be undertaken, more new species could be described to accommodate common species such as the expensive ornamental bamboo originally collected from the wild at Batan Island (North of Luzon). In this respect comparison with the Formosan bamboo flora will be useful.

For the Papua New Guinea bamboos more attention should be given to climbing *Bambusa microcephala*. Other climbing *Bambusa* species from the adjacent areas, i.e. from Indonesia (*B. cornuta* and *B. horsfieldii*) and the Philippines [*B. merrillii* and the species originally identified as *B. cornuta* by Gamble (1910) and then reidentified as *B. horsfieldii* by Santos (1986), and the climbing *Cephalostachyum mindorense*], all characterised by auricles having two horn-like appendages and a reduced number of florets in the spikelets, should be studied for comparison. Obviously the generic delimitation of Malesian *Bambusa* requires a thorough overhaul. The scrambling *Racemobambos* and *Nastus* from Papua New Guinea should simultaneously be studied with the Indonesian representatives.

With regard to Indonesia itself, there are a number of species of *Dinochloa*, the scrambling *Racemobambos*, some *Gigantochloa*, *Bambusa* and *Dendrocalamus* from Sulawesi, Borneo and Sumatra which could not be satisfactorily identified due to incomplete specimens. A study of eastern Indonesian *Bambusa atra*, *B. forbesii* and *B. amahussana* variations should be carried out in the field and ample collections should be made.

Looking ahead

From the above account it is evident that to finish a satisfactory taxonomic study, the bamboo taxonomists should get into the field personally to do field observations related to their work, so that complete collections can be built up. It will not be possible to do field work throughout Malesia in a short time, and it is suggested here to achieve this by exploring one island after the other. This field work should be annexed or correlated to other projects such as conservation or productivity of economically important bamboos. The experimental approach to support the classical taxonomy should be encouraged because it will attract more support from others who need data on biochemical properties, cytology, anatomy in relation to physical and mechanical properties and so on.

The above case histories also suggest that when studying bamboos, Malesia should be treated as one entity, because local revisions often leave significant problems unsolved. For the account in Flora Malesiana, it is suggested to do the work genus by genus, although during the work it is necessary to compare other genera to arrive at a sound generic delimitation. We should also be aware of the fact that modern Bambusoideae also include genera such as *Oryza*, *Leersia* and other herbaceous groups.

References

Brown, W.H., & A.F. Fischer. 1913. Philippine bamboos. Bull. Dept. Agric. Nat. Res. Bur. For. Philippines. 15: 9–33.

Chou, C.H., S.S. Shen & Y.H. Hwang. 1984. A biochemical aspect of phylogenetic study of Bambusaceae in Taiwan. I. The genus Bambusa. Bull. Acad. Sinica 25: 177–189.

Dali, Y. 1980. Laporan sementara penelitian pengaruh pemupukan terhadap pertumbuhan bambu di Borisallu (Sulsel). Bagian Silvikultur. Lembaga Penelitian Hutan, Bogor.

Dransfield, S. 1980. Bamboo taxonomy in the Indo-Malesian region. In: G. A. Lessard & A. Chouinard (eds.), Bamboo research in Asia: 121–130.

Dransfield, S. 1981. The genus Dinochloa (Gramineae–Bambusoideae) in Sabah. Kew Bull. 36: 613–633.

Dransfield, S. 1983a. The genus Racemobambos (Gramineae–Bambusoideae). Kew Bull. 37: 661–679.

Dransfield, S. 1983b. Notes on Schizostachyum (Gramineae–Bambusoideae) from Borneo and Sumatra. Kew Bull. 38: 321–332.

Dransfield, S. 1989. Sphaerobambos, a new genus of bamboo (Gramineae–Bambusoideae) from Malesia. Kew Bull. 44: 425–434.

Eusobio, M.A. In press. Bamboo Research and Development Project: its significance to the Philippine Development Programme and its progress. Proc. First Symp. Workshop on Bamboo, Laguna, Febr.–March 1989. In press.

Gamble, J.S. 1896. The Bambusaceae of British India. Ann. Roy. Bot. Gard. Calc. 7: 1–133.

Gamble, J.S. 1910. The bamboos of the Philippine Islands. Philipp. J. Sci. (Bot.) 5: 267–281.

Holttum, R.E. 1958. The bamboos of the Malay Peninsula. Gdns' Bull., Sing. 16: 1–135.

Holttum, R.E. 1967. The bamboos of New Guinea. Kew Bull. 21: 263–292.

Karelstschicoff, S. 1868. Die faltenformigen Verdickungen in den Zellen einiger Gramineen. Bull. Soc Imp. Nat. Moscou 41: 180–190.

Kumari, L., P.R. Reddy & C.A. Jagdish. 1985. Identification of species of Bambusa by electrophoretic pattern of peroxidase. Ind. For. 111: 603–609.

Lessard, G.A., & A. Chouinard. 1980. Bamboo research in Asia. Proc. Bamboo Workshop in Singapore, May 1980. 228 pp.

Merrill, E.D. 1916. On the identity of Blanco's species of Bambusa. Amer. J. Bot. 3: 61–62.

Merrill, E.D. 1925. An enumeration of Philippine flowering plants: 94–102. Manila.

Monod de Froideville, C. 1968. Gramineae. In: C.A. Backer & R.C. Bakhuizen van den Brink Jr, Flora of Java 3: 493–641. Noordhoff, Groningen.

Musa, N., Y.C. Sulistyaningsih & E.A. Widjaja. 1989. Morfologi, anatomi dan taksonomi Bambusa vulgaris koleksi Kebun Raya Bogor. Floribunda 1 (12): 45–47.

Rudall, P., & S. Dransfield. 1989. Fruit structure and development in Dinochloa and Ochlandra (Gramineae–Bambusoideae). Ann. Bot. 63: 29–38.

Santos, J.V. 1986. Philippine bamboos. In: Guide to Philippine flora and fauna. Natural Resources Management Centre & University of the Philippines.

Soderstrom, T.R. 1985. Bamboo systematics: yesterday, today and tomorrow. J. Amer. Bamboo Soc. 6 (1–4): 4–16.

Soeprayitno, T., T.L. Tobing & E.A. Widjaja. In press. Why do the Sundanese of West Java prefer slope-inhabiting Gigantochloa pseudoarundinacea than those growing in the valley? Proc. Int. Bamboo Workshop. Cochin, India, Nov. 1988. In press.

Stapf, O. 1904. On the fruit of Melocanna bambusoides Trin., an endospermless, viviparous genus of Bambuseae. Trans. Linn. Soc. London (Bot.) 6: 401–425.

Tabrani, G., A. Setiawan & E.A. Widjaja. 1989. Anatomi buluh jenis-jenis Schizostachyum koleksi Kebun Raya Bogor. Floribunda 1 (11): 41–44.

Widjaja, E.A. 1987. A revision of Malesian Gigantochloa (Poaceae–Bambusoideae). Reinwardtia 10: 291–380.

Widjaja, E.A., & R.N. Lester. 1987. Experimental taxonomy of the Gigantochloa atter–Gigantochloa pseudoarundinacea complex. Reinwardtia 10: 281–290.

Widjaja, E.A., & Z. Risyad. 1987. Anatomical properties of some bamboos utilized in Indonesia. In: A.N. Rao, G. Dhanarajan, C.B. Sastry (eds.), Recent Research on Bamboos: 244–249.

Wong, K.M. 1988. The growth architecture and ecology of some tropical bamboos. Paper presented at International Bamboo Conference 88, Prafance, June 1988.

5 New evidence for the reconciliation of floral organisation in Pandanaceae with normal angiosperm patterns

BENJAMIN C. STONE

Summary

New evidence is provided to support the hypothesis that floral organisation in Pandanaceae is derived from a ground plan that is not observable at the macroscopic level. When SEM micrographs of floral primordia are studied, the existence of primordia for organs which can be identified by position as tepals and bracteoles can be seen at least in *Freycinetia*. Moreover, primordia representing the non-functional sexual organs may be routinely present. It appears that suppression of subsequent development of these primordia may be responsible for the organisation of the perianthless, unisexual flowers prevalent in Pandanaceae. The floral structure as ascertained from the earliest primordial stages, however, is consistent with the normal type in Monocotyledonae, and need not be interpreted as a proto-flower or pseudoflower, a concept which has been suggested. Discussion of connected topics including sexually mosaic inflorescences, cycles of alternating sexuality, phylogenetic reconstructions, pistillodia and carpellodia, and the peculiar staminal structure termed herein the pendiculus found in some groups of *Pandanus* is presented with the aim of stimulating relevant future research.

Introduction

Uncertainty about the fundamental floral structure in the Pandanaceae has existed for a long time. Reasons for this were the highly reduced state of these flowers, their occurrence in condensed inflorescences (condensed apparently at the floral, partial inflorescence, and inflorescence levels), coupled with the inadequate materials representing early stages of development and the sparse representation of staminate inflorescences in herbaria. The difficulty of ascertaining the precise relationships of closely aggregated floral parts, the ephemeral nature of the staminate structures, and the extremely hard, dense osseous endocarps of fruits, probably contributed as well. New evidence is presented here which, in conjunction with the studies by Huynh (1982) and Cox (1981, 1982, 1983a) appear to demonstrate conclusively that flowers in certain species of the family are entirely compatible with the general construction of angiosperm, and in particular monocotyledonous, flowers. Early developmental stages of certain species of *Freycinetia* show distinct primordia of perianth parts and bracteoles,

P. Baas et al. (eds.), The Plant Diversity of Malesia, 33–55.

although they are suppressed and fail to develop beyond a microscopic size. Their position, however, strongly supports the conclusion that they represent the organs in question. Moreover, through anatomical features of the stamens, the orientation of the latter can be discerned, leading to the recognition of discrete staminal groups which can be rationally equated with definite androecia. This interpretation can be verified by the identification of pistillodes in some species. The evidence suggests that the generalised pandanaceous floral pattern is 'lilialean' with organs in triads. The hypothesis proposed here is that evolutionary developmental alterations have led in some groups to reduction, and in others to increase in organ number; and sometimes to extensive fusion (connation). The presence of 'monstrous' organs in some cases yields evidence that can be used in the interpretation of floral structure. Recently some examples of bisexual individual plants of *Freycinetia* species have been discovered, and even individual spikes of one inflorescence have been found to be mosaics of pistillate and staminate flowers, but still there has never been in the entire family a demonstrable case of bisexuality at the level of the individual flower.

Retrospective

Prior interpretation of Pandanaceous floral structure has varied from one author to another but there have been three basic notions:

1) The unisexual-derived hypothesis (Solms 1878; Warburg 1900; Stone 1968), which regards the flower as unisexual through abortion ('flores abortu unisexuales') and either lacking a perianth or, in *Sararanga,* having a rudimentary perigonium of 'undefined' character; having several to many stamens in staminate flowers; and several to many carpels in pistillate flowers. Staminodia and pistillodia may be present, but are in general rare, except in *Freycinetia.* Whether explicitly expressed or not, this hypothesis assumes that Pandanaceae are monocotyledonous angiosperms, and their ancestry is ultimately the same as that of the Angiospermae as a whole.

2) The primitively unisexual hypothesis (Meeuse 1975, 1980) in which flowers of extant Pandanaceae with staminodes or carpellodes are either merely developmental errors or signify an 'incipient stage' of a process of bisexualisation. "At least some recent groups which are diclinous, aphananthous, and anemophilous certainly cannot have had monoclinous and entomophilous ancestors, and retained the condition prevailing among cycadophytinous gymnosperms. Examples ... are ... Pandanales ... (etc.)" (Meeuse 1980). According to this hypothesis the Pandanalean flower would be an anthoid, and "... a number of magnoliophytic taxa are primarily aphananthous (e.g. palms, amentiferous groups, Urticales) and if they are zoophilous, have secondarily acquired this form of pollination e.g. by the advent of extrafloral

semaphylls (Euphorbiaceae, ... *Freycinetia, Dichromena*..." (Meeuse 1976). This hypothesis is linked to the idea that floral organisation ('the' flower) evolved during the early history of the Angiospermae and was not an existing character of the Angiospermae at their beginning.

3) An 'undecided' view, not amounting to an hypothesis, but which permits 'operational' designations of floral organs without requiring the context of a floral-structure definition. This point of view leaves unresolved the homology of organs and rejects the definition of the cupule in *Sararanga* as a form or vestige of perianth. It is based on ideas suggested or reintroduced by R. Sattler in a number of papers. From this viewpoint, Pandanalean floral structure is to be demonstrated, if at all, from new evidence, and the ancestral structure should be deduced therefrom and not accepted from earlier authorities, principles, or traditions.

New evidence

Freycinetia — In older descriptions the flowers in this genus were defined as organ groups based on the occurrence of gynoecia, which in a number of species were accompanied by staminodia (Warburg 1900: 16, fig. 7D). In the staminate spikes, the occurrence of rudimentary gynoecia (pistillodia) would similarly, but imperfectly, define the staminate flowers; imperfectly, because the technique of assessing the orientation of the stamens had not been developed, and thus the association of each stamen with a particular gynoecium rudiment was very uncertain (Warburg 1900: 26). In practice, specimens lacking staminodia and pistillodia – in fact, the usual situation – could not be used to ascertain the full definition, including meristics, of flowers *per se* in either species or genus. While variation in carpel number per gynoecium was rather readily established, the stamen number per flower generally was not.

Moreover, in the absence of any macroscopic evidence of perianth or bracts, except the major external bracts of the inflorescence proper, some doubt remained as to the status of floral make-up in general. When staminodia occur, they often lack anthers, leading to doubt as to their homology.

New sources of evidence help to elucidate the construction of the *Freycinetia* flower. Cox has revealed that in spikes examined in the very early developmental stage, primordia can be seen at high magnification (by SEM) that are consistent in position and number with incipient floral organs, and that in two species (*Freycinetia arborea* Gaudich. and *F. reineckei* Warb.) primordia of both bracteoles and tepals, as well as carpels and stamens, can be recognised. These primordia, after their appearance, with a dimension of about 100 μm, then develop differentially, with complete suppression of the bracteole and tepal primordia, and depending on the overall sex of the spike, partial or complete

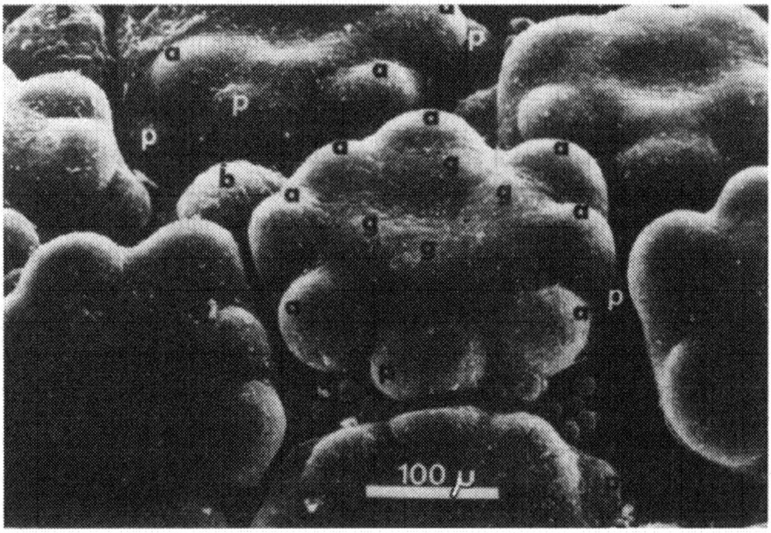

Figure 5.1 — *Freycinetia arborea.* SEM micrograph of a staminate spike showing gynoecial (pistillode) primordia (g), stamen primordia (a), tepal primordia (p), and bracteole primordia (b). From a preparation by P. A. Cox.

suppression of the androecial or gynoecial primordia. These primordial stages are shown for *Freycinetia arborea* (Figure 5.1) in SEM micrographs provided by P.A. Cox.

Further evidence in support comes from collections of a New Caledonian species, *Freycinetia spectabilis* Solms, made in 1981. This species, which is extremely similar to *F. cylindracea* Solms and closely related to both *F. coriacea* Warb. and *F. comptonii* Rendle, was found toward the summit of the Mont des Sources at c. 1000 m altitude. The staminate spikes had white bracts (flowering 14 August). Details of the flowers are shown in Figure 5.2; a and h show the leaves; in b, the spike is shown.

In Figure 5.2g, two staminate flowers, one with six stamens and one with nine stamens, and each with a distinct pistillode, are shown. No trace of perianth or bract rudiment is evident, but the relative position and absolute number of stamens relative to one pistillode is unambiguous. Figure 5.2e shows a single stamen in dorsal (abaxial) view revealing the crystal cells of the connective; these occur only on this side, thus forming a reliable indication of stamen orientation; 2f shows the adaxial side. Figure 5.2d shows a flower in top view, with seven stamens surrounding the pistillode. The external position of the connective crystal cells demonstrates that these seven stamens with their pistillode form an integral unit which must be equated with a single flower. From this evidence it appears that stamen number is variable, but the range is not large. Existing data from other specimens corroborates this and also the fact that carpel number also varies but also has a definite range.

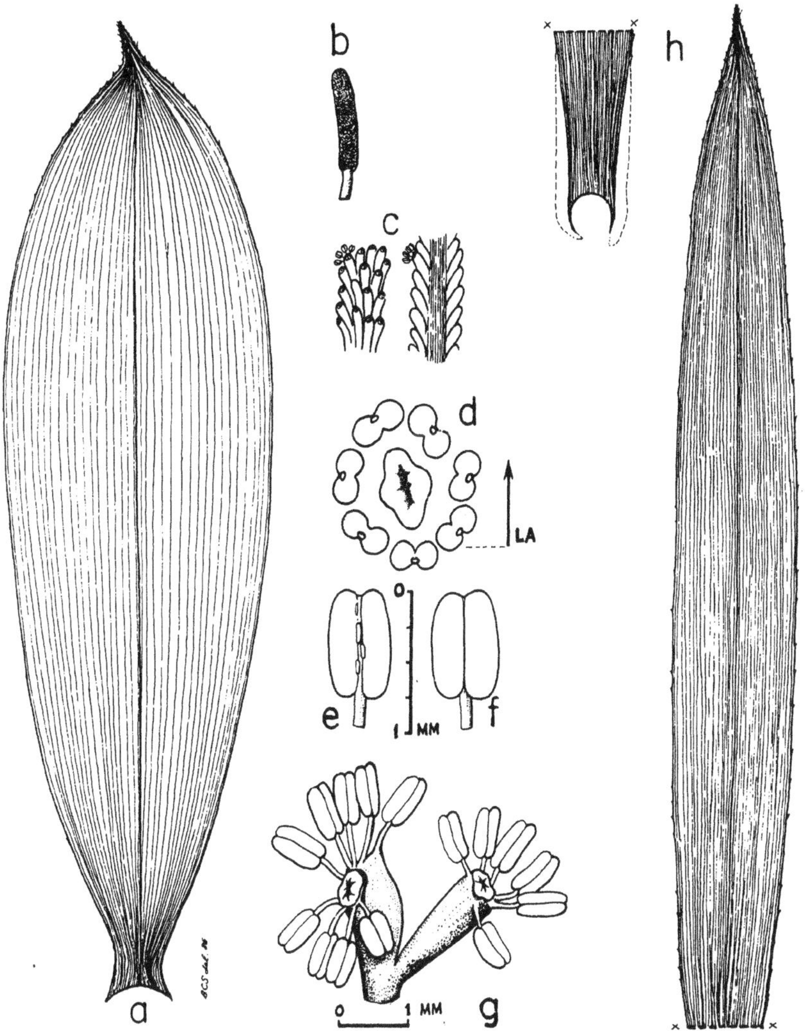

Figure 5.2 — *Freycinetia spectabilis*. a. Leaf (from *Stone 14985*); b. staminate spike; c. portion of staminate spike near base, surface view (left) and longisection view (right); d. plan of a single flower; LA = longitudinal axis of spike; e, f. stamens; abaxial view (e) shows crystal cells in connective; g. two flowers, showing central pistillodia; h. leaf (base showing auricles as dotted lines) (*Stone 14981*).

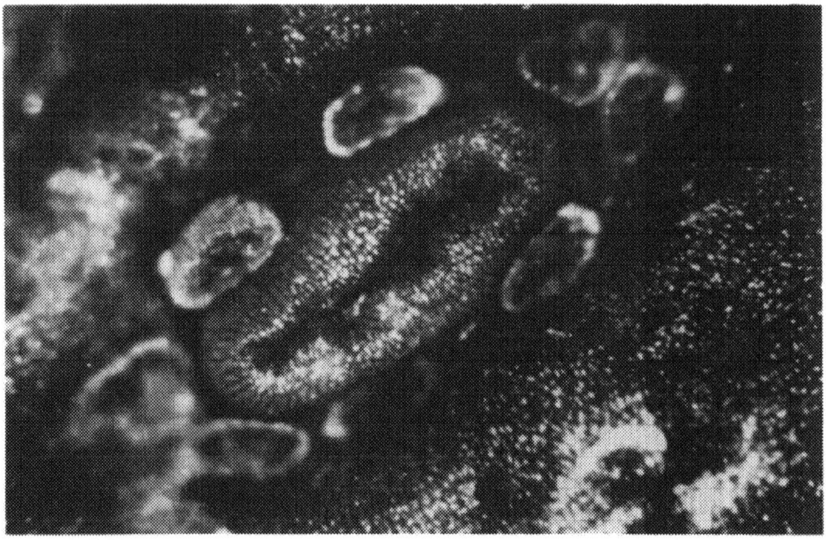

Figure 5.3 — *Freycinetia negrosensis*. LM micrograph of pistillate spike, showing developing flower; central gynoecium surrounded by 15 staminodia. Photographed by R. Sattler.

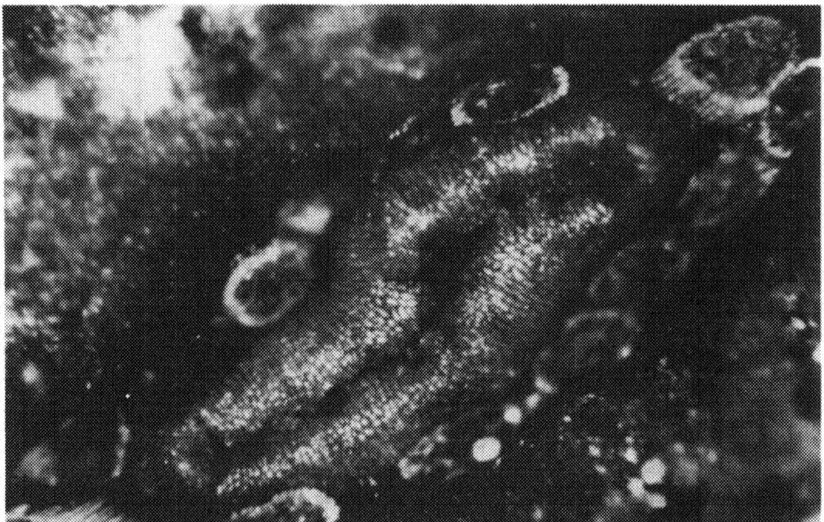

Figure 5.4 — *Freycinetia negrosensis*. LM micrograph of pistillate spike, showing developing flower with gynoecium surrounded by 8 staminodia. Photographed by R. Sattler.

In the absence of earlier developmental stages the occurrence of tepal and bract primordia cannot be recorded but on the analogy of the known primordia in *F. arborea* the floral units in *F. spectabilis* appear acceptable. Furthermore, the structures are readily equated with those in such species of *Pandanus* as *P. cominsii* Hemsl., which has been found with staminodia, and *P. barklyi* Balf. f. which has been found with pistillodia.

Inflorescences of *Freycinetia* at an early developmental stage, viewed under a high magnification with the light microscope, may clearly reveal staminodia in the pistillate spike although tepal and bracteole primordia are not visible. An example of this is shown in Figures 5.3 and 5.4. These show a developing gynoecium with adjacent staminodia from a pistillate spike of *Freycinetia negrosensis* Merr. (Sect. *Pristophyllae*) of the Philippines (voucher specimen is *Stone 12199*). These are LM micrographs (taken by R. Sattler) and show that at this stage the tepal primordia are no longer observable. However, the definite association of a specific group of androecial elements with a particular gynoecium is manifest.

Theoretical significance

The evidence described above supports the acceptance of the first hypothesis summarised earlier, that the unisexual aspect of the Pandanaceous flower is a derived condition; that in structure, the flower can be reconciled with a normal Monocotyledonous floral structure, and for that matter, with a generalised Angiosperm flower. The occurrence of primordia appears to indicate that earlier, now extinct, members of the family, could have had macroscopic perianth parts and bracteoles. The present perianthless state is neither an indication that perianth parts did not exist in earlier stages of the existence of the family (or order) nor that perianth parts cannot 'reappear' in future derivatives. In certain species, the occasional occurrence of bracteoles is known (*Pandanus cominsii*), and the occurrence of bracts on tertiary branches of the inflorescence is known in *Pandanus halleorum* Stone (in staminate plants).

Staminodia and carpellodia

One of the most interesting taxa in Pandanaceae in regard to the occurrence of staminodia and carpellodia is *Pandanus* subgenus *Martellidendron*. Huynh (1981) has recently reviewed the taxonomy and morphology of this group and described a fourth species, *P. kariangensis* Huynh (all occur in Madagascar). The noteworthy features of this subgenus are the constant occurrence of pistillodia in the staminate flowers. Moreover, the staminate flower is further characterised (at least in *P. kariangensis*) by a cycle of staminodia (Huynh terms them 'onglets') on the apex of the staminal column, within which are the normal stamens. The pistillode is generally distigmatic but sometimes has only one or from three to six stigmas. The pistillode is unilocular, but the single chamber is probably the result of suppressed growth of the tissues which, in a functional female, would develop into the endocarp, placentas, and ovules, and there seems no basis to suppose (as Huynh does) that a bilocular ovary would develop from an initially unilocular one. The fruits, in those cases where they are known, are invariably bilocular and 2-seeded. Perhaps it should be asked whether the two stigmas are really half-stigmas; if so, the pistillode is truly unilocular. The staminodes are peculiar in having stomata elevated above the epidermal surface. However, Huynh states that there are certain stamens with a structure intermediate between normal stamens and these 'onglets' which thus can be classed as staminodes, even though they totally lack vascularisation.

Unfortunately there are no collections of any species of subgenus *Martellidendron* of pistillate flowering plants at a stage that would show a pistillate flower before or at the time of anthesis. Only more or less ripe fruiting specimens are known of the pistillate plants. Consequently fine details of pistillate floral structure remain unknown. (This is in fact an extremely common situation in Pandanaceae, where the vast majority of all collections are of partly to fully ripe pistillate plants.)

Nonetheless, the flower in subgenus *Martellidendron* may approach more closely than any other example in the family to the hypothetical bisexual, tepalous condition which has been abandoned by nearly all other members of the family still extant.

Huynh (1983: 191) has proposed for usual tests of identity of staminodia and pistillodia the occurrence of certain anatomical characters:

Staminodia: presence of a single vascular bundle and absence of any fibres.

Pistillodia: presence of several vascular bundles and several strands of fibres each associated with a vascular bundle.

These anatomical criteria permit recognition of the two categories of aborted floral sexual organs. However, problems may arise when the structures are of a 'mixed nature"'or when vascular bundles and fibers are entirely absent.

If petals may originate from staminodia or from later members of the calyx cycle the 'onglets' of *Pandanus kariangensis* may be examples of the former mode of origin.

Sexual organs of 'dual' nature

Sexual organs of a dual nature may be found in staminate spikes of *Pandanus gibbsianus* Martelli, a species of Borneo. In this example, the specimen is probably an aberration or 'monstrosity'. A specimen (*T. van Niel 4400,* 17.6.1967, from Brunei) shows many carpellodia among the stamens, but they are not uniform in either form or position. Many intermediate organs also occur,

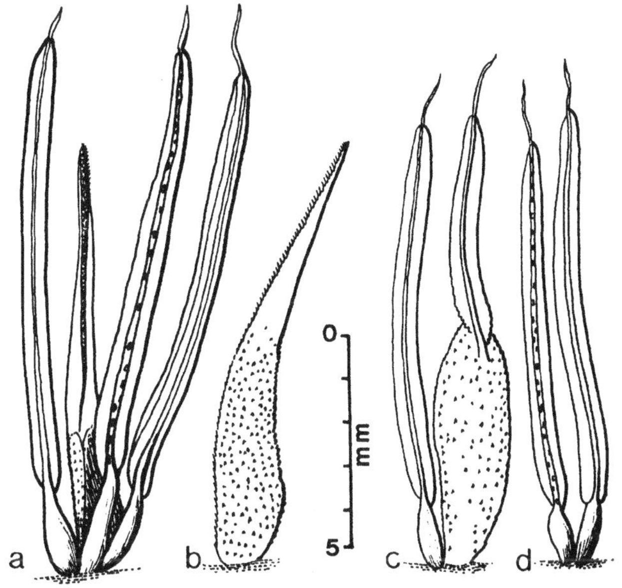

Figure 5.5 — *Pandanus gibbsianus*. Organs from a staminate spike (*van Niel 4400*). a. Staminate flower with 3 stamens and one carpellode; b. carpellode (from a) in side view (note squamulae on the lower (ovarian) portion, and the distinct stylar apex with well-developed papillose stigma); c. staminate flower with one normal stamen and one stamen of 'dual nature' with functional polleniferous apex (but shortened anther) and carpellodiform base; d. normal stamens in abaxial (left) and adaxial view.

transitional between normal stamens and 'normal' carpellodia. Figure 5.5 shows some of these organs. Figure 5.5a shows a flower with 3 normal stamens and one carpellode; the latter has no obvious trace of staminoid features. Figure 5.5b shows this same carpellode in lateral view, with its spinescent style, prominent stigmatic papillae, and the scabrid lower surface of the 'ovarian' portion (these features being characteristic of both normal carpels and carpellodia). Figure 5.5c shows another flower, but here the structure is reduced to one normal stamen and one organ with a dual nature, having a well-developed anther with connective and apiculus, and normal polleniferous thecae, but with an enlarged filament which resembles the 'ovarian' portion of a carpellode, complete with squamulae on the surface. Figure 5.5d shows normal stamens in both abaxial (left) and adaxial (right) views, demonstrating the unilateral occurrence of the raphidophorous cells along the connective. In some other examples from the same specimen, flowers with two carpellodia and one stamen, or three stamens and one carpellode, or other combinations of stamens and carpellodia, and with or without organs of a dual nature, were noted.

Despite these irregularities it is clear that the spike itself was functionally staminate. No functional gynoecia were found. Over half of the organs were normal polleniferous stamens, and less than a fifth of the organs were carpellodia. The remainder includes the 'mixed' organs of a dual nature, many of which were functionally stamens, although pollen yield ranged from normal to reduced to nil owing to the often shorter, or incomplete thecae. Even the most 'normal' of the carpellodia entirely lacked ovules. Many contained no internal chamber at all, nor any histological differentiation. Moreover the overall appearance of the inflorescence was manifestly staminate, with the open, extended spiciform plan, well separated major bracts and spikes, and normal size and position of the spikes.

This example must be contrasted with the specimens of *Freycinetia* mentioned earlier, because they showed no organs of a 'dual nature' and because in the *Pandanus gibbsianus* flowers, the carpellodia and mixed organs must be regarded as aberrant, probably not to be repeated on other individuals. But both cases further demonstrate the constraints of structure in the Pandanaceous flower, and show that it is necessary to distinguish between non-functional sexual organs which are normal components and those which are aberrations. In this way, the possibility is recognised that vestigial organs may be morphologically similar to aberrant organs yet have different implications for morphology, morphogenesis, and phylogeny.

The only other example of this kind is discussed by Huynh (1983); it occurred in a species of Reunion Island (*P. borbonicus*).

Mosaic inflorescences

Cox (1980, 1981, 1984) has shown that in *Freycinetia reineckei* the occurrence of sexually mosaic inflorescences can be shown to have an adaptive value in relation to vertebrate (bat and bird) pollinators. While such cases include no bisexual flowers, the hypothetical benefit of bisexual flowers in relation to these vertebrate pollinators is almost certainly nil. This is because of the tendency for the pollinators (especially bats) to damage and consume the staminate spikes as well as the fleshy bracts. In completely pistillate inflorescences, bract consumption occurs but much less interest is shown in the pistillate spikes. Mosaic spikes (in which zones of pistillate flowers are mixed with zones of staminate flowers on the same spike) show as expected an intermediate focus of feeding interest, and intermediate levels of seed set (dependent on intact gynoecia); but these levels are high enough to justify the repeated occurrence of such mosaics.

The occurrence of another example of mosaic inflorescences is therefore of interest, as it involves a different species and another geographic locality. *Freycinetia scabripes* Warb., a Philippine species, can now be reported as a further

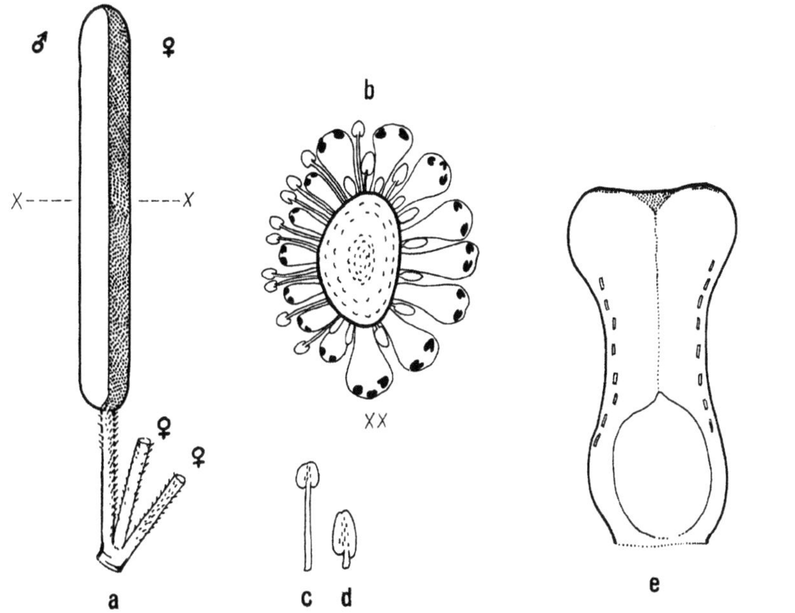

Figure 5.6 — *Freycinetia scabripes*. Inflorescence with a sexually mosaic spike. a. Inflorescence with the two fully pistillate spikes removed except for their peduncles; staminate zone of the mosaic is exterior; b. section (at x–x) of mosaic spike; staminate portion at left; c. normal stamen; d. staminode; e. pistillode. From *PNH 117425*, Samar, Philippines.

example of the phenomenon. A collection made in Samar (Mt Apoy), in May 1969, by H. Gutierrez *et al.* (*PNH 117425*) has a ternate pistillate inflorescence with two spikes completely pistillate and the third divided longitudinally into two regions, an adaxial fully pistillate zone, and an abaxial (exterior) zone of functionally staminate flowers. This spike is shown in Figure 5.6a. Note that the peduncles of the two fully pistillate spikes are bristly all round, as is normal for the species; but the peduncle of the mosaic spike is differentially bristly, its inner (adaxial) face, corresponding to the pistillate adaxial zone of the spike, being much more densely bristly than the abaxial face.

In the transverse section of the spike, Figure 5.6b, the staminate zone is seen to have short pistillodia and long stamens, while the pistillate zone has short staminodia and long, normal gynoecia.

For the present there is no evidence concerning pollinators for *Freycinetia scabripes*, but as a working hypothesis it may be suggested that the occurrence of the sexually mosaic inflorescence strongly implicates a comparable vertebrate pollination syndrome like that demonstrated for *F. reineckei*. In addition it can

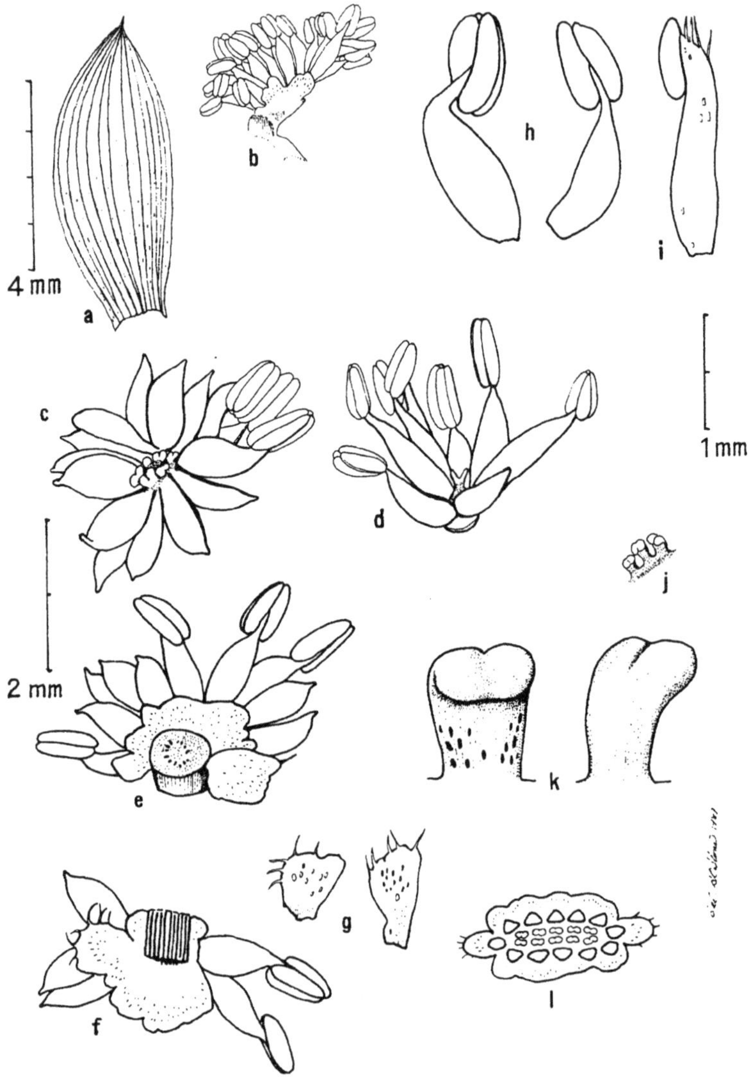

Figure 5.7 — *Sararanga sinuosa.* Details of staminate flowers. a. Bract from base of 4th order branch of inflorescence; b. a staminate flower; c. staminate flower (most anthers removed for clarity), showing central pistillodia; d. another staminate flower showing pistillodia; e. staminate flower *in situ,* showing cupule; cut end of rachis in centre; f. another view showing view of cupule from below; g. cupule lobes with raphidophorous cells and marginal spinules; h. two normal stamens; i. a stamen with the filament enhanced by dorsal spinules; j. isolated pistillodia; k. a pistillode in adaxial (l) and abaxial views; l. transverse section of a flower above the level of the cupule. Note spinules on the lobes on left and right. Pistillodia central, bilobate. Twelve stamens present. From *Whitmore BSIP 18105*, Solomon Islands.

be mentioned that the example of *Freycinetia negrosensis* Merr. (see Cox 1981; Stone 1972b) which (at least occasionally) is known to produce staminate and pistillate inflorescences on the same individual plant, probably is to be regarded as an adjunct phenomenon and related also to vertebrate pollination.

Cycles of sexuality

Poppendieck (1987) has shown that two other species, *Freycinetia cumingiana* Gaudich. (also a Philippine species) and *F. funicularis* (Lam.) Merr., under conditions of cultivation at least, definitely are capable of the production of both staminate and pistillate inflorescences from the same individual. In these cases, staminate inflorescences are associated with certain distinguishable branches, and there appears to be an alteration of staminate and pistillate states, in different years. If this phenomenon proves to be the rule for most *Freycinetia* species (a suggestion far from proven) it would perhaps be an additional generic character, as so far nothing of the kind has been found in either *Pandanus* or *Sararanga*. Such facultative or cyclic sexuality would be of considerable significance in relation to the dispersal and establishment of *Freycinetia*. In terms of floral structure, however, *Freycinetia* seems to have retained the fundamental unisexuality characteristic of the rest of the family. The inference may be drawn that the inflorescence *per se* is now the effective unit of adaptation in *Freycinetia*, and that the differential benefits of unisexuality and bisexuality in relation to pollination, fruit and seed set and dispersal, are no longer relevant at the level of the flower as a morphological unit. If this is the case in *Freycinetia*, perhaps the same question should be posed apropos of *Pandanus* and *Sararanga*. The achievement of unisexuality at the level of the flower, coupled with the overall specialisation along two different lines of the pistillate and staminate inflorescences, has perhaps bypassed any evolutionary opportunities afforded by bisexual flowers. If *Freycinetia* is exemplary, adaptations involving returns to some form of at least facultative bisexuality are not likely to invoke a return to bisexual flowers, but rather should involve various kinds of higher-level changes either within a single spike, within a particular inflorescence, or on a single individual.

Functionally the results may be equivalent, but the long term phylogenetic results will be very different.

Pistillodia in flowers of *Sararanga*

In *Sararanga sinuosa* Hemsl. (Solomon Islands), staminate flowers may have pistillodia. Investigation of these flowers has also provided some other evidence relating to stamens and the so-called 'cupule'. In Figure 5.7, some of the details are displayed. Figure 5.7b shows a single staminate flower with the irregularly lobate 'cupule' which is shown also in 7f, viewed from below, with a short segment of the peduncle axis (parallel lines); some of the individual lobes are seen to advantage also in 7e, and separately in 7g. Note that they possess raphidophorous cells and marginal spinules. Compare also the aberrant stamen in Figure 5.7i, which has spinules along the back of the filament; ordinary stamens are shown in 7h. The pistillodia have bilobed tips, shown in 7c, 7d, 7j, and 7k. Crystal cells in the epidermal tissue are found adaxially, but not abaxially (7k). The apex is bilobed but there is no stigma. In transverse section (Figure 5.7l) the staminate flower is seen to be compressed so that the stamens form a flattened ring, and the pistillodia an inner, concentric, similarly flattened ring. Comparably, the cupule is most clearly lobed at the opposite sides in the plane of compression, while the longer sides have only shallow and rather ambiguous lobation.

In describing these features, the terms 'cupule' and cupule lobes are deliberately neutral; certainly they have the position of a perianth, but their homology is still very doubtful. It seems possible that they are a modified ring of staminodia, as suggested by the similarity of the stamen of Figure 5.7i. It is also perhaps not proven that the pistillodia are homologs of gynoecia, but their position is in favour of such an interpretation. The normal gynoecia however are usually made up of a greater number of carpellary elements.

Reconstructing the phylogeny of floral structure

Some fundamental assumptions are made:

The Pandanaceous flower and inflorescence are essentially compatible with the norms of monocotyledonous Angiospermae, but perianth of any kind is fully suppressed.

There are three clearly distinguishable levels of floral organs: tepals (not clearly distinguishable as petals and sepals, and in no extant species developing beyond the primordial state); androecium (with a cycle of 6 or 3 stamens, but subject to meristic modification whether increase or reduction); gynoecium, consisting of 6 or 3 carpels, subject to meristic modification.

Because the flowers are invariably unisexual, the non-functional sexual structure is at most represented by a sterile, usually dwarfed or otherwise modified homologue, sometimes reduced to non-developing primordia.

Each flower is subtended by a bracteole, but this is normally suppressed and observable only at the stage of primordia. Bracts subtending inflorescence branches are either suppressed or reduced, except at the nodes of the primary axis.

Each carpel is monostigmatic, unilocular, and has parietal placentation; the ovules range from many to one.

The stigma is essentially an everted interior surface set with papillae; the papillae line the surface of the interior as well. In plan, the stigma is thus in the form of an arrowhead or heart, the basal notch being the point of entry into the apex of the carpel. Hence the stigma defines a symmetrical division of the carpel into two mirror-image halves, and the stigma orientation is always the indicator of the location of the placentation.

The fully formed stamen is also bilateral, and its orientation (with respect to the centre of the flower, resp. the gynoecium rudiments) is indicated by: the normally abaxial location of the crystal-cells of the connective and apiculus; the position of the endothecium in the anther; and the location of the fibre bundles along the vascular bundle of the filament. Trends in floral evolution have been along certain lines, but those commonest throughout the entire family are:

Ovules — reduction in number to one per carpel in *Pandanus* and *Sararanga,* retention of higher number or increase in number in *Freycinetia.*

Seeds — development of an indurated epidermis in seeds that remain small (*Freycinetia, Sararanga*) where the endocarp is not osseous. In contrast, in *Pandanus,* the seed epidermis is thin but the fruit has a thick endocarp.

Carpels — connation of carpels, though the locules tend to remain distinct; meristic changes in carpel number, either increase or reduction; specialisation of internal tissues (especially the development of osseous endocarp in *Pandanus,* of aerenchyma, of oil-rich cells, etc.).

Stigmas — stigmatic structure has remained basically the same even as the form and size of the stigmas has changed.

Styles — essentially lacking throughout the family.

Carpellodia / pistillodia — rarely present at macroscopic level, but probably fairly common as primordia, at least in *Freycinetia*; structurally simplified, but often with a rather complete stigma; locules devoid of ovules, usually devoid of other specialised interior tissue (cf. endocarp).

Stamens — the distinction between anther and filament can always be found, but (as discussed more fully below) there are trends to filament connation, to fasciation of stamens, to the extension of the connective both distally and proximally, and minute alterations in anatomical details.

Pollen — grains are always free; there are some alterations in form, which range from perprolate to virtually isodiametric; most changes involve the exine, with psilate, reticulate, and spinulose types commonest; the tectum may be complete or incomplete. The pore may be polar or in a few examples along the polar axis some distance away from the pole.

Staminodia — generally, reduction is the main trend; types with antherodes or without them may be found; often only a remnant of the filament remains. In some cases organs appear by position to be staminodia but are neither morphologically nor anatomically clear homologs. Occurrence of staminodia is probably more regular in species of *Freycinetia* than in the other genera.

Perianth — none in any extant species of the family; the structure in *Saranga* called the perigonium by Warburg is probably a bracteal rudiment, not homologous to a perianth. However, at the earliest developmental stages, primordia in the position of tepals may occur (cf. *Freycinetia*). The possibility that staminodia may (formerly) have been functional equivalents of a perianth, in relation perhaps to pollen vectors, is worth considering.

Calyx — none, and no rudiments or identifiable primordia; see comments above.

There are many more lines of specialisation in regard to floral organs in *Pandanus* than in the other genera. This is of course correlated with the vast number of species in this genus, and is more precisely correlated with the infrageneric taxa, especially at sectional level. A phylogenetic reconstruction of floral evolution within *Pandanus* remains to be done. In contrast, floral structure in *Freycinetia* is comparatively invariable, with differences chiefly in meristics, absolute size, ovule (seed) number, epidermal colour, and details of staminodia and pistillodia. Here the emphasis seems to be on trends involving overall characteristics of the inflorescence. Of these, the most important in some ways would be the development of fleshy bracts, including those which are essentially linear-subulate and may be non-spike-subtending. These bracts are specialised in form, colour, and particularly in tissue content with respect to nutrients and attractant chemicals, and are crucial in the role of vertebrate pollinator attractants, providing visual and olfactory clues and forming an edible reward. In contrast, the bracts in the other two genera are coriaceous, thinner (at least not fleshy), and though often brightly coloured, are not specialised as food bodies, although the youngest may be eaten by (presumably insect) predators.

Stamen morphology in *Pandanus* subgenus *Lophostigma*

(See Figures 5.8 & 5.9)

This group is singled out to display the specialised structure which is here termed the pendiculus (see Figure 5.8). This structure has been illustrated and discussed previously (see for example Huynh 1982; Stone 1974) and has been called the "basal extension of the connective." It is the extension of the connective below the anther, but above the filament, and forms a slender link between them. Like the apiculus, which is the extension of the connective slightly beyond the apex of the anther, the pendiculus is readily distinguished from the true filament by being translucent, normally whitish in dry specimens, and flexible. The apiculus usually differs from the pendiculus (and sometimes from the connective proper) by containing raphidophorous cells, usually asymmetrically located, with the dorsal (abaxial) side often solely furnished with

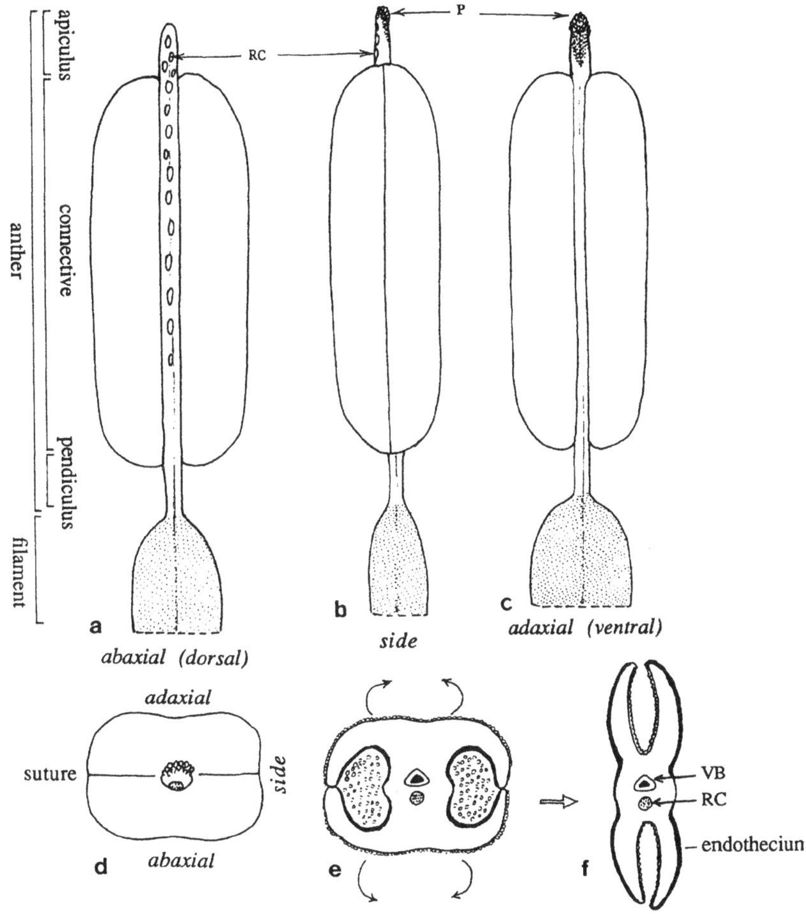

Figure 5.8 — Schematic views of a stamen of the pendicular type as found in *Pandanus* subgenus *Lophostigma,* section *Lophostigma* and section *Maysops.* a. Abaxial view of stamen, showing basal filament, pendiculus, anther and connective (the latter with crystal cells), and apiculus (with crystal cells = RC); b. the same in side view; c. the same in adaxial view; note distribution of papillae (P) on apiculus; d. top view of anther, the abaxial side on the bottom. The same, but in transverse section; pollen grains shown as small circles in pollen chambers; endothecium in black; outer anther epidermis shown as a beadlike margin; arrows show direction of reflexion of thecal lobes; e. anther after dehiscence and loss of pollen, the thecal lobes fully reflexed; note relationship of VB (vascular bundle) and RC (row of crystal cells) in the connective relative to the direction of the false chambers, and the apparently external position of the endothecium after thecal lobe reflexion.

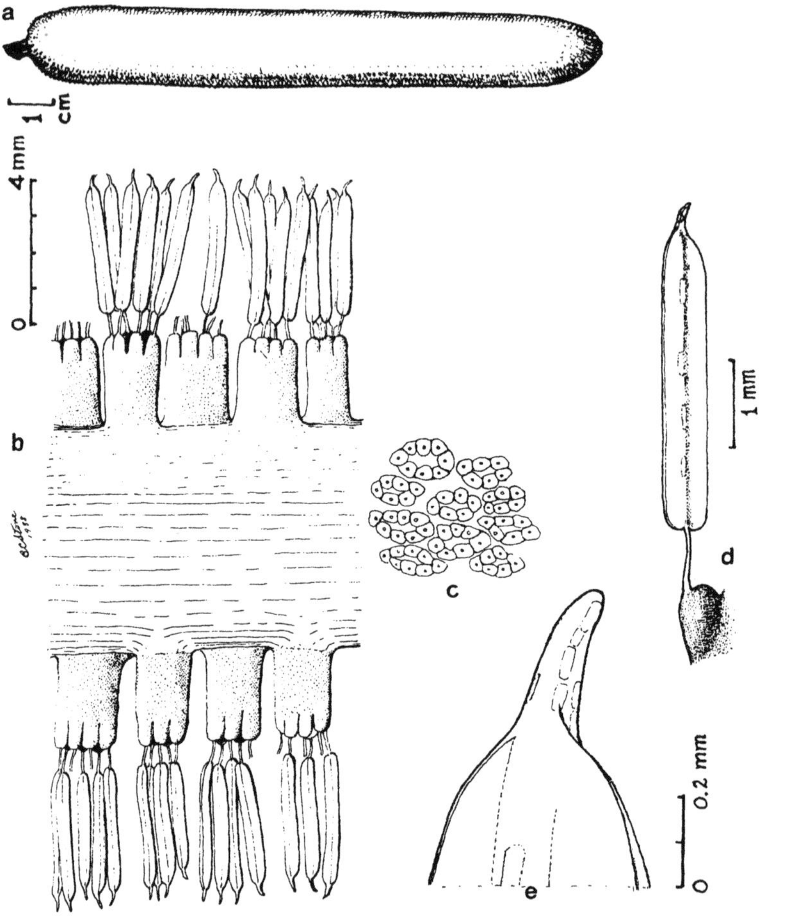

Figure 5.9 — *Pandanus krauelianus*. Staminate spike and flowers of this representative of section *Maysops*. a. One male spike; b. connate filaments and distinct pendiculus of each stamen; c. top view of staminate flowers (stamens from 6–8); d. a single stamen; note raphidophorous cells in connective and on apiculus; e. apex of stamen showing apiculus; note asymmetric distribution of raphidophorous cells (crowded on the adaxial side). From *Kairo 748*.

them; and also in some species by the occurrence of papilliform epidermal cells toward the apex. Figure 5.8 shows a stamen with an apiculus and pendiculus. Stamens of this type are known so far only in *Pandanus* subgenus *Lophostigma,* and especially in section *Maysops* of this subgenus. Often, though not always, flowers in section *Maysops* have the filaments connate into a tube, but the pendiculi are free. The stamens as a result may be easily vibrated.

The four thecae at maturity devolve into two chambers, the separating wall between members of each pair 'disappearing' and leaving only a slight ridge (usually) in the interior (Figure 5.8e, f). The anther opens by full length slits; when the lobes open widely, they may become recurved until they are in contact with the equally recurved member of the opposite pair, and as a result the post-dehiscent anther may expose the locular interiors, while the adjacent lobes in recurved position provide a 'false chamber' in which, often enough, some pollen grains may have become trapped (Figure 5.8f). The true chambers can be distinguished from the false ones by means of the endothecium, which lines the interior of the true chambers; by the orientation of the central vascular strand of the anther, which is accompanied by a fibre bundle along its ventral side; and by some raphidophorous cells scattered along its dorsal side. Thus dorsiventrality can be recognised in even a detached anther. Moreover, there is generally a difference in the epidermis/hypodermis sequence between the inner and outer face of the thecal lobes. In the unopened anther, the thecal lobes are basically parallel to the plane of dorsiventrality as defined by the sequence RC (raphidophorous cells)–VB (vascular bundle)–FB (fibre bundle), although of course the lobe ends may be inwardly curved toward each other. In a fully opened anther with reflexed lobes, the lobes lie almost perpendicular to the axis of dorsiventrality.

The orientation of individual stamens can be determined by these characters, despite distortion.

Polarisation of character states

Determining which states of a character are apomorphic and plesiomorphic should be done in the context of an outgroup. In the case of the Pandanaceae, the outgroup itself is still a controversial matter. Suggested groups include palms (Arecaceae), Cyclanthaceae, and Araceae. Cox, unpubl. (in litt.) has derived a cladogram showing putative relationships of the three genera on the basis of selected characters relating to pollination syndromes, and with Cyclanthaceae as outgroup. Still, little confidence can be placed in outgroup choice, simply because there is such a gulf (extinction) between Pandanaceae and its relatives. Cox's cladogram appears to support the derivation of the wind- or insect-pollinated genera (*Sararanga, Pandanus*) through one clade, with *Freycinetia,* the vertebrate-pollinated genus, constituting a distinct and preceding clade. A more complete cladogram, utilising all characters, not only those of the pollination syndrome, has to be carried out to assess possible alternative pathways.

A tentative selection of staminate characters with suggestions of polarisation follows:

Character	*plesiomorphic state*	*apomorphic state*
Pedicel	+	0
Stemonophore	+	0
Filament	present	suppressed
Filament connation	0	present
Filament epidermis	not papillose	papillose
Anther	elongate	ovate
Anther dehiscence	latrorse	introrse
Anther thecae	subpatent	reflexed
Connective	RC only abaxial	RC ab- & adaxial
Apiculus	present	suppressed
Apiculus papillae	absent	developed
Endothecium	3 – 4-layered	1–2-layered
Pendiculus	absent	present
Pollen tectum	incomplete	complete
Pollen exine	reticulate	psilate or spinulose
Pollen aperture	variable, along polar axis	at pole
Staminodia	present (at least as primordia)	fully suppressed

Such a listing can be, and should be, greatly extended, extending to vegetative as well as reproductive characters.

If the proposed plesiomorphic state is scored 0 and the proposed apomorphic state 1, the scores for *Pandanus, Sararanga,* and *Freycinetia* are not clearly discriminated (0–12, 5, and 6 or 7, respectively).

It may appear, however, that among species of *Pandanus,* there are some which have more plesiomorphic character states than others in respect of these selected staminate characters; and for that matter, some which have more than do species of the other two genera, and some which have more apomorphic character states. Such results may however be only artefacts of an incomplete analysis.

In passing it is worth noting the analysis of the gynoecium in Araceae by Mayo (1989). According to this author the Araceous gynoecium is 2–47-carpellate, and 1-carpellate gynoecia are unknown. Each carpel has its own stylar canal, although there is a common canal (compitum) into which each separate canal opens; the stigmatic epidermis descends into the compitum. But in some cases, the compitum is lacking, and individual stylar canals open onto the stigmatic surface. The latter condition is that found in Pandanaceae. In Pandanaceae, there is apparently no style, but if the style is defined simply as "a zone of tissue between stigma and ovary which possesses a distinct anatomy" it might be an arguable issue. However, as a morphological entity, a style is not found in Pandanaceae.

Prospective

Current evidence suggests that Pandanales (effectively the same as Pandanaceae, if fossil taxa are excluded) are plants with rather markedly derived flowers. These differ from other Monocotyledons in several respects, and from most (but not all) in the absence of macroscopic perianth parts. That the perianth may have been a feature of ancestral or preceding taxa is however suggested by the occurrence of conformably located primordia in some flowers of extant Pandanaceae. The pervasive diclinous condition extends throughout the family, with no exceptions at the level of the individual flower. However, bisexuality of a particular individual plant can be brought about in other ways, and it is now evident that *Freycinetia* at least as to some species has achieved this through two different means; one being the production of staminate and pistillate inflorescences on the same plant, usually in an alternating sequence (as in *F. cumingiana*; see Poppendieck 1987); and the other being the production of sexually mosaic spikes (as in *F. reineckei*; see Cox 1981). The latter condition seems to be a response to vertebrate pollinator behaviour; the former may be also. Such modifications are, however, quite unknown in *Pandanus* and *Sararanga*. They, however, appear to have either entomophilous or anemophilous pollination. Little is known so far, however, of the details. The available evidence suggests that insects, specifically small, rather unspecialised insects, whether beetles or thrips, are pollinators for many species of *Pandanus* (Stone 1976).

The implication of such evidence is that floral patterns in Pandanaceae are derivative from a more generalised prior Angiospermous type, and are not examples of proto-anthoids or other hypothetical stages of organisation of reproductive organs antecedent to the Angiospermae themselves.

Phylogeny, especially in terms of cladistic analysis, should be based, however, on a wide spectrum of character states drawn from vegetative as well as reproductive organs. While there are very rich data from floral structures, the fruits and seeds supply another source of data which should be made use of. The deep gulf between Pandanaceae and putatively close relatives makes a choice of outgroup difficult. However, intra-family analyses can be made. Perhaps one question that needs to be brought up again is the one relating to the integrity of *Pandanus*. In the past, botanists such as Gaudichaud, De Vriese, Kurz, Hasskarl, and perhaps others thought of more than one genus, but since the time of Solms and Warburg, all judgments have been in favour of a single large genus (Stone 1972a). Recent evidence is still in favour of one genus. However, the extraordinary diversity at the infrageneric level indicates how important it is to examine the full extent of this diversity before generalising about the genus. That this is true is emphasised by the peculiar types of floral structure and specialisation shown in such taxa as *Pandanus* subgenera *Martellidendron* and *Stonedendron*. Does the diversity of floral structure in *Pandanus* indicate a longer phylogenetic history than the other two genera have, or is it merely an adjunct of the greater number of species of *Pandanus*? The much wider geographic distribution of *Pandanus* helps to explain the higher number of species, but seems to me also to support the idea of a longer phylogenetic history. If that is true, it is difficult to envision *Freycinetia* as phylogenetically earlier. The remarkable diversity of *Pandanus* in Madagascar at species, sectional, and subgeneric rank, suggests an antiquity tracing back to an early Cretaceous time. The occurrence of subgenus *Martellidendron* in Madagascar (only), with its flowers apparently regularly containing pistillodia in the staminate flowers, is consonant with such antiquity. Yet it appears that in *Freycinetia,* staminodia and pistillodia are more common (occur in more species and with greater regularity) than is the case in *Pandanus* as a whole. Is this a real or imagined conflict of evidence?

References

Cox, P.A. 1980. Flying Fox pollination and the evolution of dioecism: prospectus of a thesis. (Privately published) 1–18.

Cox, P.A. 1981. Bisexuality in the Pandanaceae: New findings in the genus Freycinetia. Biotropica 13 (3): 195–198.

Cox, P.A. 1982. Vertebrate pollination and the maintenance of dioecism in Freycinetia. The American Naturalist 120: 65–80.

Cox, P.A. 1983a. Sex and the single flower. Natural History (New York) 92: 42–50.

Cox, P.A. 1983b. Extinction of the Hawaiian avifauna resulted in a change of pollinators for the ieie, Freycinetia arborea. Oikos 41: 195–199.

Cox, P.A. 1983c. Natural history observations on Samoan bats. Mammalia 47: 519–523.

Cox, P.A. 1984. Chiropterophily and ornithophily in Freycinetia (Pandanaceae) in Samoa. Plant Evolution and Systematics 144: 177–290.

Huynh, K.L. 1981. Pandanus kariangensis (sect. Martellidendron), une espèce nouvelle de Madagascar. Bull. Mus. Nation. d'Hist. Natur., Paris, 4e série, 3, section B, Adansonia 1: 37–55.

Huynh, K.L. 1982. La fleur male de quelques espèces de Pandanus subgen. Lophostigma (Pandanaceae) et sa signification taxonomique, phylogénique, et évolutive. Beitr. z. Biol. d. Pfl. 57: 15–83.

Huynh, K.L. 1983. Pandanus borbonicus (Pandanaceae), une espèce nouvelle de l'île de la Réunion. Candollea 38 (1): 81–103.

Mayo, S. 1989. Observations of gynoecial structure in Philodendron (Araceae). Bot. J. Linn. Soc. 100: 139–172.

Meeuse, A.D.J. 1975. Aspects of the evolution of the Monocotyledons. Acta Bot. Neerl. 24: 421–436.

Meeuse, A.D.J. 1976. Fundamental aspects of evolution of the Magnoliophyta. In: P.K.K. Nair, Glimpses in plant research 3: 82–100.

Meeuse, A.D.J. 1980. Evolution of the Magnoliophyta: Current and dissident viewpoints. In: C.P. Malik (ed.), Annual reviews of plant sciences 2: 393–442.

Poppendieck, H. 1987. Monoecy and sex changes in Freycinetia (Pandanaceae). Ann. Missouri Bot. Gard. 74: 314–320.

Solms, H. Grafen zu. 1878. Monographia Pandanacearum. Linnaea 42, n.s. 8: 1–110.
Stone, B.C. 1968. Morphological studies in Pandanaceae I. Staminodia and pistillodia of Pandanus and their hypothetical significance. Phytomorphology 18: 498–509.
Stone, B.C. 1972a. A reconsideration of the evolutionary status of the family Pandanaceae and its significance in monocotyledon phylogeny. Quart. Review of Biol. 47: 34–45.
Stone, B.C. 1972b. Materials for a monograph of Freycinetia Gaud. (Pandanaceae). XV . The Sumatran species. Fed. Mus. J. (Kuala Lumpur), n.s. (for 1970) 15: 203–207.
Stone, B.C. 1974. Toward an improved infrageneric classification of Pandanus (Pandanaceae). Bot. Jahrb. Syst. 94: 459–540.
Stone, B.C. 1976. On the biogeography of Pandanus (Pandanaceae). Comptes Rendus des Séances de la Société de Biogéographie (Paris) 45: 69–90.
Warburg, O. 1900. Pandanaceae. In: A. Engler, Das Pflanzenreich 3 (IV.9): 1–97.

6 Progress in Elaeocarpaceae

M. J.E. COODE

Summary

Progress in the preparation of an account of Elaeocarpaceae for Flora Malesiana is described. Much of the taxonomic work for the largest genus, *Elaeocarpus*, has already been done by R. Weibel of Geneva. Descriptions and some 'natural' keys exist, but need translation and putting into Flora format. The characters that define the natural groups are reliable but minute; to avert total alienation of users in the field, artificial keys, avoiding most of these minutiae, will also be needed, by islands or groups of islands. To this end, the descriptions are being recorded not just in words but in DELTA format, and Pankhurst's key-making programmes are being tested with a view to producing such keys in a reasonable time-span.

Elaeocarpaceae is a family of small to large trees; a few species are merely shrubby. The family comprises 9 genera with perhaps 500 species. The species all possess valvate sepals, a disk (which may be of various types), an ovary of fused carpels with axile ovules in two rows giving rise to endospermous seeds, and basifixed anthers. Most species have fringed, toothed or lobed petals, but a few species (in different genera) have entire petals and many species of *Sloanea* are apetalous. The family lacks stellate hairs.

Five genera occur in Malesia (Maps 6.1 & 6.2); one (*Sericolea*) is endemic. *Aceratium* is found in the Moluccas, New Guinea, the Solomons, Vanuatu and Queensland; *Dubouzetia* is known from the Moluccas, New Guinea, the Northern Territory and New Caledonia. *Elaeocarpus* is wide-ranging in the Old World but missing from the New. *Sloanea* is wider spread – from northeastern India, southern China, through to Australia, the Neotropics and Madagascar and has perhaps 130 species altogether, more than half of them in the New World.

Of the non-Malesian genera, *Peripentadenia* is a northern Queensland endemic. *Vallea* (related to *Aristotelia*) and *Crinodendron* (related to *Dubouzetia* and *Peripentadenia*) are New World genera; *Aristotelia* has a southern distribution.

These distributions suggest a family origin in Gondwanaland, although no species have survived on the African mainland. The often quoted report from the London Clay needs re-assessment.

P. Baas et al. (eds.), The Plant Diversity of Malesia, 57–61.

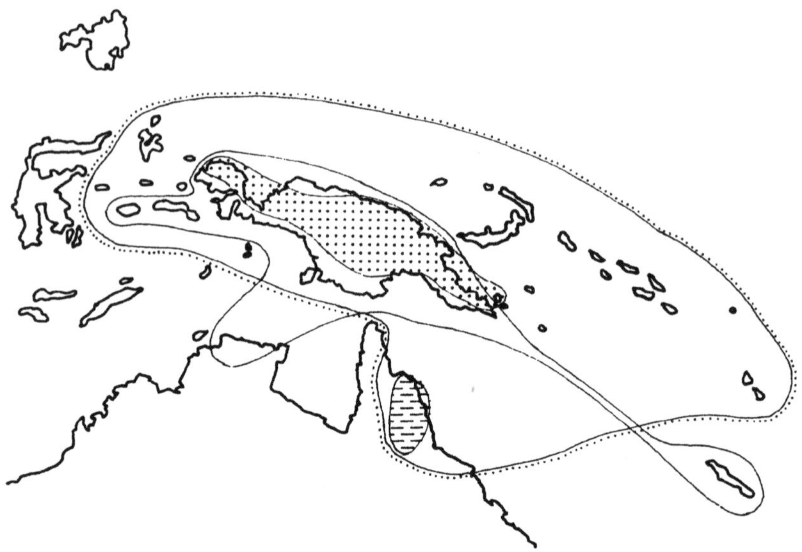

Map 6.1 — Approximate distributions for *Aceratium* (line with dots); *Dubouzetia* (single line); *Peripentadenia* (horizontal dashes); *Sericolea* (dots).

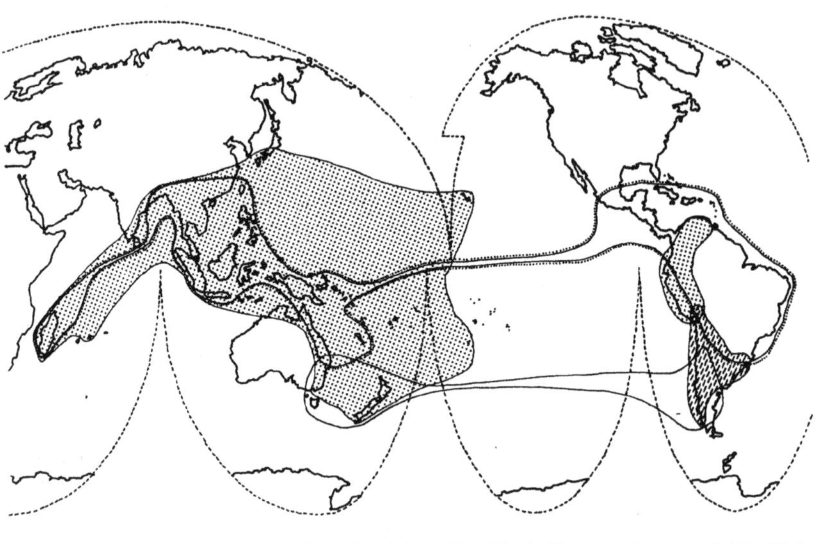

Map 6.2 — Approximate distributions for *Aristotelia* (single line, southern amphi-Pacific); *Crinodendron* (oblique dashes); *Elaeocarpus* (dots – Old World); *Sloanea* (line with dots, equatorial amphi-Pacific); *Vallea* (dots – New World).

Map 6.3. Approximate distribution of *Elaeocarpus* with some species numbers for outlying areas; total number of species above, number of endemics below the oblique lines.

The largest genus is *Elaeocarpus*, with about 350 species (Map 6.3). It can be best divided, perhaps into subgenera, on seed type. This is unfortunate since only about half the specimens have fruit, and less than half of these have properly formed seeds. Most species have straight embryos with broad cotyledons (Figure 6.1, left – this is the type found in all the other genera except *Sericolea* which has somewhat curved seeds); roughly 40% of Malesian *Elaeocarpus* species have strongly curved embryos with narrow cotyledons (Figure 6.1, right; see also Weibel 1968). All 29 species from New Caledonia have straight embryos; Madagascar's 8 have curved embryos (Tirel 1982, 1985) but the two Mauritian endemics have straight ones. The picture is confused by a few seeming intermediates in Australia (Coode 1984).

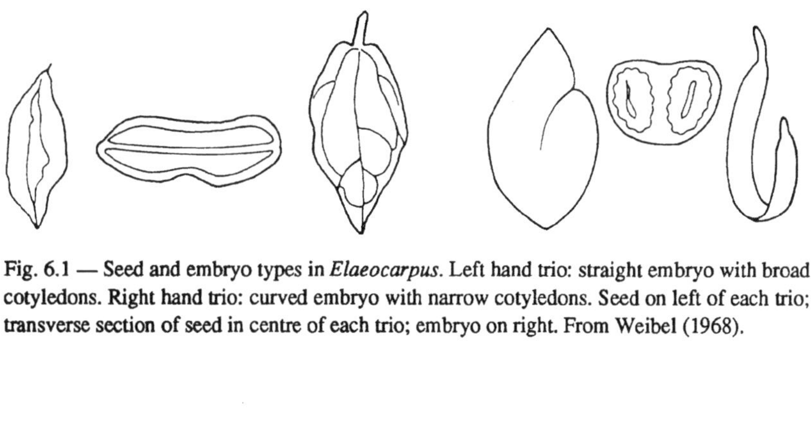

Fig. 6.1 — Seed and embryo types in *Elaeocarpus*. Left hand trio: straight embryo with broad cotyledons. Right hand trio: curved embryo with narrow cotyledons. Seed on left of each trio; transverse section of seed in centre of each trio; embryo on right. From Weibel (1968).

Since the seed characters are shown by relatively few specimens, to place an unknown specimen it is necessary to establish the next most important character – number of ovules per ovary loculus. This, too, is unfortunate, since it can scarcely be determined in the field. In Malesia, those species with curved seeds always have more than 6 ovules per loculus; those with straight seeds can have the whole range possible – 2, 4, 6, 8, or 10–14 ovules per loculus. In Madagascar, however, there are species with curved seeds developed from 3–4-ovulate loculi.

Most species fall fairly easily into a section or group. Some groups are distinctive in both flower and fruit; others are defined on floral characters only.

My involvement with the family began when the late J. S. Womersley asked me to work on them while I was in Lae in the late 1960's. When Marius Jacobs died, I agreed to take over his plan to help Raymond Weibel of Geneva to produce an account of *Elaeocarpus* for Flora Malesiana, while I did *Sloanea* and *Dubouzetia* and M.M.J. van Balgooy *Sericolea* and *Aceratium*. I have published an account of *Sloanea* in the Old World (Coode 1983) and van Balgooy a revision of *Sericolea* (van Balgooy 1982). *Dubouzetia* and *Aceratium* are covered in Coode (1981), so the compilation of Flora Malesiana accounts for these genera should not take too long.

Elaeocarpus is a different story. Weibel has worked on this genus since the war, and has accumulated notes, descriptions and keys, in Latin and French. He has published papers describing new species, and his concepts of known species can be understood from his determination labels. But the rest, particularly the sectional arrangement, remains locked up in his manuscripts. In spite of all this careful, thorough work, the final account for the Flora is likely to be somewhat provisional. In Malesia, there are about 230 species of which about 30 are represented by material inadequate for naming but good enough to demonstrate that they are new. Another c. 30 names are uninterpretable; their types are lost or represented by scraps, and their descriptions inadequate. There are also several complexes or species-groups that defy resolution in the herbarium and need study in the field.

Helping Weibel entails, initially, translation of his descriptions of species, selection of notes suitable for publication in Flora Malesiana and translation of keys to sections and to species within sections, where he has attempted them. I have so far translated about 3/4 of the descriptions and keys. Then, a general bringing up to date with the last ten years' material, well under way, and the gathering of specimen data for the generation of a determination list (for which I am using dBase III+ and have about 3500 records so far). Finally, an overall edit to bring everything into consistency.

But most important contribution will be the provision of artificial keys, island by island, since keys in which the early leads require accurate floral dissection would be impractical for the majority of users.

It took 3 almost full-time years to write keys to the 75-odd New Guinea species in the traditional way and so it seemed probable that it would take too long to write a similar key to all the Malesian species. Therefore I decided to build up a DELTA matrix on a portable computer and use R.J. Pankhurst's KCONI program to construct artificial keys. However, experience now suggests that if the sole aim is to build up keys for traditional publication in a book, DELTA is unsuitable. It takes too long to fill out a matrix and it is less time-consuming to learn the key characters of the species.

However, I am not regretting the decision to use DELTA, since it is useful for identification of unknown or incomplete specimens (using Pankhurst's Onlin5). This must be the way to go in future – the straightjacket of a key printed on paper will be found, in comparison, to be so restricting that such keys will eventually die out.

Also, in time, the intention is to use the matrix for cladistics, after very considerable changes, but here is not the place for consideration of the suitability of DELTA and the currently available programs for identification, key-writing, generation of descriptions and phylogenetic studies.

References

Balgooy, M.M.J. van. 1982. A revision of Sericolea Schlechter (Elaeocarpaceae). Blumea 28: 103–141.

Coode, M.J.E. 1981. In: E.E. Henty (ed.), Elaeocarpaceae in Handbooks of the Flora of Papua New Guinea 2: 38–185. Melbourne.

Coode, M.J.E. 1983. A conspectus of Sloanea (Elaeocarpaceae) in the Old World. Kew Bull. 38: 347–427.

Coode, M.J.E. 1984. Elaeocarpus in Australia and New Zealand. Kew Bull. 39: 509–586.

Tirel, C. 1982. Eléocarpacées. Flore de Nouvelle-Calédonie et Dépendances 11: 3–124.

Tirel, C. 1985. Eléocarpacées. Flore de Madagascar et des Comores, Fam. 125: 3–51.

Weibel, R. 1968. Morphologie de l'embryon et de la graine des Elaeocarpus. Candollea 23: 101–108.

7 Correlations in the germination patterns of Santalacean and other mistletoes

J. KUIJT

Summary

Within a framework of the emerging ecological (rather than merely systematic) content of the concept 'mistletoe', various observations are presented on embryo and seedling structure of the Santalacean mistletoes, i.e., *Phacellaria* and the *'Henslowia'* complex of genera. An attempt is made to interpret, and to extrapolate from, observations on gamocotyly and embryo reduction in the light of certain parallel evolutionary developments in Eremolepidaceae, Loranthaceae, and Viscaceae. It is suggested that cryptocotyly and primary haustoria, and consequently unitunicate radicular apices and adventitious shoot generation from the haustorial disk and/or endophytic system, have evolved in a number of Santalacean mistletoes.

Introduction

The systematic meaning of the word 'mistletoe' has undergone significant changes in the last few decades. Originally extended from European species to all members of what was until recently regarded as a single family, Loranthaceae s.l., the concept became fragmented as workers began to realise that at least Viscaceae and Loranthaceae have distinct evolutionary pedigrees with respect to their branch-parasitic habit. We now recognise that this same ecological niche was also invaded independently by Misodendraceae and some Asian Santalaceae. The typical mistletoe habit thus evolved separately in a number of places in the Santalalean complex; in fact, we cannot close our eyes to the possibility of it having originated more than once in any one natural family of that order. As a result of these shifts in thinking, the concept 'mistletoe' has become transformed from a taxonomic one to an ecological one: it has come to refer largely to an ecological niche, even though its use remains limited to the order Santalales.

In Santalaceae, this niche is occupied by the highly reduced genus *Phacellaria* of continental south-east Asia, and by a complex of genera originally called *Henslowia*, reaching from India to New Guinea and the Solomon Islands and even to north-eastern Queensland (Stauffer 1966; Smith 1951). Two outstanding European systematists concerned themselves with this difficult complex; tragically, both died prematurely, and in each case their last, unfinished papers on the subject were published posthumously (Danser 1940, 1955; Stauffer

P. Baas et al. (eds.), The Plant Diversity of Malesia, 63–72.

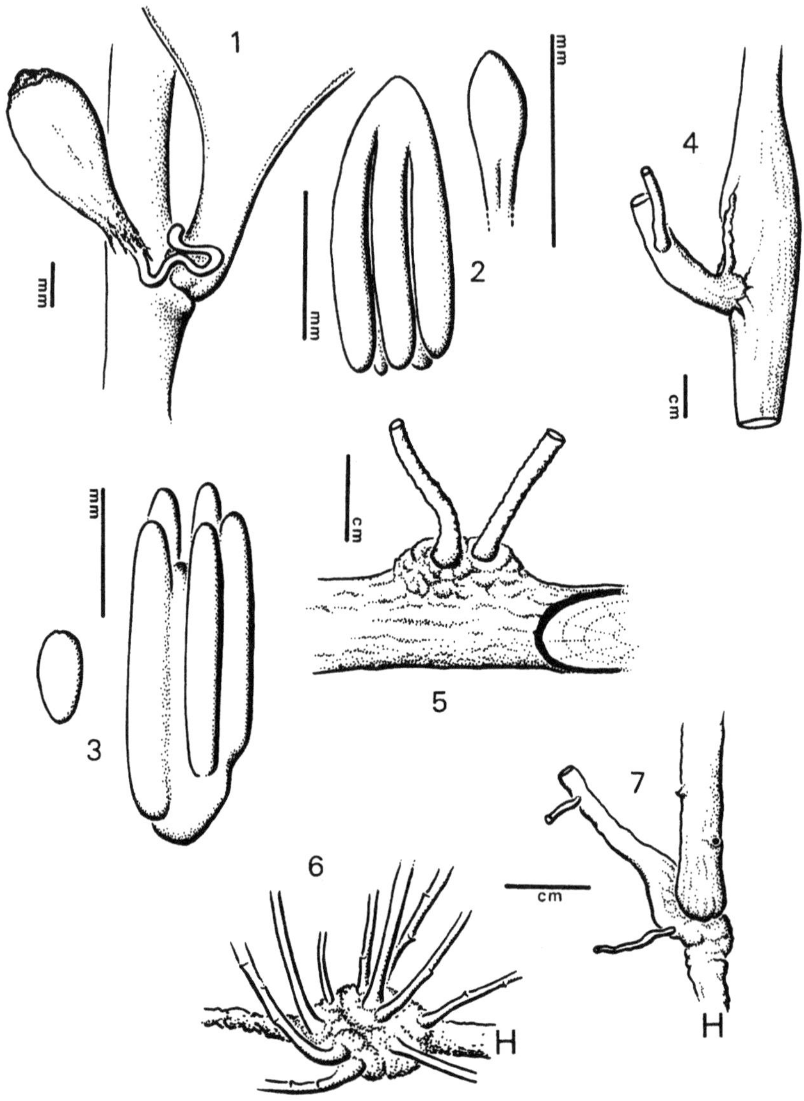

Figures 7.1–7.7 – 1: *Cladomyza dendromyzoides* Stauffer: germinating seedling. *Sleumer & Vink 4484* (L). – 2: *Cladomyza cuneata* Danser: endosperm body (left) and embryo (right). *Stauffer 5633* (Z). – 3: *Phacellaria malayana* Ridley: endosperm body (right) and embryo (left). Malaysia: Gunung Ulu Kali, 1800 m, 8 Jan. 1987. – 4: *Cladomyza uncinata* Danser: basal attachment on host stem, with emergent cortical strand. *Stauffer et al. 5545* (L). – 5: *Cladomyza cuneata* Danser: two stems attached to host stem. *Van Royen 18246* (L). – 6: *Cladomyza erecta* Danser: multiple stem emergence. H = host stems. *Lam 1974* (L). – 7: *Daenikera corallina* Hürlimann & Stauffer: basal attachment to host root (H). *Stauffer et al. 5793* (L).

1966). Since the distinction between the segregate genera accepted by both workers rests on anatomical characters of the fruit, one is left with an uneasy feeling that the systematics of the group is partly unresolved. My focus here, however, is not systematic in the usual sense. Rather, I should like to present a few original observations on seedlings and germination, and explore the extent to which the available information on other mistletoes allows certain extrapolations.

I should first like to submit that the genus *Phacellaria,* notwithstanding its advanced mistletoe habit, may not be closely related to the *Henslowia* complex. The genus seems to be more closely related to *Anthobolus* and *Exocarpos* as indicated especially by its (sometimes branching) spicate, squamate inflorescence (see Stauffer 1959). While we may marvel at the extreme convergence between *Phacellaria* and the Viscacean genus *Arceuthobium,* both in appearance and in the perhaps exclusively adventitious shoot production from the endophytic system (see below), *Phacellaria* does not seem to be out of place in Santalaceae.

No seedling of any Santalacean mistletoe has, to my knowledge, been described or illustrated, although a sketch of a juvenile plant of *Dendrotrophe umbellata* (Blume) Miquel was included in van Steenis' (1933) note [van Steenis claimed that this species is non-parasitic, a statement later drawn in doubt by both Danser (1940) and Stauffer (1966)]. The seedlings of terrestrial Santalaceae generally are poorly known (Kuijt 1978; Stauffer 1959) but may be either epigaeus or hypogaeus.

Observations

The first Santalacean mistletoe to which I want to refer is *Cladomyza dendromyzoides* Stauffer, the seedling of which is radically different from that of terrestrial ones (Figure 7.1). The 'seed' adheres laterally to the host branch by means of a tissue which is fibrous when dry. The radicle elongates to about 4 mm and is sinuous, following the surface contours of the host.

In *Cladomyza cuneata* Danser, secondly, I had available only mature fruits, not seedlings. The seed is completely contained in a nearly spherical mass of hard sclerenchyma which, like the sclerenchyma mentioned for *Phacellaria* below, is morphologically perhaps referable to the mesocarp as in other Santalaceae (see, for example, Bhatnagar 1960, 1965). Within this sclerenchyma, there are 5 elongated, parallel grooves or cavities. These spaces are occupied by as many narrow, fleshy lobes of the endosperm, the lobes being united below by a thickened portion (Figure 7.2). Thus the endosperm body as a whole, which is 2 mm long, has a remarkable but spurious resemblance to many *Psittacanthus* (Loranthaceae) embryos (Kuijt 1967, 1970). Embedded in the thickened part is a minute embryo about 0.7 mm long. Its main portion is ellipsoid to broadly fusiform, but one end extends into two distinct, narrow cotyledons. I have not been able to ascertain whether the cotyledonary tips are completely separate, but they seem to be.

The germinating seeds of *Dendromyza reinwardtiana* which were available to me, thirdly, were not individually attached to the host, but rather formed a tangled mass interconnected by the acicular, fibrous vascular bundles of the ovary wall which, in this genus, take the place of viscin (see Danser 1940: plate 14, fig. 2, and my Figure 7.8). The radicle of some seedlings had reached a length of 3 mm, but could probably have undergone further growth. At its upper end is a slight constriction presumably marking the transition to the hypocotyl. The latter is about 0.3 mm long at this time. Above it, a pair of fused cotyledons is present, marked by a shallow groove on each side of the seedling. Even the 1 mm long tip of the cotyledons is completely fused into a smooth, sharply acute, conical structure lacking any surface grooves.

In *Phacellaria,* where again I had only mature fruits at my disposal, the living portion of the seed is invested in a thin layer of sclerenchyma, and consists largely of an endosperm body which is simple and blunt at the portion turned toward the fruit base (Figure 7.3). Somewhat higher, five (4–6 according to Danser 1940) prominent longitudinal, rounded ridges exist which terminate in blunt, spur-like extensions. The entire structure is about 3 mm long. The embryo is situated in the depression between the spurs, and is 0.7 mm long. It is ellipsoid and completely undifferentiated, each end being smooth and rounded, and is completely submerged in endosperm tissue. It must be emphasised that the fruit is fully mature at this stage, being some 6 mm long.

Discussion

The above observations immediately suggests some further questions in the light of recent observations on other mistletoe seedlings (Kuijt 1982a, 1986, 1989b).

Cryptocotyly and gamocotyly

There can be little doubt that cryptocotyly is an advanced feature in mistletoes and elsewhere. It is not, of course, a consequence of the mistletoe habit in itself, as demonstrated by its existence in innumerable terrestrial plants, including the Santalaceous *Comandra umbellata* (Kuijt 1978) and *Santalum album* (Bhatnagar 1965). Among other mistletoes it exists in the majority of Loranthaceae in the Old World (Docters van Leeuwen 1954), in *Tristerix* and *Ligaria* (Kuijt 1982a), and in transitional forms in *Psittacanthus sonorae* (Watson) Kuijt (Kuijt 1973) and *Desmaria mutabilis* (Poepp. & Endl.) van Tieghem (Kuijt 1985) (see also *Ixocactus* in Kuijt 1987).

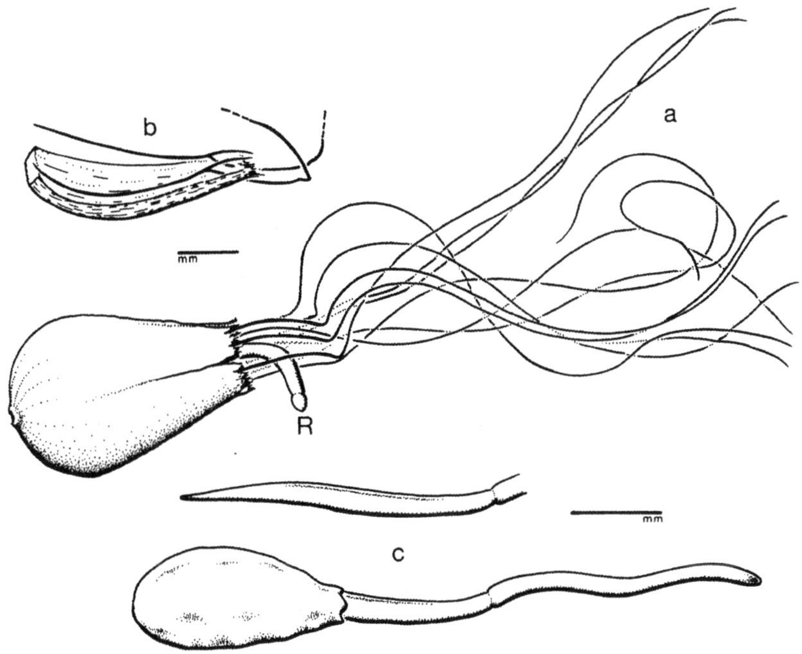

Figure 7.8 — *Dendromyza reinwardtiana* (Blume) Danser: a. germinating fruit (R = radicle); b. section of fruit wall, showing bristle attachment; c. germinating seedling and attached embryo, the fused cotyledons shown separately above. *Stauffer 5618* (Z).

Cryptocotyly as an evolutionary phenomenon in the mistletoes has, in a number of instances, led to further structural changes in the cotyledons. These changes are of two entirely separate kinds. The first possible change is an extreme reduction of the cotyledons as seen especially in *Arceuthobium* (Kuijt 1960) but also in *Viscum minimum* (Kuijt 1986) and *Phoradendron californicum* (Kuijt 1989b). It is a notable fact that this trend, among mistletoes, has so far been documented only in Viscaceae. We cannot say, however, that such cotyledonary reduction is always linked to cryptocotyly, as is seen in other, apparently phanerocotylous mistletoes such as the squamate *Korthalsella* (Stevenson 1934) and *Eubrachion* (Bhandari & Indira 1969) (see also *Lepidoceras peruvianus* Kuijt in Kuijt 1988).

The second possible evolutionary development following cryptocotyly is fusion of cotyledons, either completely or in part. The latter is seen in most Old World Loranthaceae, and elsewhere in *Tristerix* and possibly in *Tripodanthus flagellaris* (Venturelli 1983; but see Kuijt 1982a, where the seedling is described and illustrated as having separate cotyledons). Fusion apparently begins at the cotyledonary tips, but one advanced instance is known of complete fusion, *Tristerix aphyllus* (DC.) Barlow & Wiens (Kuijt 1982a).

From this short survey we can extrapolate with reasonable certainty that *Cladomyza, Dendromyza,* and *Phacellaria* are also cryptocotylar; the observed seedling of the first genus leaves no doubt of this (Figure 7.1) and, as far as the others are concerned, no other gamocotylar mistletoes are known which are not also cryptocotylar. It is, indeed, difficult to visualise how a pair of fused cotyledons would withdraw from the endosperm and lead to any further developments. The fact that the *Phacellaria* seedling is so rudimentary that in reality we cannot speak of fused cotyledons does not affect this argument. These considerations, in turn, suggest yet another avenue of thought.

Implicit in the three known instances of complete cotyledonary fusion (*Tristerix aphyllus,* and the instances of *Dendromyza* and *Phacellaria* here reported) is the fact that there is no epicotyl; at least, the shoot apical meristem is completely missing. What this inevitably means is that the aerial shoot(s), which will be formed later, must originate adventitiously either from the haustorial disk or from the endophyte itself. The former is known elsewhere from several Viscaceae and from *Ixocactus* of Loranthaceae (Kuijt 1986); the latter is most strikingly exemplified in *Arceuthobium* (Viscaceae: Kuijt 1960) but also in *Phacellaria* itself (see photographs in Danser 1939). The generation of adventitious shoots from the endophyte is also seen, of course, in some species with normally functioning epicotyls, as in *Viscum album* L. and most *Tristerix* species. But in *Tristerix aphyllus, Dendromyza,* and *Phacellaria,* such adventitious shoot generation has *replaced* the epicotyl, a phenomenon representing an extremely advanced condition.

Yet another line of thought revolves around the evolution of primary haustoria. It is generally accepted, not only for mistletoes but for parasitic angiosperms generally (Kuijt 1969; Weber 1980), that primary haustoria represent a late evolutionary acquisition. I have argued previously (Kuijt 1969) that, among mistletoes, the acquisition of primary haustoria has paralelled the evolutionary 'ascent of the tree'. In many Loranthaceae, and in some Eremolepidaceae, secondary haustoria are later formed in addition to the primary haustorium from the epicortical roots growing out from the base of the plant and/or its upper shoots (Kuijt 1964, 1982b, 1989a). All cryptocotylar mistletoes known to me (see above; this includes the transitional species mentioned) have also evolved primary haustoria. The extrapolation would thus appear to be justified that *Dendromyza* and *Phacellaria* will also be found to have primary haustoria, even though some of the species in the '*Henslowia* complex' are known also to generate secondary haustoria from shoot-borne epicortical roots (Figure 7.9; Stauffer 1966), a situation completely parallel to *Struthanthus* and certain other Loranthaceae. The fact that *Cladomyza* has also evolved primary haustoria is

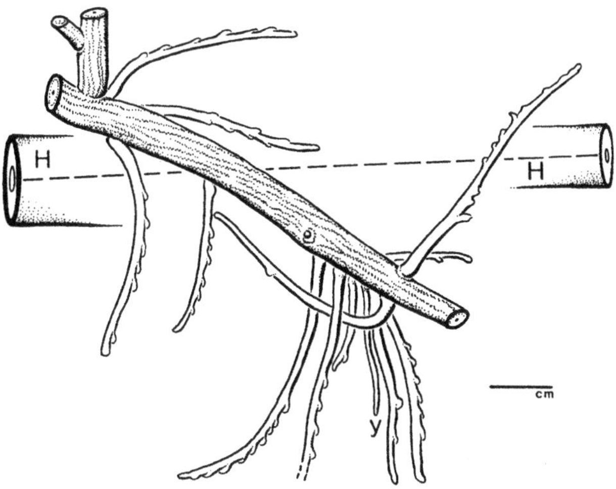

Figure 7.9 — *Dendrotrophe* sp.: portion of a twining stem with haustoria-bearing stem roots drawn as if unwrapped from host stem (H). All root tips except y were sunken in the host tissues. *Sayers 21255* (L).

strongly indicated, though not proven, by the simple attachment of *C. uncinata* Danser on a host stem (Figure 7.4), an attachment identical to that of most Viscaceae.

Nearly all mistletoes which develop primary haustoria have unitunicate radicular apical meristems lacking a rootcap. Possibly the only exception is present in *Psittacanthus* (Kuijt 1967), a specialised situation of no consequence here. This correlation leads me to make one further, very tentative suggestion: it is very likely, although by no means certain, that the radicular apices of *Cladomyza, Dendromyza,* and *Phacellaria* will be found to be unitunicate or nearly so. The striking similarity of the *Cladomyza* seedling illustrated in Figure 7.1, especially to those of *Arceuthobium* (Kuijt 1960: fig. 4) and some *Phoradendron* (Kuijt 1969: figs. 2–28c) makes such a conjecture a reasonable one.

The rare monotypic Santalaceous genus *Daenikera,* endemic to New Caledonia, poses special problems in relation to the above discussion, for two reasons: it has a simple, graft-like haustorial connection, and it may parasitise either underground parts (Figure 7.7) or the erect, above-ground stem of the host (Stauffer 1966: figs. 3 & 4). It is extremely difficult to visualise how such a plant can locate and parasitise a host root when equipped with only a primary haustorium, but the existence of the latter seems to be demonstrated in the mature attachment, which lacks secondary roots completely. No other terrestrial parasitic Santalales are known to have primary haustoria. Yet it is equally difficult to understand how a parasite lacking a primary haustorium (and specialised adhesive tissue) can develop a single-stemmed attachment on an upright host stem as shown in Stauffer's photographs. Clearly, this interesting paradox can be resolved only by careful field studies. Perhaps similar puzzles are posed by *Dendrotrophe varians* (Blume) Miquel which, in South Borneo, is said to be sometimes terrestrial (Hambali 1977); and in *Exocarpos pullei* Pilger, which may parasitise either stems or roots (Lam 1945; Stauffer 1959). Secondary haustoria, however, are known to occur in the latter species.

It needs to be mentioned that some of the above observations on Santalacean mistletoes disagree with those of Hambali (1977). Hambali believes that Santalacean mistletoes are deficient in specialised adhesive tissues around the seed. Indeed, moist viscin tissue as known in Loranthaceae seems to be absent; however, Danser (1940) already described various elaborate hair-like tissues which serve the same function. Furthermore, the present Figure 7.1 leaves no doubt that such tissues in *Cladomyza dendromyzoides* are effective; and the instances mentioned above implying establishment of parasites on vertical host surfaces can scarcely be interpreted otherwise. Hambali's intriguing idea of what we might call, following Atsatt (1965) in a very different situation, 'coordinated dispersal' of mistletoe seeds, must be accorded separate standing. Hambali observed that bird excretions may contain seeds of Loranthaceae and Viscaceae mixed with those of *Hylomyza,* and suggests that the less glutinous seeds of the latter may have a better chance of becoming established when assisted in this fashion by the more efficient viscin of the other mistletoes. This suggestion might well have validity, not only in the coordination of present-day dispersal of mistletoe seeds, but also in the evolution of the mistletoe habit in the past.

Multiple shoot emergence

The close association of two or more mistletoe shoots theoretically may be due to either multiple emergence or conjoint seed dispersal. Certainly the latter is a common phenomenon (see Docters van Leeuwen 1954; Restrepo 1987). In reality, confusion exists only when there are very few mistletoe shoots, as in Figure 7.5 (*Cladomyza cuneata* Danser). When many shoots emerge together, as in Figure 7.6 (*Cladomyza erecta* Danser), the likelihood is that we are concerned with multiple emergence, for a host branch is not likely to survive the simultaneous attack by larger numbers of germinating seedlings, and these seedlings themselves frequently compete strongly among each other. If *Cladomyza* or others indeed are capable of multiple emergence, as in *Phacellaria,* we should characterise this behaviour as a highly advanced feature. However, the meager observations before us can do little more than suggest specific field observations to collectors.

Acknowledgements

The Natural Sciences and Engineering Research Council of Canada supplied continuing financial support. Mr. Gregori Hambali is acknowledged for indispensable support in collecting *Phacellaria*. I am also grateful to Prof. C.D.K. Cook of Zürich for making available materials collected by the late H.U. Stauffer.

References

Atsatt, P.R. 1965. Angiosperm parasite and host: coordinated dispersal. Science 149: 1389–1390.

Bhandari, N.N., & K. Indira. 1969. Studies in the Viscaceae. IV. Embryology of Eubrachion (Hook. et Arn.) Engl. Bot. Notiser 122: 183–203.

Bhatnagar, S.P. 1960. Morphological and embryological studies in the family Santalaceae. IV. Mida salicifolia A. Cunn. Phytomorphology 10: 198–207.

Bhatnagar, S.P. 1965. Studies in angiospermous parasites. No. 2. Santalum album – the Sandalwood Tree. Bull. Nat. Bot. Gard. 112: 1–90.

Danser, B.H. 1939. A revision of the genus Phacellaria (Santalaceae). Blumea 3: 212–235.

Danser, B.H. 1940. On some genera of Santalaceae Osyrideae from the Malay Archipelago, mainly from New Guinea. Nova Guinea, n.s., 4: 133–150.

Danser, B.H. 1955. Supplementary notes on the Santalaceous genera Dendromyza and Cladomyza (with pictures of these genera and Hylomyza). Nova Guinea, n.s., 6: 261–277.

Docters van Leeuwen, W.M. 1954. On the biology of some Javanese Loranthaceae and the role birds play in their life-history. Beaufortia, Misc. Publ. 4: 105–207.

Hambali, G.G. 1977. On mistletoe parasitism. Proc. 6th Asian-Pacific Weed Sci. Soc.: 58–66.

Kuijt, J. 1960. Morphological aspects of parasitism in the dwarf mistletoes (Arceuthobium). Univ. Calif. Publ. Bot. 30: 337–436.

Kuijt, J. 1964. Critical observations on the parasitism of New World mistletoes. Can. J. Bot. 42: 1243–1278.

Kuijt, J. 1967. On the structure and origin of the seedling of Psittacanthus schiedeanus (Loranthaceae). Can. J. Bot. 45: 1497–1506.

Kuijt, J. 1969. The biology of parasitic flowering plants. Univ. of California Press, Berkeley & Los Angeles.

Kuijt, J. 1970. Seedling establishment in Psittacanthus (Loranthaceae). Can. J. Bot. 48: 705–711.

Kuijt, J. 1973. Further evidence for the systematic position of Psittacanthus sonorae (Loranthaceae). Madroño 22: 177–185.

Kuijt, J. 1978. Germination of Comandra (Santalaceae). Madroño 25: 202–204.

Kuijt, J.. 1982a. Seedling morphology and its systematic significance in Loranthaceae of the New World, with supplementary comments on Eremolepidaceae. Bot. Jahrb. 103: 305–342.

Kuijt, J. 1982b. Epicortical roots and vegetative reproduction in Loranthaceae (s.s.) of the New World. Beitr. Biol. Pflanzen 56: 307–316.

Kuijt, J. 1985. Morphology, biology, and systematic relationships of Desmaria (Loranthaceae). Plant Syst. Evol. 151: 121–130.

Kuijt, J. 1986. Observations on establishment and early shoot emergence of Viscum minimum (Viscaceae). Acta Bot. Neerl. 35: 449–456.

Kuijt, J. 1987. Miscellaneous mistletoe notes, 10–19. Brittonia 39: 447–459.

Kuijt, J. 1988. Monograph of the Eremolepidaceae. Syst. Bot. Monographs 18: 1–60.

Kuijt, J. 1989a. Additional notes on the parasitism of New World Loranthaceae. Beitr. Biol. Pflanzen 64: 115–125.

Kuijt, J. 1989b. A note on the germination and establishment of Phoradendron californicum (Viscaceae). Madroño 36: 175–179.

Lam, H.J. 1945. Fragmenta Papuana. Sargentia 5: 1–196.

Restrepo, C. 1987. Aspectos ecologicos de la diseminación de cinco especies de muérdagos por aves. Humboldtia 1: 65–116.

Smith, L.S. 1951. Henslowia – an addition to the Australian genera of Santalaceae. Queensl. Natural. 14: 62–65.

Stauffer, H.U. 1959. Revisio Anthobolearum. Mitt. Bot. Mus. Univ. Zürich 213: 1–260.

Stauffer, H.U. 1966. Santalales-Studien. X. Amphorogyneae, eine neue Tribus der Santalaceae. Viertelj. Naturf. Ges. Zürich 114: 49–76.

Steenis, C.G.G.J. van. 1933. Het geslacht Henslowia op Java. Trop. Natuur 22: 97–99.

Stevenson, G.B. 1934. The life history of the New Zealand species of the parasitic genus Korthalsella. Trans. Proc. Roy. Soc. New Zeal. 64: 175–190.

Venturelli, M. 1983. Estudos embriologicos em Loranthaceae: genero Tripodanthus. Kurtziana 16: 71–90.

Weber, H.C. 1980. Evolution of the secondary haustoria to a primary haustorium in the parasitic Scrophulariaceae/Orobanchaceae. Plant Syst. Evol. 156: 127–131.

8 Tabernaemontana (Apocynaceae): discussion of its delimitation

A. J. M. LEEUWENBERG

Summary

Delimitation of the genus *Tabernaemontana* (Apocynaceae–Plumerioideae–Tabernaemontaneae), which occurs in the tropical belt around the world, changed many times during its history and contributed considerably to the list of synonyms. The present paper gives a survey of the history of this delimitation. Especially the authors who changed the genus concept by making segregates or who reduced generic names to synonyms are mentioned. Afterwards some critical remarks are given.

History of the genus

The genus *Tabernaemontana* was already known to Linnaeus (1753) who based it on two Caribbean species, *T. citrifolia* (confused with *T. divaricata*) and *T. laurifolia,* and the Indian *T. alternifolia* (= *T. divaricata*). Subsequently species were described in *Pandaca* Du Petit Thouars (1806), *Rejoua* Gaudichaud (1826) and *Conopharyngia* G. Don (1837), the first and the third based on the Madagascan *T. retusa* and the second on *T. aurantiaca* from near Irian Jaya, Bawak. G. Don was also the author of the tribe Tabernaemontaneae.

The dispute about the delimitation started with A. De Candolle (1844) and it lasts up to the present (Allorge 1985; P.T. Li 1986; Leeuwenberg 1976, 1984, 1988). De Candolle founded two segregates on American species, viz. *Bonafousia* (*T. undulata*) and *Peschiera* (*T. echinata, T. hystrix* and *T. muricata*). *Tabernaemontana echinata* as well as *T. muricata* were based on mixtures.

Miers (1878) added several species to the two segregates mentioned and to *Tabernaemontana sensu stricto*. Moreover, he founded several more segregates: *Anartia, Codonemma, Merizadenia, Phrissocarpus,* and *Taberna,* all based on American species. However, his book contains many errors, as several of the names given with these genera refer to species not even belonging to the tribe Tabernaemontaneae. He compared these genera with *Stemmadenia,* a good genus of this tribe and two other genera no more belonging there, *Geissospermum* (Plumerieae at present) and *Rhigospira* (Ambelanieae at present).

P. Baas et al. (eds.), The Plant Diversity of Malesia, 73–81.

Baillon (1889) made *Ochronerium* based on the Madagascan *T. humblottii*. Schumann (1896) published the African genus *Gabunia*. Stapf (1902) founded *Ervatamia* on Asian species and *Pterotaberna* on an African one. Furthermore he added several species to *Conopharyngia* and *Gabunia*, and he also raised the tribe to the rank of subfamily.

Markgraf (1935, 1937, 1938) contributed considerably to the list of generic and specific names, firstly for Asia with *Oistanthera, Pagiantha* and *Testudipes* and with new species and new combinations in *Ervatamia* and *Rejoua* and secondly for America with *Stenosolen* and *Taberna* (the latter homonymous with that of Miers; no species in common) and also new species and new combinations in *Anacampta, Anartia, Bonafousia, Peschiera,* and *Tabernaemontana sensu stricto*. He compared the Asian genera with *Eucorymbia* (at present Apocynoideae-Ichnocarpeae) and *Voacanga* and the American genera with *Macoubea* (at present Macoubeeae) and *Stemmadenia*. He reduced Miers's *Codonemma* and *Phrissocarpus* to synonyms of *Anacampta*. A few years later (1941) he published *Domkeocarpa* with *D. pendula* from Cameroun which turned out to be a new synonym of *T. ventricosa*. Woodson erected some months later in the same year *Quadricasaea* with two species from Colombia [both reduced to synonyms of *Bonafousia muelleriana* by Allorge (1985) = *T. macrocalyx*].

Pichon (1948) described three new genera from Madagascar, *Hazunta* (several species, almost all synonyms of *T. coffeoides*), *Muntafara* (only *T. sessilifolia*) and *Pandacastrum* (only *P. saccharatum* = *T. ciliata*). In the same paper he reduced *Peschiera, Bonafousia, Pandaca* and *Gabunia* to the rank of subgenus and he made four new subgenera: *Pupala* (based on *T. calcarea* from Madagascar), *Lepidosiphon* (= genus *Oistanthera*), *Leptopharyngia* and *Sarcopharyngia*. In 1949 Pichon listed *Pandaca, Conopharyngia, Bonafousia, Peschiera, Anacampta, Phrissocarpus, Codonemma, Merizadenia, Anartia, Ochronerium, Gabunia, Oistanthera, Stenosolen,* and *Domkeocarpa* as synonyms of *Tabernaemontana*.

In the same paper he maintained the four Madagascan genera he described in the preceding year and the segregated *Pagiantha* (of which he moved the section *Corymbosae* to *Ervatamia*), *Rejoua, Ervatamia* and *Pterotaberna*. Furthermore he included *Callichilia* in the tribe with the two segregates *Ephippiocarpa* and *Hedranthera* (at present synonyms of *Callichilia*), *Daturicarpa* (a synonym of *Tabernanthe*; Leeuwenberg 1988), *Carvalhoa, Tabernanthe, Schizozygia, Stemmadenia, Crioceras, Calocrater, Voacanga* and with an interrogation mark *Eucorymbia*. He left out *Macoubea*. Moreover, he reduced the level of the subfamily Tabernaemontanoideae again to that of a tribe.

Markgraf (1970) reinstated *Pandaca* to which he assigned *Ochronerium* as synonym and on that occasion he combined all species described in *Tabernaemontana* from Madagascar by Pichon (1948) with this genus. Markgraf (1972) published a fifth genus for Madagascar, *Capuronetta*.

Boiteau in Boiteau & Allorge (1976) completed the long list of generic names all based on African species formerly treated in *Conopharyngia* and *Gabunia*: *Camerunia, Leptopharyngia, Sarcopharyngia,* and *Protogabunia.* The three first were known as infrageneric taxa to Pichon (1948).

Markgraf (1976) made a key to the five genera he kept up in this tribe for Madagascar. Leeuwenberg (1976, 1984) reduced the names which did not yet have the status of synonyms according to any of the earlier authors, *Camerunia, Capuronetta, Hazunta, Leptopharyngia, Muntafara, Pandacastrum, Protogabunia,* and *Sarcopharyngia*; later (1988) he also reduced *Pterotaberna* to a synonym. Independently P.T. Li (1986) did the same with *Ervatamia.*

Allorge (1985) made *Anacampta* and *Quadricasaea* synonyms of *Bonafousia,* maintained *Anartia, Bonafousia, Peschiera,* and *Stenosolen,* mentioned *Stemmadenia,* brought back *Macoubea,* added *Woytkowskia* as a new member and raised the tribe again to the level of subfamily. Leeuwenberg (1984, 1988) preferred to follow G. Don (1837) as for the level of the tribe, accepted *Woytkowskia* here, but not *Macoubea,* nor *Eucorymbia* (respectively Plumerioideae-Macoubeeae and Apocynoideae-Ichnocarpeae at present).

Not discussed are the species once placed in *Tabernaemontana,* but at present classified in other genera assigned to the tribe or other tribes.

Discussion of the segregation

Efforts were undertaken to understand why the above cited authors drew their conclusions as they did. Understanding is hampered by the facts that their observations were partly wrong due to the poor condition of the specimens they studied, and that they frequently mentioned species with the correct specific epithet in one segregate and under a synonym in another, e.g. Miers (1878) gives *T. cymosa* under *Merizadenia, Peschiera* and *Taberna* and *T. flavicans* under *Anartia, Bonafousia* and *Taberna*; Markgraf (1935) *T. pandacaqui* under *Ervatamia, Pagiantha* and *Rejoua,* (1938) *T. muricata* and *T. undulata* under *Anacampta* and *Bonafousia,* and (1976) *T. calcarea* under *Hazunta* and *Pandaca,* and finally *T. ciliata* under *Pandaca* and *Pandacastrum.*

Miers (1878) considered the presence of a disk-like thickening at the base of the ovary known from several American *Tabernaemontana* species a character to keep up the segregates, but his observations were not correct. A true disk separate from the ovary is known from *Tabernanthe* and *Voacanga,* but not from *Tabernaemontana sensu lato.* Furthermore the height of the insertion of the stamens and the shape of the corolla were used.

Markgraf (1935) considered the more or less clear auricles at the base of the corolla lobes useful for distinguishing *Pagiantha* and *Rejoua* from *Ervatamia* and *Testupides.* The clear intrapetiolar stipules inspired him to separate *Pagiantha* from *Rejoua* and the relatively long sepals to keep *Testudipes* apart from

Ervatamia. In Markgraf's key (1938) to the American genera maintained in the Tabernaemontaneae distinctions are very slight. Pichon (1949) restricted *Pagiantha* to the species with relatively thick fleshy corollas and relatively thick leaves. Markgraf (1976) produced a key to the Madagascan genera, *Capuronetta, Hazunta, Muntafara, Pandaca* and *Pandacastrum* which gives partly other characters for them than are mentioned in the descriptions.

When Boiteau described four new genera in Boiteau & Allorge (1976), he raised three infrageneric taxa of Pichon (1948) and distinguished the fourth as completely new. For the distinction he mainly copied Pichon's observations which are partly erroneous, e.g. only one of the species he assigned to *Gabunia* lacks colleters in the calyx, the size of the pistil head may vary more, the sepals show other variation, the indumentum does not show the regularity in variation he supposed and the relative length of petioles and peduncles has not the mentioned correlation.

Allorge (1985) undertook every effort to maintain *Anartia, Bonafousia, Peschiera* and *Stenosolen*. With these efforts the first error was that she supposed that the modules of Tabernaemontaneae may bear one, two or several leaf pairs. Observations of living plants in the field and of specimens preserved in the herbaria show that they may bear one to about ten leaf pairs. The apical pair consists in usually unequal leaves with a more or less gradual variation. In the other pairs the size of the leaves may also be unequal. *Tabernaemontana heterophylla* from America, as the name aptly implies, has a very large size discrepancy between the two leaves of the apical leaf pairs. The same is true for *T. letestui* from Africa. The number of leaf pairs is relatively small in some American species, e.g. in *T. undulata* and *T. disticha* mostly 1 or 2. In most African species the variation is much larger, e.g. in *T. eglandulosa* and *T. glandulosa* they may vary between 2 and 10 in a single branch.

The species the present author keeps in *Tabernaemontana* might be placed in reasonable distinguishable sections based on characters of anthers, pistil head, texture of corolla tube and lobes, texture of leaves, shape of fruit and thickness of fruit wall.

The Asian species (formerly in *Ervatamia* and *Pagiantha*), the African *T. inconspicua* (formerly *Pterotaberna*) and the Madagascan *T. sessilifolia* (formerly *Muntafara*) have a subglobose or cylindrical pistil head without a basal ring and oblong basally cordate anthers. The other species (all American, the African except one, and the Madagascan except one) have a more or less lampshade-shaped pistil head with five more or less pronounced lobes at the apex, five more or less pronounced longitudinal grooves and an entire, undulate or lobed basal ring, and narrowly triangular basally sagittate anthers. *Tabernaemontana aurantiaca* has the pistil head of the first group and the anthers of the second.

Thick fleshy corollas are known from the species placed in *Conopharyngia* and some of *Bonafousia, Pagiantha* and *Pandaca*. The corollas are thin in e.g.

Capuronetta, Hazunta, Peschiera, Stenosolen and most of the species of *Ervatamia*. In this respect they are intermediate in e.g. *Gabunia, Muntafara* and several species of *Anartia, Ervatamia* and *Pandaca.*

Thick leaves are the rule in e.g. *Capuronetta, Conopharyngia* and *Pagiantha,* but may occur in *Bonafousia* and *Pandaca.* Thin leaves are mainly represented in e.g. *Ervatamia, Hazunta, Peschiera, Protogabunia, Stenosolen,* and *Tabernaemontana sensu stricto.* Intermediate quality leaves are present in e.g. species of *Anartia, Bonafousia, Camerunia, Gabunia, Rejoua,* and *Tabernaemontana sensu stricto.*

Subglobose fruits with thick walls are known from *Bonafousia, Conopharyngia, Pagiantha, Pandaca,* and *Rejoua.* Pod-like thin-walled fruits occur in *Ervatamia, Hazunta, Muntafara, Peschiera, Stenosolen,* and *Tabernaemontana sensu stricto.* In both characters intermediate fruits are known from for instance *Anartia, Bonafousia* (*sensu* Allorge), *Camerunia, Gabunia, Pandaca,* and *Peschiera.*

The revision of the tribe Tabernaemontaneae may help to stabilise the delimitation of the genera and consequently also the names of the species.

A subdivision of the tribe

The tribe Tabernaemontaneae comprises 10 genera: 1) *Calocrater* (1 species, Gabon); 2) *Callichilia* (inclusive of *Ephippiocarpa* and *Hedranthera*: 7 species in Africa); 3) *Carvalhoa* (1 species in southeastern Africa); 4) *Crioceras* (1 species in Gabon to Angola); 5) *Schizozygia* (1 species in Central and East Africa); 6) *Stemmadenia* (12–13 species in tropical America); 7) *Tabernaemontana* (about 110 species, worldwide tropics); 8) *Tabernanthe* (inclusive of *Daturicarpa*; 2 species in Central Africa); 9. *Voacanga* (12 species, 7 in tropical Africa and 5 in the Flora Malesiana area); 10) *Woytkowskia* (2 species in tropical South America).

Tabernaemontana has 30 synonyms: *Pandaca* (1806), *Conopharyngia* (1837), *Rejoua* (1829), *Bonafousia* (1844), *Peschiera* (1844), *Taberna* Miers (1878), *Anacampta* (1878), *Phrissocarpus* (1878), *Codonemma* (1878), *Merizadenia* (1878), *Anartia* (1878), *Ochronerium* (1889), *Gabunia* (1896), *Ervatamia* (1902), *Pterotaberna* (1902), *Pagiantha* (1935), *Oistanthera* (1935), *Testudipes* (1935), *Stenosolen* (1937), *Taberna* Mgf. (1938), *Domkeocarpa* (1941), *Quadricasaea* (1941), *Hazunta* (1948), *Muntafara* (1948), *Pandacastrum* (1948), *Capuronetta* (1972), *Sarcopharyngia* (1976), *Camerunia* (1976), *Leptopharyngia* (1976), and *Protogabunia* (1976). Nine of these were reduced by the present author.

The genus *Tabernaemontana* may be subdivided into some weakly defined sections (see Table 8.1). The synonymy given is based exclusively on the type species.

Table 8.1 — Sections of *Tabernaemontana* characterised.

	Tabernaemontana	*Peschiera*	*Bonafousia*	*Ervatamia*
Leaves	mostly thin	mostly thin	mostly rather thick	mostly thin
Inflorescence	mostly lax and long	mostly lax and long	mostly congested and short	mostly lax and long
Corolla	thin	thin	rather thin	thin
Tube	almost cylindrical	almost cylindrical	almost cylindrical	almost cylindrical
Stamens	inserted in upper half of tube, exserted or not	inserted in lower half of tube, deeply included	inserted in upper half of tube, included	inserted on various levels, included
Anthers	sagittate at the base	sagittate at the base	sagittate at the base	cordate at the base
Pistil head	various, mostly with basal ring and with 5 lobes around stigmoid apex	various, with basal ring and with 5 lobes around stigmoid apex	± lampshade-shaped, with basal ring, often with 5 longitudinal grooves and 5 lobes around stigmoid apex	mostly subglobose, without basal ring, not grooved nor lobed
Mericarps	mostly pod-like, thin-walled and dehiscent, smooth	rather thick-walled and dehiscent, mostly prickly or warty, obliquely ellipsoid	rather thick-walled and dehiscent, smooth, mostly obliquely ellipsoid	mostly pod-like, thin-walled and dehiscent, smooth
Pulp	absent	absent	absent	absent

(Table 8.1 continued)

	Pagiantha	*Pandaca*	*Rejoua*
Leaves	mostly thick	mostly thick	rather thick
Inflorescence	rather lax, mostly long	lax	lax
Corolla	mostly thick and fleshy	thick and fleshy or not	thin
Tube	almost cylindrical	almost cylindrical	narrowly infundibuliform
Stamens	inserted on various levels, included	mostly inserted in lower half of tube, deeply included	inserted in lower half of tube, included
Anthers	cordate at the base	sagittate at the base	sagittate at the base
Pistil head	mostly subglobose, without basal ring, not grooved nor lobed	mostly lampshade-shaped, with basal ring, with 5 longitudinal grooves and 5 lobes around stigmoid apex	cylindrical, without basal ring, with 5 longitudinal grooves mostly and often with 5 lobes around stigmoid apex
Mericarps	mostly thick-walled and dehiscent, smooth, mostly subglobose	mostly very thick-walled, subglobose or ellipsoid, mostly dehiscent, mostly smooth	subglobose to pod-like and indehiscent, smooth, spongy
Pulp	absent	absent	spongy, not separable from fruit wall

1. *Tabernaemontana* (including *Capuronetta, Hazunta, Leptopharyngia, Oistanthera, Protogabunia* and *Taberna* Miers).
2. *Peschiera* (including *Stenosolen*).
3. *Bonafousia* (including *Anacampta, Anartia, Camerunia, Codonemma, Merizadenia, Phrissocarpus, Quadricasaea* and *Taberna* Mgf.).
4. *Ervatamia* (including *Muntafara, Pterotaberna* and *Testudipes*).
5. *Pagiantha.*
6. *Pandaca* (including *Conopharyngia, Domkeocarpa, Gabunia, Ochronerium* and *Pandacastrum*).
7. *Rejoua.*

The characters (e.g. shape of pistil head, anthers, height of insertion of stamens and texture of leaves and corolla) often used to distinguish segregates of *Tabernaemontana* cannot be used at genus level, but may reasonably well be valid at section level.

References

Allorge, L. 1985. Monographie des Apocynacées – Tabernaemontanoïdées Américaines. Mém. Mus. natn. Hist. Nat. sér. 2, sér. B, Botanique 30: 1–216.

Baillon, H. 1889. Ochronerium. Bull. Soc. Linn. Paris 1: 774–775.

Boiteau, P., & L. Allorge. 1976. Sur le statut des Conopharyngia au sens de Stapf. Adansonia sér. 2, 16: 259–281.

De Candolle, A. 1844. Apocynaceae. Prodromus Systematis Naturalis regni vegetabilis 8: 317–489. Fortin, Masson & Soc., Paris.

Don, G. 1837. General System 4. Apocynaceae: p. 68–105. Rivington et al., London.

Du Petit Thouars, L.-M.A.A. 1806. Genera nova Madagascariensia: p. 10, Pandaca. Paris.

Gaudichaud-Beaupré, C. 1826–1830. Voyage autour du monde ... exécuté sur les corvettes de S.M. l'Uranie et la Physicienne. Louis de Freycinet, Paris. Rejoua on p. 450–451. 1829.

Leeuwenberg, A.J.M. 1976. The Apocynaceae of Africa. I. Tabernaemontana L. Adansonia sér. 2, 16: 383–392.

Leeuwenberg, A.J.M. 1984. In: T.A. van Beek, R. Verpoorte, A. Baerhein Svendsen, A.J.M. Leeuwenberg & N.G. Bisset, Tabernaemontana L. (Apocynaceae): A review of its taxonomy, phytochemistry, ethnobotany and pharmacology. J. Ethnopharmacology 10: 1–156.

Leeuwenberg, A.J.M. 1988. Series of revisions of Apocynaceae XXI. Notes on Tabernaemontaneae. Agricultural Univ. Wageningen Papers 87-5: 1–32.

Li, P.T. 1986. Revision of Chinese and Thai Tabernaemontana. Guihaia 6: 167–174.

Linnaeus, C. 1753. Species Plantarum (Tabernaemontana): 210–211. Salvius, Stockholm.

Markgraf, F. 1935. Die Gliederung der asiatischen Tabernaemontanoideen. Notizbl. Bot. Garten Berlin 12: 540–552.

Markgraf, F. 1937. Apocynaceae. In: A.A. Pulle (ed.), Flora of Surinam 4.1. Additions and corrections: 443–467.

Markgraf, F. 1938. Die amerikanischen Tabernaemontanoideen. Notizbl. Bot. Garten Berlin 14: 151–184.

Markgraf, F. 1941. Ein neue Apocynaceen-Gattung aus Kamerun. Notizbl. Bot. Garten Berlin 15: 421–422.

Markgraf, F. 1970. Nouveaux taxons d'Apocynacées malgaches. Adansonia sér. 2, 10: 23–33.
Markgraf, F. 1972. Capuronetta, genre nouveau d'Apocynacées malgaches. Adansonia sér. 2, 12: 61–64.
Markgraf, F. 1976. Flore de Madagascar et des Comores. 169e Famille. Apocynacées. Muséum national d'histoire naturelle, Paris.
Miers, J. 1878. On the Apocynaceae of South America. Williams & Norgate, London.
Pichon, M. 1948. Classification des Apocynacées: VI, Genre 'Tabernaemontana'. In: H. Humbert (ed.), Notulae Systematicae 13: 230–254.
Pichon, M. 1949. Classification des Apocynacées: IX, Rauvolfiées, Alstoniées, Allamandées et Tabernémontanoidées. Mém. Mus. natn. Hist. Nat. sér. 2, 27: 153–252 & pl. 10–20 ('1948').
Schumann, K. 1896. Apocynaceae africanae. In: A. Engler (ed.), Bot. Jahrb. 23: 219–231.
Stapf, O. 1902. Apocynaceae. In: W.T. Thiselton-Dyer (ed.), Flora of Tropical Africa 4. 1: 24–231. L. Reeve & Co., Ashford, Kent.
Woodson, R.E. 1941. Miscellaneous new Asclepiadaceae and Apocynaceae from tropical America. Ann. Missouri Bot. Gard. 28: 271–286.

9 Pollinaria of some tropical Asian Asclepiadaceae

M. A. RAHMAN

Summary

Pollinaria of 35 tropical Asian asclepiad species from 21 genera were studied, and a comparative account of 11 morphological characters was tabulated. The variation is of taxonomic significance and provides diagnostic characters for separating some genera and species within the family.

Introduction

The Asclepiadaceae (Robert Brown 1809) are separated from the Apocynaceae by the formation of distinctive waxy pollen masses, called pollinia. A pair of pollinia (subfamily Asclepiadoideae) is attached to a horny pollen carrier, called corpusculum, either directly or by its translator arms (caudicle + retinaculum) forming a pollinarium (see diagram on page 86). Each corpusculum is attached to the margin of the stigmatic head between the anthers, and the adjoining pollinaria are either pendulous (tribes Secamoneae and Cynancheae) or erect/horizontal (tribes Marsdanieae, Ceropegieae and Stapelieae) and each situated in alternative anther loculi. The subfamily Secamonoideae, where all four (not two, as in Asclepiadoideae) loculi of an anther are fertile (Safwat 1962), are characterised by the presence of 10 pairs of pollinia (two pairs in each pollinarium). Therefore, the pollinaria play a central taxonomic role in the family. Nevertheless, they have been studied by only a few workers since Robert Brown (1809). El-Gazzar *et al.* (1973, 1974) have studied the asclepiad pollinaria of 89 species of more than 20 genera from Africa. The only other comparative studies of pollinaria have been made by Schill & Jakel (1978), Rao (1984: Indian *Asclepias curassavica*) and Rahman & Wilcock (1989: Asian *Calotropis* spp.). The present study deals with the pollinaria of 35 Asian species from 21 genera.

Material and Methods

Tropical Asian (Malaysia, Hong Kong, Thailand, Burma, Bangladesh, India, Bhutan, Nepal, Sri Lanka, and Pakistan) herbarium material of Asclepiadaceae was borrowed from ABD, BM, CAL, DACB, E, K, and L. Mature

P. Baas et al. (eds.), The Plant Diversity of Malesia, 83–92.

flowers were reconstituted by boiling in water for 3–4 minutes and kept in separate labelled 5 ml vials. Two pollinaria were extracted intact from the anther sacs of each softened flower and placed on microscope slides. Permanent slides each with a single pollinarium were mounted in Euparal.

The list of specimens examined is as follows:

Toxocarpus acuminatus (2): Bangladesh: Sylhet, Wallich Ascl. no 127 (E) & Cat. no 8242 (K-W).

Toxocarpus himalensis (3): Bangladesh: Sylhet, Rahman & Hossain 57 (ABD, DACB). India: Assam, Jenkins 261 (K); Khasia hills, Griffith 3759 (K).

Genianthus laurifolius (5): Bangladesh: Chittagong Hill Tracts, King's coll. 140 (CAL). East Himalaya: Labdah, Cave s.n. (E). India: Hort. Bot. Gard. Calcutta, Wall. no 1141 (BM); Mysore, Salda. & Rama. 1164 (E). Thailand: Chiang dao, Hui Hia, Sarrett 1234 (ABD, E).

Oxystelma esculentum (6): Bangladesh: Dhaka, C.B. Clarke 7704 (K); Sundarbans, Gamble 10109 (K). Burma: Pakokku, Gamble 11 (K). India: Goalpara, Hamilton 772 (E). Sri Lanka: Thwaites 2837 (BM). Thailand: Bangkok, Kerr 3777 (ABD).

Calotropis acia (10): Bangladesh: Dinajpur, Beldanga, Das s.n. (DACB); Rangamati, Bandarban, Rahman & Khan 54 (ABD); Tangail, Madhupur, C.B. Clarke 7757 (K), 7840 (K). India: Pilibhit, Inayat 22173a (K), Gamble 3233a (K); Mahhope, Inayat 25944 (K); Sikkim, J.D.Hooker s.n. (K); Gwalior, Maries 26 (CAL). Nepal: Sibgonja-Mahara Bahara, Murata & Togashi 6302664 (K).

Asclepias curassavica (7): Bangladesh: Chittagong, Barabkundu, Rahman & Hossain 52 (ABD, BM); Cox's Bazar, Matamari, Cowan 20 (E). Bhutan: Shamchi, Deo Pani khola, Grieson & Long 3503 (E). India: Assam, Prain s.n. (ABD); Sikkim, Jeffrey s.n. (E). Pakistan: Karachi Univ. Campus, Qureshi s.n. (KUH). Thailand: Doi Sutep, Kerr 874 (ABD).

Pentabothra nana (2): India: Camrup, Hamilton 764 (E); Philibhit, Gara, Inayat 22162 (K).

Raphistemma pulchellum (5): Bangladesh: Chittagong, Kodala hills, Badal Khan 38 (CAL). Burma: Ruby mines, Lace 5985 (E). India: Goalpara, Hamilton 749 (E). Malay Peninsula, King's coll. 10483 (E). Thailand: Bangkok, Kerr 4418 (ABD, E).

Pentatropis capensis (4): Bangladesh: Satkhira, Khan et al. 6647 (DACB). India: 24-Parganalis, C.B. Clarke 21750A (BM); Madras, Palicut, Wight Cat. no 1545 (E, K). Sri Lanka: Hambantota, Huber 32 (BM).

Pergularia daemia (4): Bangladesh: Magura, Hassan 14 (DACB). India: Sarajkund, Burtt 6790 (E). Pakistan: Karachi, Abedin 5820 (KUH). Sri Lanka: Buttawa beach, Comanor 870 (E).

Holostemma ada-kodien (4): Bhutan: N. Tashingang, Dangme chu, Grieson & Long 235 (E). Burma: Yamethin, Lace 4276 (E). India: Sambalpur, Haines 5729 (ABD). Nepal: Teku Nala, Kanai et al. 721200 (E).

Cynanchum callialata (4): Bangladesh: Dhaka, Rahman 62 (ABD). Burma: May myo, Lace 4349 (E). India: Gengetica, Hamilton 767 (E), Wallich Ascl. no 85 (E).

Cynanchum corymbosum (4): Bangladesh: Sylhet, Wallich Ascl. no 81 (E) & Cat. no 8222 (E, K-W). India: Assam, F.K. Ward 20210 (BM). Thailand: Songkla, Hut Yai, Kerr 13650 (ABD).

Cynanchum wallichii (4): Bangladesh: Sylhet, Wallich Ascl. no 80a (E). Burma: Kachin State, N. Triangle, F.K. Ward 21275 (E). India: Khasia hills, Lobb s.n. (CGE), Hooker & Thomson 1702 (K).

Sarcolobus carinatus (5): Bangladesh: Cox's Bazar, Ring bong mangroves, Rahman & Hossain 42 (ABD, K); Sundarbans, C.B. Clarke 33382 (BM), Khan 544 (DACB). India: Samulcotta, Hyne s.n. (LIV). Thailand: Trang Sikao, Kerr 19021 (ABD).

Sarcolobus globosus (7): Bangladesh: Sundarbans, C.B. Clarke 33416 (BM); Lace 2321 (E). India: Calcutta, C.B. Clarke 21608 (BM); Delhi, Wallich 789 (BM). Malesia: Java, Horsfield 7 (BM); Pulau Penang, Fox 12710 (BM); Philippines, Wenzel 1684 (BM). Thailand: Pak nam, Kerr 16115 (ABD).

Pentasacme caudatum (6): Bangladesh: Sylhet, Wallich Cat. no 8234A (BM, K-W). Burma: Kachin State, Sumprabun, Keenan 3839 (K). Hong Kong: Mt Gauge, Bod 123 (E). India: Khasia, Hooker & Thomson s.n. (BM, K). Malay Peninsula: Perak, Ridley 2885 (BM, K).

Pentasacme wallichii (4): Bangladesh: Sylhet, Wallich Ascl. no 74 (K). India: Kumaun, Sarju valley, Duthie 3147 (E), Inayat 24687 (CAL). Nepal: Bim Khola, 6500 ft, Stainton et al. 272 (BM, E).

Gymnema acuminatum (6): Bangladesh: Chittagong, Karerhat, Rahman & Hossain 35 (ABD, DACB); Sylhet, Wallich Ascl. no 65 (E). Burma: Tavoy, Wallich Ascl. no 69 (E, K). India: Bot. Gard. Calcutta, Wallich Cat. no 8187B (BM, K-W). Malay Peninsula: Malacca, Griffith s.n. (K). Thailand: Sam Roi, Yawt, Kerr 10962 (ABD).

Gymnema sylvestris (4): India: Bot. Gard. Calcutta, Wallich Ascl. no 72a (E); Penins. Ind. Or., Wight Cat. no 1530 (BM); Sambalpur, Mooney 1367 (E). Sri Lanka, Macrae 562 (CGE).

Bidaria indica (2): India: Anaimali hills, Beddome 278 (K); Nilghiri hills, Wight s.n. (K).

Bidaria inodorum (4): Bangladesh: Sylhet, Wallich Ascl. no 66c (E). India: Munger hills, Hamilton 754 (E); Darjeeling, Cowan s.n. (E). Thailand: Surat, Kao Meo, Kerr 12467 (ABD).

Gongronema nepalense (6): Bangladesh: Chittagong, Karerhat, Rahman & Hossain 36 (BM); Sylhet, Jaintapur, Huq et al. 6196 (DACB). Burma: May Myo, Rodger 600 (E). India: Khasia, Neing Klao, C.B. Clarke (BM). Nepal: Pokhara, Madikhola, Stainton et al. 6456 (E); Wallich Ascl. no 73a (CGE, E).

Gongronema thomsonii (3): E Himalaya: Ghoom, Cave s.n. (E). India: Darjeeling, Lace 2304 (E); Sikkim, Raugbee, C.B. Clarke 12126B (BM).

Marsdenia hamiltonii (2): India: Neterhat, Haines 4342 (ABD); Suknagar, Hamilton s.n. (E).

Marsdenia royle (3): India: Mussoori, Bahugema s.n. (ABD); Nainital, A.A. Khan s.n. (ABD); Simla, Watt 9930 (E).

Marsdenia thyrsiflora (4): Bangladesh: Dhaka, Jinjira, Soejarto 4727 (DACB); Magora, C.B. Clarke 21800F (BM). India: Assam, Master s.n. (K); Khasia, Griffith 3771 (K).

Telosma pallida (8): Bangladesh: Jessore, Keshobpur, Huq 599 (DACB); Norail, Bura Bhaduri, Rahman & Hossain 19 (ABD, K). Burma: Nayingyan, Lace 4884 (E). India: Palaman, Banashatpur, Haines 2345 (ABD); NW Himalaya, Chamba State, Lace 1764 (E). Nepal: Kali Gandak, Stainton 582 (E). Pakistan: Rawalpindi, Topi Park, Stewart 11063 (KUH). Thailand: Chieng Mai, Kerr 397 (ABD, E).

Wattakaka volubilis (8): Bangladesh: Dhaka, Sonargoan, Huq 3487 (DACB). Burma: Rangoon, Roxburgh s.n. (BM). Hong Kong: Ford 21728 (BM). India: Dharmapur, Haines 5732 (ABD); Hassan Dist., Saldanha 13021 (E). Nepal: Birat Nagar:, Stainton 9 (E). Sri Lanka: Hambantota, Jayasuria & Austin 2250 (K). Thailand: Sanders s.n. (K).

Dischidia bengalensis (7): Bangladesh: Sylhet, Chattak, C.B. Clarke 8390 (K), Rahman & Hossain 58 (DBD, DACB). Bhutan: E. Tashingang, Grieson & Long 2082 (E, K). India: Assam, Hooker f. s.n. (K). Malay Peninsula: Phang State, Reod 16236 (K). Nepal: Stainton 5287 (BM). Thailand: Adang, Kerr 14031 (ABD).

Dischidia major (3): Burma: Shan State, Robertson 291 (K); Bhams, Kaukkeoe valley, Lace 6289 (K); Tenasserim, Beddome s.n. (BM). Singapore: Jahore, Herb. Hort. Bot. Singapore s.n. (ABD).

Hoya fusca (5): Bangladesh: Sylhet, Wallich Ascl. no 30b (E, K). Bhutan: Latu la, Cooper 4678 (E). India: Khasia, Mont., Hooker & Thomson s.n. (BM), C.B. Clarke 6226 (K). Nepal: Sheopuri, Schilling 672 (K).

Hoya parasitica (8): Bangladesh: Chittagong, Ichamati, Das & Alam 3911 (DACB); Rangunia, Rahman & Hossain 17 (ABD, E). Burma: Pegu, Kurz 225 (K). India: Assam, Parry 207 (K); Darjeeling, Gamble 2342 (K). Malay Peninsula: Perak, Wray 2273 (K); Singapore, Cluny road, Sinclair 5106 (E). Thailand: Surat, Kerr 12711 (ABD).

Ceropegia longifolia (5): Burma: May Myo, Lace 5966 (E). India: Shillong, C.B. Clarke 44490A (K); Khasia Mont., J.D. Hooker s.n. (K). Nepal: Mugu Karnali, Polunin 5181 (BM); Madikhola, Stainton 8773 (E).

Ceropegia tuberosa (2): India: Orissa, Baruni hills, Haines 4251 (ABD); Madura, Pulney hills, Anglade 1066 (ABD).

Results and Discussion

Table 9.1 shows the variation in shapes and sizes of pollinia, corpuscles and translator arms, attachment of pollinia to caudicle, translator arms to corpuscles, and the orientation of pollinia (see Diagram).

From Table 9.1 and Figures 9.1–9.14 it is clear that there is a wide range of variation in size, shape of pollinia, T-arms, corpuscles and their attachment within the family. The largest pollinia (1000–1642 μm) occur in *Asclepias curassavica* (Figure 9.5), *Oxystelma esculentum* (Figure 9.6), *Calotropis acia* and *Holostemma ada-kodien* – all members of tribe Cynancheae. On the other hand the smallest pollinaria occur in *Toxocarpus* (Figure 9.1) and *Genianthus* (Figure 9.2), the representatives of tribe Secamoneae. The 19 species from tribe Marsdenieae and two species from tribe Ceropegieae are intermediate in size of pollinaria. The shape of pollinia ranges from ovate or lanceolate to ovate- or oblong-lanceolate, with oblong to pyriform or reniform types. The size and shape of corpuscles also vary greatly from species to species; length ranging from 66 μm in *Toxocarpus acuminatus* to 688 μm in *Holostemma ada-kodien*; the shape ranges from ovate (Figures 9.3, 9.4) or triangu-

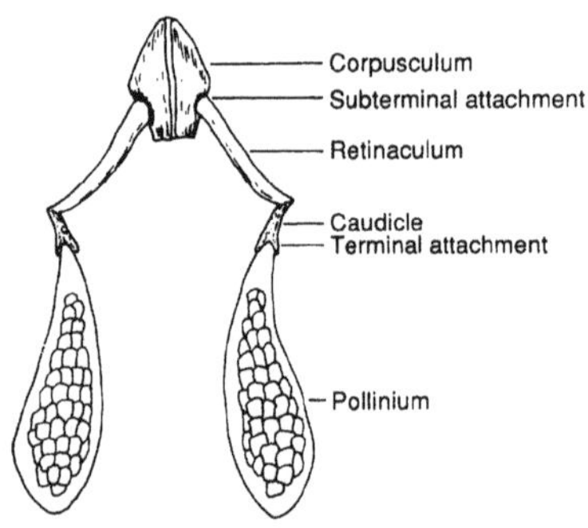

Diagram — The pollinarium of *Asclepias curassavica.*

lar (Figure 9.8) to spindle-shaped (Figures 9.7, 011). The attachment of pollinia and corpuscles to the caudicle is also a characteristic feature in some species, ranging from terminal or subterminal to median.

This study reveals that the pollinaria provide some potential diagnostic features which are taxonomically useful at levels of tribes, genera and species:

Characters diagnostic of tribes

Tribes	Number of pollinia in a pollinarium	Orientation of of pollinaria	Pollinia sessile/ caudicular
Secamoneae	4	pendulous	sessile
Cynancheae	2	pendulous	caudicular
Marsdenieae	2	erect	caudicular
Ceropegieae	2	erect	caudicular
Stapelieae*	2	horizontal	caudicular

* From El-Gazzar *et al.* (1973, 1974).

The species of Secamoneae so far examined all have sessile pollinia which are pyriform and are the smallest in size within the family. These characters also may prove to be diagnostic for the tribe.

Characters diagnostic of genera

A number of diagnostic characters at the generic level are now known. For instance, the genus *Sarcolobus* (all 4 species; Rintz 1980) is characterised by the presence of twisted translator arms and corpuscles longer than their pollinia. Pollinia with beaks at the tips and median attachment to caudicles are characteristic of the genus *Pentasacme*. The flattened and wide tip of the corpusculum and the median attachment with its translator arms in a pollinarium are diagnostic characters of the genus *Ceropegia*.

Legend of Table 9.1 (on pages 88 & 89):

Measurements (µm): average (and range); pollinia (length × breadth*), corpuscles (length × breadth*), T arms (length × breadth*).

FC = Forms of caudicle; 1 = club-shaped, 2 = filiform/simple, 3 = more or less triangular, 0 = absent.
PC = Attachment of pollinia to caudicle; 1 = T arms absent, 2 = terminal, 3 = subterminal, 4 = median.
TC = Attachment of translator arms to corpuscles; 1 = basal, 2 = subbasal, 3 = median, 4 = sessile (T arms absent).
OP = Orientation of pollinia; 1 = pendulous, 2 = erect.
SP = Shapes of pollinia; 1 = ovate, 2 = lanceolate, 3 = ovate-lanceolate, 4 = oblong, 5 = oblong-lanceolate, 6 = pyriform, 7 = reniform.

* Measurement at the broadest region.

Table 9.1 — Comparative account of 11 morphological characters of the pollinaria of 35 species of Asclepiadaceae from Asia. See legend on page 87.

Tribes / Genera / Species	Measurements in μm: Pollinia	Corpuscles	Translator Arms	FC	PC	TC	OP	SP
Tribe Secamoneae								
Toxocarpus								
acuminatus	156 (148–168) × 90 (80–100)	66 (62–68) × 49 (46–52)	0 × 0	0	1	4	1	6
himalensis	156 (144–161) × 82 (78–86)	74 (72–76) × 90 (84–98)	0 × 0	0	1	4	1	6
Genianthus								
laurifolius	140 (115–149) × 90 (74–98)	115 (107–127) × 98 (92–108)	0 × 0	0	1	4	1	6
Tribe Cynancheae								
Oxystelma								
esculentum	1367 (1196–1442) × 369 (344–393)	379 (360–400) × 262 (238–277)	66 (57–74) × 49 (41–57)	2	2	2	1	5
Calotropis								
acia	1642 (1410–1721) × 596 (541–656)	496 (443–451) × 160 (148–164)	497 (410–574) × 61 (49–65)	2	2	2	1	5
Asclepias								
curassavica	1035 (1004–1074) × 367 (344–385)	439 (410–458) × 283 (270–295)	585 (576–598) × 140 (131–148)	2	2	2	1	2
Pentabothra								
nana	746 (738–754) × 246 (239–253)	229 (220–238) × 131 (123–139)	393 (387–399) × 98 (95–101)	2	2	2	1	2
Raphistemma								
pulchellum	991 (983–997) × 590 (583–598)	656 (647–665) × 606 (596–615)	754 (744–765) × 410 (401–420)	1	3	1	1	1
Pentatropis								
capensis	295 (279–321) × 180 (172–190)	270 (262–277) × 197 (186–203)	98 (92–105) × 82 (74–87)	1	3	1	1	1
Pergularia								
daemia	656 (639–704) × 295 (282–303)	262 (249–270) × 246 (239–254)	115 (108–125) × 66 (60–70)	2	2	1	1	3
Holostemma								
ada-kodien	1640 (1590–2000) × 344 (320–410)	688 (656–738) × 229 (213–240)	980 (900–1300) × 148 (115–180)	2	2	1	1	2
Cynanchum								
callialata	439 (413–478) × 195 (167–213)	320 (311–326) × 197 (180–213)	123 (115–132) × 151 (132–164)	1	3	1	1	1
corymbosum	406 (377–426) × 213 (205–229)	347 (336–360) × 204 (180–246)	287 (279–295) × 107 (98–115)	1	3	1	1	4
wallichii	574 (550–606) × 213 (188–221)	385 (377–393) × 254 (246–262)	164 (148–180) × 148 (140–164)	1	3	1	1	4

Tribe Marsdenieae								
Sarcolobus								
carinatus	197 (172–229) × 115 (98–123)	213 (200–224) × 74 (69–84)	262 (249–278) × 66 (62–72)	2	2	1	2	3
globosus	320 (295–334) × 156 (138–164)	385 (374–793) × 115 (107–123)	598 (557–629) × 66 (62–72)	2	2	1	2	7
Pentasacme								
caudatum	426 (410–443) × 205 (182–216)	205 (197–229) × 57 (49–66)	49 (46–54) × 33 (30–36)	1	4	2	2	3
wallichii	361 (344–369) × 229 (226–234)	213 (208–220) × 66 (62–72)	49 (42–56) × 33 (30–38)	1	4	2	2	3
Gymnema								
acuminatum	393 (377–410) × 188 (167–200)	172 (150–184) × 131 (118–136)	82 (75–90) × 49 (46–54)	3	3	1	2	3
sylvestris	172 (167–180) × 82 (78–90)	115 (105–120) × 66 (62–74)	82 (78–88) × 41 (39–49)	3	3	1	2	5
Bidaria								
indica	598 (590–606) × 205 (197–213)	393 (387–403) × 188 (180–197)	115 (108–125) × 66 (59–73)	3	3	2	2	4
inodorum	514 (475–533) × 180 (164–188)	246 (235–262) × 128 (118–132)	49 (36–57) × 43 (33–46)	3	3	2	2	5
Gongronema								
nepalense	336 (311–377) × 96 (90–102)	121 (107–131) × 93 (82–102)	264 (246–275) × 60 (51–64)	2	2	1	2	2
thomsonii	279 (270–282) × 164 (157–172)	197 (190–205) × 148 (141–152)	82 (77–85) × 49 (46–54)	2	2	1	2	1
Marsdenia								
hamiltonii	402 (385–415) × 197 (193–201)	115 (108–121) × 82 (79–85)	246 (238–254) × 82 (79–85)	2	2	2	2	3
royle	426 (400–443) × 188 (164–196)	311 (287–318) × 180 (172–186)	164 (157–167) × 49 (46–52)	2	2	1	2	5
thyrsiflora	295 (279–303) × 131 (115–138)	85 (80–88) × 74 (66–78)	197 (180–205 × 106 (88–110)	1	2	1	2	5
Telosma								
pallida	730 (672–819) × 235 (213–278)	412 (369–475) × 197 (164–229)	167 (148–197) × 106 (98–115)	3	3	2	2	5
Wattakaka								
volubilis	601 (557–671) × 195 (172–245)	325 (303–344) × 109 (98–125)	102 (90–115) × 51 (48–66)	1	2	1	2	5
Dischidia								
bengalensis	262 (249–275) × 115 (107–122)	115 (108–125) × 49 (42–59)	311 (295–325) × 131 (118–139)	3	2	1	2	4
major	639 (623–648) × 213 (203–221)	344 (331–357) × 156 (149–161)	344 (336–351) × 164 (160–166)	1	2	1	2	4
Hoya								
fusca	803 (786–885) × 279 (262–295)	377 (360–393) × 131 (115–139)	33 (29–38) × 49 (42–55)	1	3	3	2	5
parasitica	721 (688–754) × 238 (205–250)	311 (287–320) × 148 (139–157)	82 (75–90) × 82 (65–85)	1	3	2	2	5
Tribe Ceropegieae								
Ceropegia								
longifolia	330 (295–336) × 221 (212–230)	270 (262–277) × 107 (98–115)	82 (65–90) × 33 (31–36)	1	3	3	2	3
tuberosa	295 (287–303) × 197 (188–206)	156 (148–164) × 98 (92–104)	58 (52–64) × 33 (29–37)	1	3	3	2	3

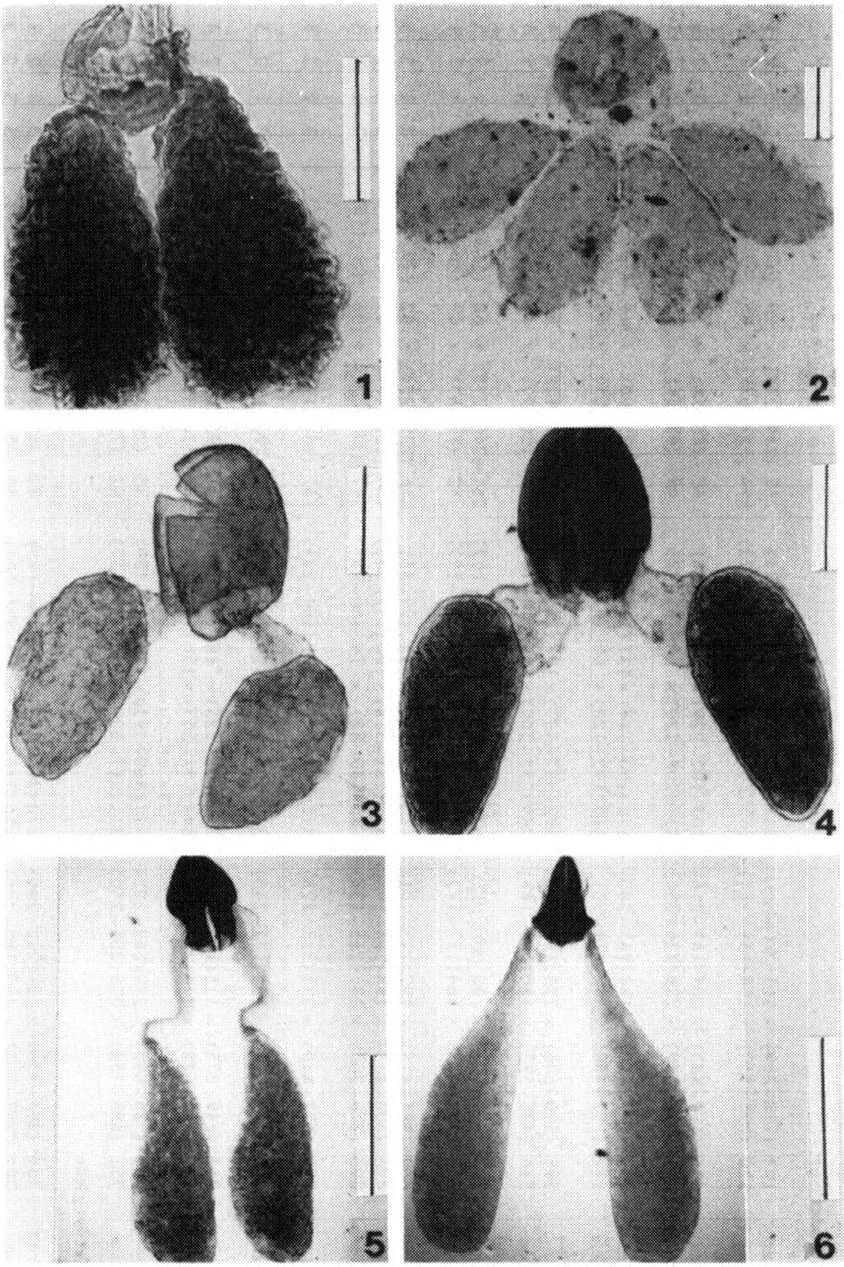

For the legends of Figures 9.1–9.6 and 9.7–9.14, see page 92.

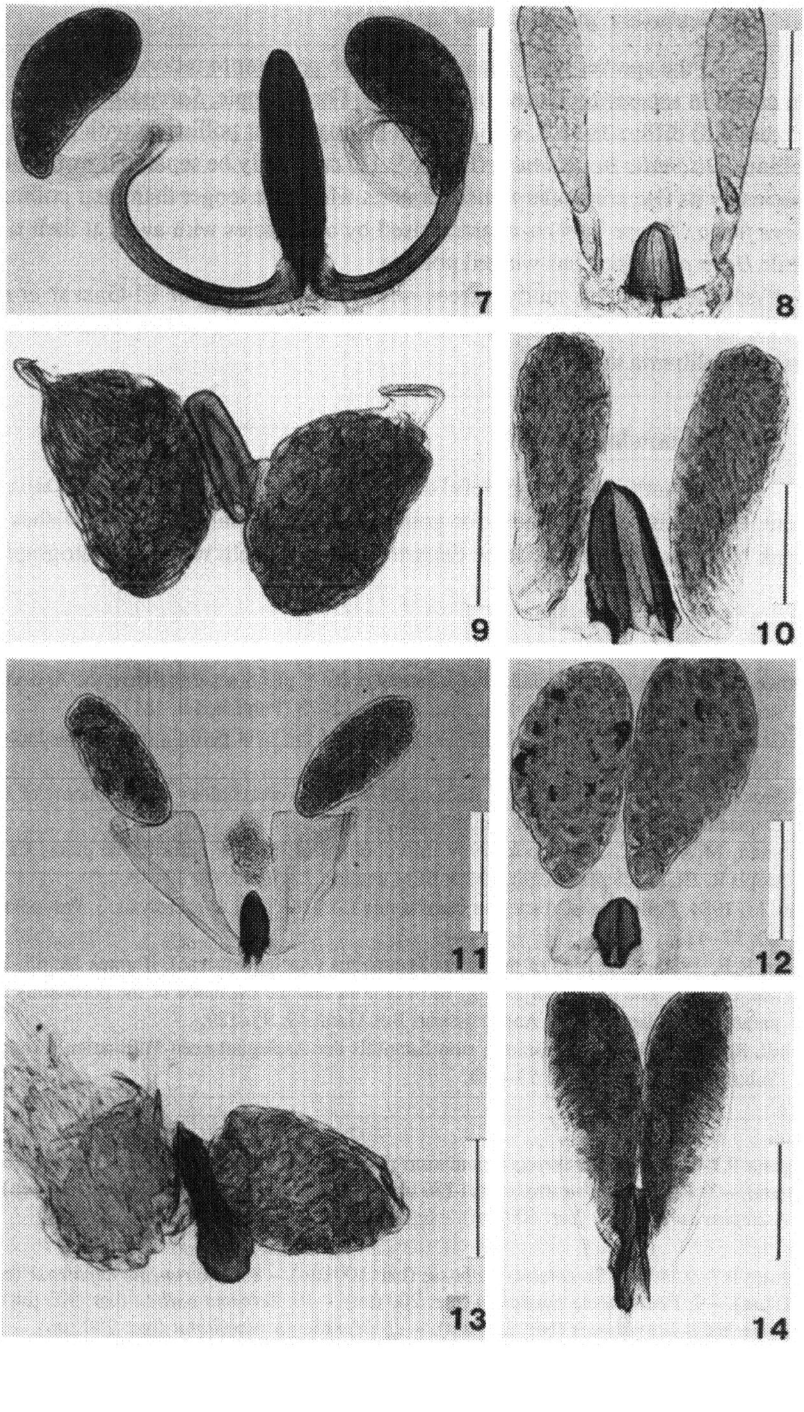
7
8
9
10
11
12
13
14

Characters diagnostic of species

At the species level some features are peculiar to individual species or are useful in separating closely related taxa. For example, *Sarcolobus globosus* (Figure 9.8) differs from *S. carinatus* in having larger pollinaria with reniform pollinia. *Dischidia bengalensis* (Figure 9.11) can easily be separated from other species by its flat, triangular translator arms which are longer than their pollinia. *Hoya fusca* (Figure 9.14) is characterised by corpuscles with awns at their tips while *Hoya parasitica* has winged pollinia.

Evidence from this study agrees with the conclusion of El-Gazzar *et al.* (1973, 1974) that many genera and even species can be distinguished on the basis of pollinaria alone.

Acknowledgements

The author is most grateful to Dr. C.C. Wilcock (Plant Science Department, University of Aberdeen) for going through the manuscript and wishes to thank Norman Little of the same department for his help with the photographs.

References

Brown, R. 1809. On the Asclepiadaceae, a natural order of plants separated from the Apocynaceae of Jussieu. Mem. Wern. Nat. Hist. Soc. 1: 12–78 (Published in 1811).

El-Gazzar, A., & M.K. Hamza. 1973. Morphology of the twin pollinia of Asclepiadaceae. Pollen et Spores 15: 459–470.

El-Gazzar, A., M.K. Hamza & A.A. Badawi. 1974. Pollen morphology and taxonomy of Asclepiadaceae. Pollen et Spores 16: 227–238.

Rahman, M.A., & C.C. Wilcock. 1989. Morphology of the pollinaria in the genus Calotropis R. Br. (Asclepiadaceae): LM & SEM studies. Asklepios 48: 37–40.

Rao, M. 1984. Pollinaria of Asclepias curassavica L.: LM, SEM & acetolysis. J. Palynology 20: 37–41.

Rintz, R.E. 1980. A revision of the genus Sarcolobus (Asclepiadaceae). Blumea 26: 65–79.

Safwat, F. 1962. The floral morphology of Secamone and the evolution of the pollinating apparatus in Asclepiadaceae. Ann. Missouri Bot. Gard. 49: 95–129.

Schill, R., & U. Jakel. 1978. Beitrag zum Kenntnis der Asclepiadaceen-Pollinarien. Trop. u. Subtrop. Pflanzenwelt 22: 53–170.

←←

Figures 9.1–9.6. – 1: *Toxocarpus acuminatus* (bar: 60 µm). – 2: *Genianthus laurifolius* (bar: 50 µm). – 3: *Pentatropis capensis* (bar: 150 µm). – 4: *Cynanchum callialata* (bar: 180 µm). – 5: *Asclepias curassavica* (bar: 600 µm). – 6: *Oxystelma esculentum* (bar: 600 µm).

←

Figures 9.7–9.14. – 7: *Sarcolobus globosus* (bar: 100 µm). – 8: *Gongronema nepalense* (bar: 180 µm). – 9: *Pentasacme caudatum* (bar: 200 µm). – 10: *Telosma pallida* (bar: 300 µm). – 11: *Dischidia bengalensis* (bar: 250 µm). – 12: *Marsdenia hamiltonii* (bar: 200 µm). – 13: *Ceropegia longifolia* (bar: 250 µm). – 14: *Hoya fusca* (bar: 350 µm).

10 Alkaloids and cyanogenic glycosides of Malesian plants as taxonomic markers

R. HEGNAUER

Summary

Buitenzorg (now Bogor) was one of the first centres of intensive phytochemical investigations of tropical plants, and Flora Malesiana is the first flora project of global importance which began in 1960 to incorporate phytochemical paragraphs in its family monographs. Starting with the phytochemical investigations performed by M. Greshoff in the years 1888–1892 in Java an attempt is made to demonstrate the importance of some aspects of secondary metabolism for modern plant taxonomy. A short discussion of *secondary metabolism* is given, and two classes of *secondary metabolites* or *natural products*, alkaloids and cyanogenic glycosides, are defined. To illustrate the taxonomic meaning of chemical characters the following taxa were chosen:

Alkaloids: Alangiaceae, Amaryllidaceae, Apocynaceae, Lauraceae, Rubiaceae, and *Ancistrocladus* and *Daphniphyllum*.

Cyanogenic glycosides: Araceae and Flacourtiaceae.

Introduction

Many features of plant metabolism are worthwhile of being considered in taxonomic work. This is especially true of the so-called secondary metabolism which is responsible for the enormous chemical diversity in the plant (and animal!) kingdom. There are many classes of secondary metabolites which, collectively, are also called Natural Products, because a great deal of technical and medicinal uses of plants are conditioned by their accumulation in certain taxa. For a long time most plant taxonomists were reluctant to make use of chemical characters since they were badly acquainted with them. It is one of the many merits of Kees van Steenis to have fully appreciated the taxonomic potentialities of non-form-based characters. Accordingly he successively began to include discussions and evaluations of many kinds of characters in the general chapters of family monographs. Phytochemistry was taken into account since 1960. As far as I am aware, Flora Malesiana was the first big flora project which started to deal with phytochemical and chemotaxonomic aspects of plant families. Perhaps it is not much surprising that, besides his many other pioneering activities, Kees also pioneered in the chemical approach to plant taxonomy. As everybody

P. Baas et al. (eds.), The Plant Diversity of Malesia, 93–104.

knows, Java had several laboratories of world fame devoted to botanical research at the molecular levels. Enthusiastic and highly gifted scientists such as Treub, Greshoff, van Romburgh, Boorsma and many others had been working hard and successfully in the famous laboratories of the Botanical Garden of Buitenzorg. Their contributions to the chemical knowledge of tropical plants – and in the case of Greshoff to chemotaxonomy – were outstanding. Perhaps nowhere during the past turn of the century more favourable conditions for combining plant taxonomy with phytochemistry existed than in Buitenzorg.

MEDEDEELINGEN

UIT

'S LANDS PLANTENTUIN.

VII.

CHEMISCH-PHARMACOLOGISCH LABORATORIUM

EERSTE VERSLAG

VAN HET

ONDERZOEK NAAR DE PLANTENSTOFFEN

VAN

NEDERLANDSCH-INDIË

DOOR

M. GRESHOFF.

BATAVIA
LANDSDRUKKERIJ
1890.

Figure 10.1 — Title page of M. Greshoff's first report on his phytochemical investigations performed in the laboratories of the Botanical Garden Buitenzorg (Bogor), Java.

It is my intention to demonstrate – starting with the work of Greshoff (1890, 1898) – that phytochemistry has indeed something to offer to plant taxonomy. To do this I shall confine myself to vascular plants and to two classes of Natural Products, alkaloids and cyanogenic glycosides.

Secondary metabolism

Functionally plant metabolites can roughly be classified in three categories:

1. *Essential compounds* which are everywhere in plants and animals. They are involved in the fundamental processes of life. Examples are a number of amino acids (e.g. alanine, valine, leucine, isoleucine, ornithine, phenylalanine, tyrosine, tryptophan and some 17 others), sugars (e.g. ribose, deoxyribose, glucose, galactose and others), organic acids (e.g. citric, aconitic, isocitric, acetic, mevalonic acids and others) and biopolymers such as nucleic acids and certain proteins (e.g. many enzymes) and lipids.
2. *Structural compounds* which give strength to the body such as cellulose and lignin in vascular plants and a number of proteins and lipids shaping cellular architecture.
3. *Non-essential or secondary metabolites* which occur erratically in the plant and animal kingdoms. They are not needed for the fundamental metabolism of life nor for the architecture of cells and organisms. Nevertheless, collectively, they are everywhere. Secondary metabolites are products of catabolic processes or of derailments of primary metabolic routes. Because they seem not to fulfil functions essential to life as such, selection pressure on secondary metabolism is assumed to be far less strong than it is on primary metabolism. Mutations are tolerated if new metabolites are not noxious. This may explain the overwhelming chemical diversity of secondary metabolites (Mothes 1973).

For reviews of secondary metabolism see, for example, Zenk 1967; Geissman & Crout 1969; Hegnauer 1975; Luckner 1984.

Alkaloids

Alkaloids are N-containing, usually heterocyclic organic bases which are water-soluble as salts and rather lipophilic in the free-base-form. Most alkaloids are biologically active compounds and toxic to man and other mammals in rather low concentrations. From the biogenetic point of view alkaloids in the broad sense are a highly heterogeneous class of natural products. For taxonomic purposes pathways of alkaloid synthesis in the taxa under study should always be taken into account (Hegnauer 1988). Because alkaloids are the active

INHOUD.

Figure 10.2 — Table of contents of Greshoff's first report (Figure 10.1). Note that most attention was paid to three families, Leguminosae, Apocynaceae and Lauraceae, and to one class of secondary metabolites, cyanogenic compounds.

principles of many medicinal plants and sometimes cause large-scale intoxications of livestock or may produce addiction in man (e.g. caffeine, nicotine, morphine, cocaine), this group of phytoconstituents attracted the attention of chemists since the start of scientific phytochemical investigations of useful and noxious plants in the beginning of the past century.

In August 1888 the Dutch government created a research position for the study of medicinal plants in Indonesia. Greshoff was the scientist appointed to start this project connected with the Botanical Garden of Buitenzorg (now Bogor). From the beginning Greshoff deviated in a significant way from the directions he had received. I like to give you his own motivation for this rather unusual behaviour (Greshoff I: 1–2):

> "Aan het chemisch onderzoek der Indische plantenstoffen werd de botanische verwantschap der gewassen ten grondslag gelegd. Hoewel aan de als geneeskrachtig bekend staande planten een bijzondere belangstelling wordt gewijd, worden dus geenszins de inlandsche geneesmiddelen achtereenvolgens en uitsluitend onderzocht, maar wel de Indische plantenfamiliën systematisch geanalyseerd en de planten zooveel mogelijk in haar natuurlijk verband ter hand genomen. Op die wijze ontvangt het onderzoek een eigen wetenschappelijken grondslag en kunnen de elders reeds verkregen ervaringen ook dienstbaar gemaakt worden aan de phytochemische studie der Indische gewassen."

[Transl.: The chemical research of Indian (= Indonesian) plant substances was founded on botanical affinity. Although special attention is paid to plants known for their medicinal properties, studies are by no means limited to the screening of native medicines. As much as possible Indonesian plant families are systematically analysed, and natural relationships of the plants are taken into account. In this manner the research has its own scientific basis and experience obtained elsewhere can be applied to the phytochemical study of Indonesian plants.]

It is clear from these introductory remarks to the first report of his phytochemical work in Buitenzorg that Greshoff started large-scale chemotaxonomic research in the tropics. This is astonishing for the time concerned and shows that Maurits Greshoff was an eminent scientist. The following examples illustrate how his pioneering work grew out in the course of time to a remarkable sound chemical knowledge of a number of taxa, and in fact to a chemotaxonomy of vascular plants (Hegnauer 1962–1989).

Alangiaceae

(Greshoff II: 89–92; Chemotaxonomie III: 78, 676; VII: 321; VIII: 25–28, 33)

Occurrence of bitter and significantly toxic alkaloids was reported by Greshoff for leaves and barks of four species of *Alangium s.l.* In 1964 the first alangiaceous alkaloids could definitely be identified as emetine-type bases. Later tubulosine-type alkaloids became also known from *Alangium salviifolium*. Alkaloids derived from one secologanin and two phenylethylamines (emetines) or from one secologanin, one phenylethylamine and one tryptamine (tubulosines) indicate biochemical relationships with Icacinaceae and Rubiaceae.

Amaryllidaceae

(Greshoff II: 190–191)

Greshoff reported one species of *Crinum* and one species of *Haemanthus* as alkaloid-bearing. The alkaloid from *Crinum asiaticum* was obtained in crystalline form and shown to be weakly toxic to chickens. Research with amaryllidaceous plants was continued later in Buitenzorg by Gorter who isolated the main alkaloid, meanwhile called lycorine, from bulbs, rhizomes, roots and seeds of 12 plant samples representing 10 genera of the family. Today a large number of lycorine-type alkaloids, i.e. C_6-C_1-N-C_2-C_6 bases, is known. Every amaryllidaceous plant seems to produce and store alkaloids of this type which derive from tyrosine (tyrosine → *p*-hydroxyphenylacetaldehyde [C_6-C_2]) and phenylalanine (phenylalanine → cinnamic acids → benzaldehydes [C_6-C_1]) and are variously called belladine-type, lycorine-type or amaryllidaceous alkaloids. The metabolic pathway producing such alkaloids is a key character of true Amaryllidaceae and seems to be lacking in modern Agavaceae, Alstroemeriaceae, Haemodoraceae and Hypoxidaceae (Chemotaxonomie II: 53–70, 475; VII: 573–580).

Apocynaceae
(Greshoff I: 45–76; II: 118–135)

Greshoff paid much attention to this family which was renowned already in his time for the many toxic plants it harbours. He investigated 46 species belonging to 35 genera. He remarked that most alkaloid-bearing taxa belong to Plumerioideae *sensu* Schumann and most non-alkaloidal genera, including those accumulating highly toxic strophanthin-like cardenolides, belong to Echitoideae. In his time the precise nature of the alkaloids demonstrated to be present in these plants was not yet known; therefore he proposed, probably erroneously, that the alkaloid-containing echitoid genera *Chonemorpha* and *Rhynchodia* should be transferred to Plumerioideae. Today five alkaloidal types are known from the family (Chemotaxonomie VIII: 48–61, 698):

1. Complex indolic alkaloids in which a tryptamine unit is combined with a secologanin unit. These alkaloids, of which reserpine and yohimbine are examples, were demonstrated to be present in hundreds of species belonging to 50 to 60 genera that all belong in modern treatments to four tribes of Plumerioideae, namely Carisseae, Tabernaemontaneae, Plumerieae and Rauvolfieae (Leeuwenberg 1980; Kisakürek *et al.* 1983).
2. Simple piperidine- and pyridine-type basic monoterpenoids derived from iridoid or secoiridoid glycosides. They accompany the mother glycosides in many plants and are sometimes artifacts of isolation (if NH_3 is used during the search for alkaloids). In Apocynaceae they are mainly known from Plumerioideae and Cerberoideae (included by Schumann in the former).
3. Spermidine alkaloids: in Apocynaceae only known from *Oncinotis* (Echitoideae).
4. Pyrrolizidine alkaloids: known from the genera *Alafia, Anodendron, Parsonsia* and *Urechites,* all belonging to Echitoideae.
5. Basic C_{21}-steroids (amino- and iminopregnanoids): with one exception all genera known to produce this type of pseudo-alkaloids belong to Echitoideae. The exception is *Holarrhena* (Plumerioideae-Alstonieae) which contains amino- and iminopregnans together with cardenolides and seems to form a link between Plumerioideae and Echitoideae. Kisakürek *et al.* (1983) proposed to transfer *Holarrhena* to Echitoideae.

In conclusion it can be said that roughly speaking Greshoff was right when he accentuated the importance of alkaloids on the one hand and glycosidic bitter principles (cardenolides; iridoid glycosides) on the other for the natural classification of Apocynaceae.

Lauraceae
(Greshoff I: 77–101; II: 160–168)

Greshoff investigated 37 species representing 20 genera for alkaloids and demonstrated their nearly universal presence in the family. Moreover, he showed that a characteristic alkaloid with tetanising properties and called lauro-tetanine by him occurs probably in at least ten genera. Alkaloids in general, and more specifically laurotetanine, were shown by this investigation to represent a remarkable feature of the whole family. In the meantime the structure of laurotetanine became known. It is a phenolic aporphine-type base and belongs therefore to the phenylalanine- and tyrosine-derived class of benzylisoquinoline alkaloids which is not only characteristic of Lauraceae (Chemotaxonomie IV: 350–381, 478–481; V: 292; VIII: 633–664, 717), but of the whole order or superorder or subclass Polycarpicae (= Magnoliiflorae = Annoniflorae = Magnoliidae *sensu* Cronquist). At the same time Greshoff demonstrated universal presence of alkaloids in Hernandiaceae (*Gyrocarpus asiaticus, Hernandia ovigera* and *sonora, Illigera pulchra* [bark probably contains lauro-tetanine]) and herewith procured arguments for biochemical relationships between Hernandiaceae and Lauraceae.

Rubiaceae
(Greshoff II: 92–105)

Greshoff tested material from 30 species representing 21 genera and both subfamilies of K. Schumann, i.e. Cinchonoideae and Coffeoideae, of this huge family. He demonstrated presence of alkaloids in all investigated species of 14 genera and their absence in seven species representing seven genera. No clear-cut differences between Cinchonoideae and Coffeoideae were observable, but a strong tendency for accumulation of alkaloids had been demonstrated for Rubiaceae by these investigations. Today it is easy to understand why far-reaching chemotaxonomic conclusions were impossible in Greshoff's days. On the one hand Schumann's classification of the family is considered to be rather artificial and on the other the alkaloidal chemistry of the family is very diverse (Chemotaxonomie VI: 130–174, 730–734, 794; Leeuwenberg 1980; Hemingway & Phillipson 1980; Kisakürek *et al.* 1983; Hegnauer 1988). The following alkaloidal types are presently known from Rubiaceae:

1. Complex indole alkaloids similar to those of Loganiaceae and Apocynaceae. They can be arranged in three main types: a) With non-rearranged indole and secologanin units. b) With rearranged secologanin unit and intact indole unit. c) With the indole unit rearranged to a quinoline heterocyclic system (quinine and related alkaloids).

From Rubiaceae only subtypes a and c are known. Moreover, synthesis and accumulation seem to be realised mainly in members of Cinchonoideae-Naucleeae (a) and -Cinchoneae (a, c). In Rubioideae such alkaloids occur only sporadically. In several rubioid genera they are replaced by secologanin-containing alkaloids of the emetine-type. In emetine and related alkaloids tryptamine is replaced by phenylethylamine or tyramine. For a chemist these alkaloids are different because they contain two isoquinoline systems in place of one or two indole systems. Emetine-like isoquinolines were also reported from a few genera of Cinchonoideae but many observations concerning the occurrence of emetine-like alkaloids are in need of confirmation by modern analytical methods. Tubulosine (see *sub* Alangiaceae) was isolated from *Pogonopus tubulosus* (Cinchonoideae).

2. Simple tryptamine derivatives. They range from protoalkaloids like N-methyl-tyramines to ß-carboline-bases such as harmane and the more complex tarennine (= dihydroelaeocarpidine) from *Tarenna bipindensis* and occur erratically in Rubiaceae and many other families of dicotyledons.
3. Dimeric to pentameric tryptamines like the calycanthine-, chimonanthine-, hodgkinsine- and psychotridine-type bases which are known from the genera *Hodgkinsonia, Palicourea* and *Psychotria*. The calycanthine and chimonanthine subtypes are more characteristic of Calycanthaceae than of Rubiaceae.
4. The monomeric and dimeric borrerine-type alkaloids in which one tryptamine or two tryptamines are combined with a hemiterpene unit (e.g. borrerine) or a monoterpene unit (e.g. borreverine). Such alkaloids are known from species of *Borreria (Spermacoce)* of Rubiaceae and from *Flindersia laevicarpa* of Rutaceae.
5. Macrocyclic peptide alkaloids resembling those of Rhamnaceae are known from *Canthium euryoides*. The main oxytocic principle of *Oldenlandia affinis* was found to be a different peptide; the latter is accompanied by the protoalkaloids 5-hydroxytryptamine (= serotonin) and tetramethylputrescine (Gran 1973).
6. Papilionaceae-type chinolizidine alkaloids were reported from *Readea membranacea* from Fiji.
7. N-Methylated purines, especially caffeine, occur amply in seeds of species of *Coffea*. Caffeine was also reported to be present in species of *Genipa* and *Oldenlandia*, but these reports are in need of confirmation.
8. Chromanoid alkaloids in which a 2-methylchromone unit is combined with nicotinic acid occur in large numbers in taxa of the genus *Schumanniophyton*.

It becomes clear from this short résumé that many alkaloidal types occur in the family. The only ones which seem to be taxonomically rewarding are those mentioned *sub* 1a and c. They strongly indicate biochemical relationships with Loganiaceae and Apocynaceae, and suggest a link of this alliance with Alangiaceae (emetine- and tubulosine-type alkaloids) and Icacinaceae (tubulosine-type alkaloids: Chemotaxonomie IV: 494; VIII: 572–574).

Short notes on Daphniphyllaceae and Ancistrocladaceae, taxa incertae sedis
(Greshoff II: 169–172; Boorsma 1899)

Greshoff reported that seeds, leaves and bark of *Daphniphyllum bancanum* contain alkaloids. From the leaves and from bark a probably pure alkaloid could be isolated and crystallised from ether; it was called daphniphylline. Pharmacological investigations performed by Professor Plugge in Groningen showed that daphniphylline is a rather toxic compound with a paralysing action on the heart and on respiration.

Daphniphyllaceae are now known to produce iridoid glycosides and a series of alkaloids which represent a totally new type. *Daphniphyllum* alkaloids are aminated squalene derivatives, i.e. they are of triterpenoid origin. This alkaloidal type seems to be restricted to Daphniphyllaceae which are presently assumed by several authors to have their closest relatives in Hamamelidaceae *s.l.* (Chemotaxonomie VIII: 376–379).

Boorsma (1899) isolated a crystalline alkaloid from the leaves of *Ancistrocladus vahlii* and showed that it arrests respiration and produces convulsions in frogs. Today we know that five species of *Ancistrocladus* contain several acetogenic isoquinoline alkaloids together with acetogenic naphthoquinones (Chemotaxonomie VIII: 39–42). Similar constituents are present in Dioncophyllaceae and, to a lesser extent, in Nepenthaceae and Droseraceae. In terms of secondary metabolism these taxa are more or less intimately related.

Cyanogenic compounds

Natural products which are able to release prussic acid (= hydrocyanic acid) are called cyanogenic compounds. For reviews see Hegnauer 1977; Nahrstedt 1987; Chemotaxonomie VII: 345–374. Apart from cyanolipids of Sapindaceae and the rather rare compounds with a free cyanohydrin group such as *p*-glucosyloxybenzaldehydecyanohydrin of *Nandina domestica* and *Goodia lotifolia* they are cyanohydrins stabilised by glycosylation of the hydroxyl in the cyanohydrin group. Spontaneous cyanogenesis only occurs if appropriate enzymes for the cleavage of the glycosidic bond are present at the same time. In some instances plant parts can contain a large amount of cyanogenic glycosides without being spontaneously cyanogenic, because they lack enzymes splitting the glycosidic bond; an example is *Aster ptarmicoides*. The biogenesis of the cyanohydrin part of cyanogenic glycosides proceeds from proteinogenic (phenylalanine, tyrosine, valine, isoleucine, leucine) or non-proteinogenic (cyclopentenylglycine, nicotinic acid) amino acids resulting in six presently known families of cyanogenic plant constituents with some fifty individual compounds. These families of chemical constituents represent taxonomically useful characters.

Araceae

(Greshoff I: 106–109)

Greshoff showed that several Indonesian plants belonging to the genera *Lasia* and *Cyrtosperma* have more or less strongly cyanogenic spadices and leaves, and that leaves of *Amorphophallus campanulatus* and three species of *Homalonema* did not release prussic acid. At the same time he demonstrated that Araceae contain a new type of cyanogenic constituents in which hydrocyanic acid is loosely bound. Today we know (Chemotaxonomie VII: 583) that cyanogenesis occurs in many araceous plants, that they contain triglochinin, and that the instability of their cyanogenic compound reported by Greshoff is conditioned by the presence of extremely active enzymes and not by loose binding of HCN. Storage of triglochinin is a character of a large number of Araceae.

Flacourtiaceae

(Greshoff I: 109–115; II: 30–34; Chemotaxonomie VIII: 487–496, 715)

Greshoff showed that all parts of *Pangium edule* are strongly cyanogenic and that a specific kind of cyanogenic compounds must be present in *Pangium, Hydnocarpus* (incl. *Taraktogenos*), *Gynocardia odorata* and two species of *Ryparosa* and that cyanogenic compounds are lacking in *Bixa orellana, Cochlospermum gossypium,* Flacourtiaceae-Scolopieae, -Flacourtieae, -Casearieae and in *Alsodeia haplobotrys* (Violaceae). He considered this special type of cyanogenic compounds to represent an attribute which characterises Pangieae. Curiously enough he did not detect cyanogenesis in *Turnera ulmifolia,* now known to contain cyanogenic glycosides; this may have been caused by a lack of splitting enzymes in the plant material investigated by him. At present we know that cyanogenic glycosides derived from cyclopentenylglycine (gynocardin, barterin, deidaclin) are present in all members of Flacourtiaceae-Pangieae, most members of -Oncobeae and in many members of Achariaceae, Malesherbiaceae, Passifloraceae (Olafsdottir *et al.* 1989) and Turneraceae, and possibly also occur in Caricaceae. These natural products are to be considered as an important character of part of Violales *sensu* Cronquist.

Concluding remarks

In the nearly hundred years which elapsed since Greshoff was working in Buitenzorg an undreamt explosive increase of our chemical and biological knowledge of natural products was achieved. The fact that a number of Greshoff's early results and theories gloriously stood the test of time pays tribute to this enthusiastic and gifted scientist. In my feeling it was mainly Greshoff's pioneering work which induced Kees van Steenis to pay attention to chemical characters in his monumental taxomic work.

References

Boorsma, W.G. 1899. Nadere resultaten van het onderzoek naar de plantenstoffen van Nederlandsch-Indië, Meded. uit 's Lands Plantentuin, no XXXI: 4–5. Batavia.

Chemotaxonomie II–VIII, see Hegnauer 1962–1989.

Geissman, T.A., & D.H.G. Crout. 1969. Organic chemistry of secondary plant metabolism. Freeman, Cooper & Co. San Francisco, California.

Gran, L. 1973. Oxytocic principles of Oldenlandia affinis. Isolation of tetramethylputrescine from Oldenlandia affinis. Lloydia (J. Nat. Prod.) 46: 174–178, 209–210.

Greshoff, M. 1890 & 1898. Eerste en Tweede verslag van het onderzoek naar de plantenstoffen van Nederlandsch-Indië, Meded. uit 's Lands Plantentuin, no VII & XXV. Batavia. — Cited in the text as Greshoff I and Greshoff II.

Hegnauer, R. 1962–1989. Chemotaxonomie der Pflanzen, vol. I–VIII. Birkhäuser Verlag, Basel, Boston, Berlin. — Cited in the text as Chemotaxonomie II to VIII.

Hegnauer, R. 1975. Secondary metabolism and crop plants. In: O.H. Frankel & J.G. Hawkes (eds.), Crop genetic resources for today and tomorrow: 249–265. Cambridge Univ. Press, Cambridge etc.

Hegnauer, R. 1977. Cyanogenic compounds as systematic markers in tracheophytes. In: H. Kubitzki (ed.), Flowering plants. Evolution and classification of higher categories. Plant. Syst. Evol., Suppl. 1: 191–209.

Hegnauer, R. 1988. Biochemistry, distribution and taxonomic relevance of higher plant alkaloids. Phytochemistry 27: 2423–2427.

Hemingway, S.R., & J.D. Phillipson. 1980. Alkaloids of Rubiaceae. In: J.D. Phillipson & M.H. Zenk (eds.), Indole and biogenetically related alkaloids: 63–90. Acad. Press, London etc.

Kisakürek, M.V., A.J.M. Leeuwenberg & M. Hesse. 1983. A chemotaxonomic investigation of the plant families of Apocynaceae, Loganiaceae, and Rubiaceae by their indole alkaloid content. In: S.E. Pelletier (ed.), Alkaloids: Chemical and biological perspectives. Vol. 1: 211–376. John Wiley and Sons, New York etc.

Leeuwenberg, A.J.M. 1980. The taxonomic position of some genera in the Loganiaceae, Apocynaceae, and Rubiaceae, related families which contain indole alkaloids. In: J.D. Phillipson & M.H. Zenk (eds.), Indole and biogenetically related alkaloids: 1–10. Acad. Press, London etc.

Luckner, M. 1984. Secondary metabolism in microorganisms, plants, and animals. Second revised and enlarged edition. Springer-Verlag, Berlin etc.

Mothes, K. 1973. Pflanze und Tier, ein Vergleich auf der Ebene des Sekundärstoffwechsels, Sitzungsber. Österr. Akad. der Wissensch., Sonderheft: 1–37 zu Abt. I, 181, Jahrg. 1972.

Nahrstedt, A. 1987. Recent developments in chemistry, distribution and biology of the cyanogenic glycosides. In: K. Hostettmann & P. J. Lea (eds.), Biologically active natural products: 213–234. Annual Proc. Phytochem. Soc. Europe. Clarendon Press, Oxford.
Olafsdottir, E. S., J.V. Andersen & J.W. Jaroszewski. 1989. Cyanohydrin glycosides of Passifloraceae. Phytochemistry 28: 127–132.
Zenk, M.H. 1967. Biochemie und Physiologie sekundärer Pflanzenstoffe. Ber. Dtsch. Bot. Ges. 80: 573–591.

11 The systematics of the Polypodiaceae

E. HENNIPMAN

Summary

Concepts and methods of monographic systematic research on ferns have changed considerably during the last decades. Recently Haufler and collaborators studying *Polypodium vulgare sensu lato* have suggested that apart from divergent speciation in sexual diploids and hybrid speciation (e.g. allopolyploids), genetical diploidisation of hybrid species through gene silencing is an intrinsic speciation mechanism that allows the polyploid to return to diploid expression. The monographer working with herbarium material faces the problem of recognising these different types of evolutionary species and speciation mechanisms occurring in ferns. It appears that micromorphological and ultrastructural character states can help to trace evolutionary species in the herbarium. This questions the application of a broad morphological species concept in tropical floras.

The study of insufficiently known features such as venation patterns, paraphyses and sterile frond indument, improve the ability to recognise inferred monophyletic groups within the family. The recognition of monophyletic genera in the Polypodiaceae is exemplified with a discussion of genera of Platycerioideae (*Platycerium, Pyrrosia*) and the ant-ferns. Character analyses of selected features using traditional and new techniques show that characters of the exospore and the perispore are most liable to change. An example shows that character states of the spore may express themselves in submicroscopical and ultrastructural characteristics in a similar way in quite different evolutionary lines.

Introduction

During the last 10 years work on ferns in the Netherlands has concentrated on the systematics of the Polypodiaceae, one of the most dubious groups of relationships in ferns (Holttum 1972). The family is here considered in a restricted sense, that is, with the exclusion of the Grammitidaceae and the Loxogrammaceae. The Polypodiaceae thus construed are a cosmopolitan family with about 500 species which are mainly concentrated in the lowlands and lower montane forests of the American and Asian tropics. The systematic position of the family amongst homosporous leptosporangiate ferns and the delimitation of species and genera by different authors, show little congruence.

The Polypodiaceae project was initiated in 1977 at the Rijksherbarium, Leiden, in order to execute monographic research of the whole family. Since

P. Baas et al. (eds.), The Plant Diversity of Malesia, 105–120.

1980 it has been the principal research project on ferns also at the University of Utrecht. Monographic research is here regarded as the best starting point for generating natural classifications. Such classifications represent testable scientific hypotheses and are at the same time starting points for research on historical biogeography and comparative ecology. Inferred monophyletic groups are traced by character analysis of known and insufficiently known features and especially at the microscopic and ultrastructural level.

Cladistic analysis is used as a tool to trace phylogenetic relationships between divergent species.

This paper illustrates the changing concepts and methods of systematic research on ferns during the last decades. It deals especially with some basal activities of systematic research such as the delimitation of species and higher taxa, speciation mechanisms, and the exploration of new characters. It is expected that these new developments will have a strong bearing on our ability to recognise and describe fern taxa in Polypodiaceae and other mainly tropical ferns by herbarium taxonomists.

Species delimitation

The herbarium taxonomist traditionally uses morphological criteria for delimiting species. When adopting the evolutionary species concept, however, the morphology allows one only to draw a first hypothesis for recognising such species. Such hypotheses are open to testing.

Manton (1950) and Wagner (1954) started testing the traditional morphological species concept in temperate ferns introducing characters from karyology and genetic systems. They greatly improved our perception of fern species by putting a number of features of ferns into a new perspective. These features include:

- fern species have high chromosome numbers;
- fern species have two free-living generations, including a tiny bisexual microscopic gametophyte apart from the variously developed asexual sporophyte;
- fern species can realise various breeding systems, e.g. inbreeding, outcrossing, or both;
- natural hybrids between species of the same genus (or of different genera) occur on an extensive scale;
- fern species and hybrids commonly form unreduced, diploid spores, which produce functional diploid gametophytes with diploid gametes.

As a consequence ferns in general can express themselves as diploids, polyploids and notoriously reticulate complexes. The situation with the *Polypodium vulgare* L. alliance is a case in point.

The Polypodium vulgare alliance

Karyological characters — The karyological results together with increasing knowledge about breeding systems in ferns had a profound effect on the delimitation of species. The original circumscription of *Polypodium vulgare* was retained by most authors for almost two centuries. However, Manton (1950) found three different cytotypes, 2n, 4n and 6n, within the European specimens of *Polypodium vulgare*, and suggested that each cytotype should be regarded as a different species. The idea of using differences in genetic systems as species markers generated much debate. Wagner (1951) thought it better to interpret the three cytotypes in *P. vulgare s.l.* as varieties based on an inferred original diploid species. At that time most pteridologists were of the opinion that even if minor morphological characters were found, specific recognition of these cytotypes was not warranted.

Manton's ideas about the species composition of *Polypodium vulgare s.l.* were tested by Shivas (1961a, b) who produced two elegant papers dealing with the karyology, micromorphology and meiotic behaviour in natural and synthesised hybrids.

Shivas convincingly demonstrated the true nature of the cytotypes discovered by Manton as separate species. She formalised Manton's view in recognising the allotetraploid as *Polypodium vulgare* L., the diploid as *P. australe* Fée and newly described the hexaploid *P. interjectum* Shivas. The herbarium taxonomist nowadays has little difficulty in properly identifying these different species as a number of microscopical characters such as the presence and shape of the sporangiasters are associated with the karyological differences.

Isozymic characters — In recent years Haufler and collaborators (for references see Barrington *et al.* 1989) have been studying the *Polypodium vulgare* alliance on a global scale using eletrophoresis. Haufler and Windham (1989) discovered that the representatives of the widespread *P. vulgare* alliance outside Europe are most heterogeneous. Thus the diploid Japanese populations earlier referred to *P. virginianum* are not conspecific with that species in North America. They belong to *P. sibericum* Siblivinsky, a recently described species which now appears to occur sporadically also in eastern Asia and North America, and possibly also in Scandinavia. This species and diploid *P. glycyrrhiza* from North America are the parents of tetraploid *P. vulgare*. It is further likely that *P. sibericum* also hybridised with diploid *P. virginianum* from North America to give rise to tetraploid *P. virginianum* from eastern North America. *Polypodium sibericum* also hybridised with other species, including the Japanese *P. fauriei*, the latter being an unexpected new member of the *P. vulgare* alliance. The position of the European *P. australe* is not yet elucidated; the original idea of a closest relationship of this species with allotetraploid *P. vulgare* is not supported.

The new evidence accumulated by Haufler and collaborators studying isozymic characters results in a much more resolved taxonomy of the *P. vulgare* alliance worldwide. This alliance comprises a group of c. 10 notoriously reticulating mainly diploid and tetraploid evolutionary species, several of which are still undescribed. However, most of them are quite distinct in micromorphological characters apart from isozymic markers, and can be recognised also from herbarium material.

The changing concepts about speciation in ferns

The discovery that quite a few of the temperate ferns show much hybridisation, polyploidisation, hence reticulation, started much discussion on the evolutionary significance and historical fate of reticulating species. It has been suggested (Wagner 1951) that, at least in the ferns, high chromosome numbers have a high survival rate, and that, when polyploidy is common in a group, the low-numbered species will die out first. Later Wagner (1970) questioned the value of studying the variations of karyology and of genetic systems as if they were a kind of 'evolutionary noise' (Wagner 1970: 146). The main pillars of plant evolution were regarded to be populations of 'normal' species, i.e. diploid, sexual and outbreeding. However, others have argued that the significance of hybridisation in generating new variability and new evolutionary lines in plants is grossly underestimated. They suggested that the interaction of divergent genomes may have a profound positive effect on the fitness of hybrids and their polyploid derivatives, and afford novel variation on which selection may act. Allopolyploidisation might be even advantageous over natural selection in diploids in generating novel variation (Levin 1970).

It has also been suggested that polyploidisation and hybridisation is largely confined to temperate areas which endured dramatic changes in climate in recent geological history. However, this cannot explain the karyological diversity in ferns. Manton and her collaborators, and especially Walker (see for references Walker 1985) found an essentially similar situation in tropical fern floras. Hennipman (1977) traced notoriously reticulate complexes in tropical *Bolbitis*. Barrington, when studying tropical *Polystichum* (in Barrington *et al.* 1989) in montane Central America, found that reticulation in the montane tropics progresses much the same as in temperate areas. A high incidence of polyploidy is known from all the major evolutionary lines of the pteridophytes (Lovis 1977). Recently Hickey *et al.* (1989) reported a 56% incidence of polyploidy in the historical genus *Isoëtes*.

The karyological studies executed during the past decades include more than 20% of the total fern flora of the world. It is estimated that the fern floras of temperate and of tropical regions alike are for 50–60% of polyploid origin. Walker further remarked that the percentage of hybrids in a flora, e.g. 27% for

North America and 34% for Britain, is a function of the intensity of biological sampling rather than their true absence (Walker 1985).

Diploidisation of polyploid genomes through gene silencing; Haufler's approach

Haufler and collaborators have recently discovered esential new data regarding the mechanisms of fern evolution using the *Polypodium vulgare* alliance on a global scale as a model. As a result of their studies the idea is developing that the situation in genetically diploid ferns with polyploid chromosome numbers is not an original condition but has evolved along with diversification. It was further suggested (Haufler & Soltis 1986) that a random process called gene silencing is the mechanism operating in ferns that causes the genetical diploidisation of polyploids. Haufler and Sweeney (1989) studied isozymic markers in individual diploid gametophytes arising from the sexual allotetraploid hybrid species *Polypodium vulgare s.s.* with heterozygous banding patterns. They found that part of the marker loci of their original diploid progenitors do not express themselves anymore in the sexual tetraploid hybrid species. As a result the tetraploid *P. vulgare* can loose expression of part of its marker loci of their parental diploids and get diploidised. These authors thus validated the hypothesis of Werth and Windham (in Haufler & Sweeney 1989) which says that reciprocal gene silencing may promote divergence and perhaps speciation at the polyploid level.

Like polyploidisation the process of gene silencing can be regarded an intrinsic capacity, hence an ongoing process. It allows the polyploid to return to the advantages of diploid expression, and also permits subsequent divergent speciation (Barrington *et al.* 1989. Haufler 1987).

Diploidised hybrid species meet the requirements of evolutionary species as they have spatio-temporal limits, their own identity and their own evolutionary tendencies and historical fate. However, they originate from convergent speciation, and are not necessarily unique events as this type of speciation may occur again and again in different geographical sites and at different times.

Associated with this is the observation in the complexes studied so far that the allotetraploid derivatives have a far wider distribution than the diploid progenitors, if at all present anymore. Obviously allotetraploids have in general a better fitness and may outcompete their diploid parents in time. The problems with species concepts in ferns have been recently discussed in detail in America (see Haufler 1989a, b).

In conclusion, the emerging idea is that, apart from divergent speciation in sexual diploids (primary speciation) and hybrid speciation (secondary speciation; e.g. allopolyploidisation), diploidisation of hybrid species through gene silencing (tertiary speciation; Haufler in Barrington *et al.* 1989) is an intrinsic speciation mechanism. If so, the everlasting problem of the understanding of the high incidence of polyploidy in ferns is put into a completely new perspective.

Evolutionary species, speciation mechanisms, and the herbarium taxonomist

With the likelihood that hybrid species are similar to divergent species in evolutionary potential (Barrington *et al.* 1989), the herbarium monographer of tropical ferns may further want to intensify his ability to trace inferred hybrid species using the (micro)morphological information from herbarium material. This kind of information may be considered in relation to geographical and ecological characteristics which appear to be more often correlated with speciation events than earlier understood.

The herbarium taxonomist faces the problem of tracing the various types of evolutionary species mentioned above in the group under study, and to recognise hybrids. He may fail to do so when dealing with groups like *Adiantum* where virtually indistinguishable cryptic species occur (Paris *et al.* 1989), or with reticulate complexes.

On the other hand it has been shown in the past that, in many cases, hybrid species can be inferred or recognised successfully from herbarium material. Certainly one needs a keen eye for the presence of so-called structural irregularities (Wagner 1962) in the expression of macro- and micromorphological characters. Besides, one should not refrain from studying essential (sub)microscopical features such as shape and size of stomates, paraphyses and the shape and quality of the spores, and the number of spores per sporangium (Hennipman 1977, Barrington *et al.* 1986).

It should further be remarked that the general application of a broad morphological species concept in tropical floras in the past did not allow the recognition of local species even if minor morphological differences correlated with geographical and ecological characteristics were present. It seems likely that this practice cannot be upheld, not only for theoretical reasons as pointed out above but for practical reasons as well. As a result it is likely that future hypotheses on evolutionary species may often comprise much narrower species descriptions. As a consequence it seems worthwhile to reconsider the taxonomic status of several so-called local species which were earlier referred to (reticulated) species complexes characterised by macro- (and micro)morphological differences.

Evolutionary species may also better meet the demands of (applied) botanists who often have great difficulty in the implementation of a 'broad' species in the field situation in the tropics.

The implementation of these new views in herbarium practice is also challenging when one considers the complete lack of consensus about the use of the rank of species or lower ranks in ferns. This is nicely illustrated by Yatskievych and Moran (1989) who perused fifty fern monographs published during the last fifty years. They found that in some cases it was impossible to determine why the monographer used an infraspecific instead of the specific category. They also showed that the term 'variety' has lost its meaning since it is defined in so many ways that it is impossible to know *a priori* what kind of variation was intended to be denoted (l.c.: 43).

Our increasing perception of the speciation mechanisms that operate in temperate and tropical ferns, and our improvement in recognising evolutionary species is a major achievement. It is expected that the new concepts indicated above may lead to the improved delineation of evolutionary species also in tropical floras. The recognition of evolutionary species which often are much more narrowly circumscribed than the traditional macromorphological species coincides with the increasing need for a species concept that duly recognises ecological and geographical factors apart from (micro)morphological character states. The search for evolutionary species may also lead towards a more consistent formal taxonomy which is a promising perspective.

The recognition of monophyletic genera

Scientific classifications should primarily be natural and secondarily convenient. Classifications should be regarded as scientific theories that other people should be able to test. Nevertheless, modern pteridologists differ as far as the recognition of natural taxa and especially genera is concerned. As regards the Polypodiaceae, Pichi Sermolli (1977) recognised 63 genera which he referred to 14 natural 'groups' of the same rank. A provisional classification of the family (Hennipman *et al.* 1990) enumerates only 29 genera which are accommodated into two subfamilies. The latter classification mainly results from monographic work in our project applying cladistic analysis (Bosman 1986; Hennipman & Roos 1982; Hennipman & Hetterscheid 1984; Hovenkamp 1986; Ravensberg & Hennipman 1986; Rödl-Linder 1990; Roos 1985). The following examples illustrate part of the problems involved.

.The generic subdivision of the Platycerioideae

Stellate hairs have been regarded as the most primitive type of the polypodiaceous indument by Holttum (1947) who at that time accepted Bower's idea (1928) that the evolution of peltate scales from stellate hairs is a 'simple' one. Jarrett (1981) stated just the reverse. Holttum (1954) also suggested that stellate hairs similar to those found in the platycerioids may also occur in the drynarioids and in *Phymatodes* (= *Phymatosorus*). Pichi Sermolli (1977), who recognised 14 distinct groups of genera within the Polypodiaceae *s.s.*, recognised two different groups with stellate hairs, the group formed by *Pyrrosia, Drymoglossum* and *Saxiglossum* on the one hand, and the group formed by *Platycerium* on the other. He did not regard these two groups as closely related: "anyhow, if a relationship exists, it is very remote." (Pichi Sermolli 1977: 377). The distinctness of *Platycerium* was also recognised by Ching (1940), who gave *Platycerium* family rank.

The problems encountered thus are:

- the homology of the stellate hairs;
- the classification of the species with homologous stellate hairs.

From a detailed study on the ontogeny of the stellate hairs as occurring in the platycerioids, together with a study of the frond indument in other Polypodiaceae (Hennipman & Roos 1982; Hovenkamp 1986) it is shown that:

- Stellate hairs occur exclusively in the platycerioids; those reported for other Polypodiaceae are morphologically simple scales.
- Stellate hairs of the platycerioids are morphologically connected by means of all possible intermediates with two-celled glandular hairs as found on the gametophytes and the juvenile sporophytes of the platycerioids (as well as the fronds of most of the Polypodiaceae), thus confirming the observations made by Strazewski as early as 1915.
- Neither stellate hairs nor an indument similar to them have been found in the ontogenetic series of the scales of other Polypodiaceae. The idea that scales should be regarded as a progressive development of stellate hairs, or *vice versa,* therefore could not be confirmed.
- Ontogenetic studies show that scales and stellate hairs result from differences in the progressive development of similar few-celled uniseriate or branched glandular hairs that are found on the rhizome apices and the juvenile fronds respectively.
- Stellate hairs are the only type of indument on the fronds of the platycerioids. Scales are never found on the fronds, neither are stellate hairs present on the rhizomes. This divergence of indument is unique in the Polypodiaceae, as mentioned already by other authors and including Jarrett (1980).

Parallel to these studies of the indument, the systematics of the platycerioids as described by these unique stellate hairs were studied with the following results:

- The monophyly of *Platycerium* could be indicated; its taxonomic status as a natural genus thus being confirmed. It also appeared that the presence of dichotomous fronds is a derived (apomorphic) character state, thus not representing an original condition as advocated recently by others.

- The species referred to *Drymoglossum* even when this genus is taken in the restricted sense as delineated by Copeland (1947), form a paraphyletic group. *Drymoglossum* with five species and *Saxiglossum* with one species are distributed throughout different monophyletic groups within *Pyrrosia*. Therefore, the position taken by some modern authors, who recognise *Drymoglossum* and *Saxiglossum* separate from *Pyrrosia*, is not confirmed by the results obtained by Ravensberg and Hennipman (1986) and by Hovenkamp (1986).

However, at the same time Hovenkamp's work on *Pyrrosia* proper showed an unexpected and most surprising result, namely that the monophyly of *Pyrrosia* could not be demonstrated. The basal node in his cladogram of the platycerioids is a trichotomy consisting of *Platycerium*, the *Pyrrosia africana* group with two species, here abbreviated as *'Pyrrosia–2'*, and the remaining species of *Pyrrosia* with 49 species, here indicated as *'Pyrrosia–1'*. See Figures 11.1 and 11.2. All three groups can be described by unique character states, and consequently may be treated as three different monophyletic groups. The three possible fully resolved cladograms are corroborated when using character states as provided by Hovenkamp (1986).

- *Hypothesis 1* which says that *Platycerium* and *Pyrrosia–2* are mutually more related to each other than either one to *Pyrrosia–1* is corroborated by the shared possession of polo- and anomocytic stomata. This support, however, is weak as this type of stomata is the type commonly found outside the group of which the monophyly is tested (Sen & Hennipman 1981).
- *Hypothesis 2* which says that *Pyrrosia–1* and *Pyrrosia–2* are closest is supported by the shared presence of entire fronds. Entire fronds, however, are considered the original condition in the platycerioids, a view supported by the heteroblastic frond studies performed on *Platycerium* species.
- *Hypothesis 3* which says that *Platycerium* and *Pyrrosia–1* are more related to each other than to *Pyrrosia–2* is supported by the shared possession of 'massive sclerification' of the rhizome.According to Hovenkamp such sclerifications are not found elsewhere in the Polypodiaceae. Using outgroup comparison as the criterion for determining polarity of character states within the Polypodiaceae, the shared possession of massive sclerification should be considered a synapomorphy for *Platycerium* and the group of *Pyrrosia–1*.

If this character state is indeed a synapomorphy, the group of *Pyrrosia–2* should be given generic recognition within the Platycerioideae. The rather complex venation types as found in both *Platycerium* and *Pyrrosia–1* then should be interpreted as progressive developments of a venation pattern with an indistinct differentiation between secondary and tertiary veins like that retained in the group of *Pyrrosia–2*. The hypothesis put forward that *Pyrrosia* should be divided

	Characters: 1. sclerification of rhizome; 2. stomata pericytic; 3. stomata polo-/anomocytic; 4. 'woolly' stellate hairs; 5. fronds dichotomous; 6. fronds entire; 7. spores colliculate; 8, 9, 10. venation types									
	1	2	3	4	5	6	7	8	9	10
PY-1	1	1	0	1	0	1	0	0	0	1
PY-2	0	0	1	0	0	1	1	1	0	0
PLAT	1	0	1	0	1	0	0	0	1	0
OUTGR	0 (1)	0	1	0	0	1/0	0	0	0	0
	*	+	*	+	+	*	+	+	+	+

Figure 11.1 — Datamatrix of character states used to discuss generic relationships in the Platycerioideae. PY-1 = *Pyrrosia* p.p., PY-2 = *P. africana* group, PLAT = *Platycerium*, OUTGR = outgroup. * = character state supporting dichotomous cladogram in Figure 11.2; + = autapomorphy.

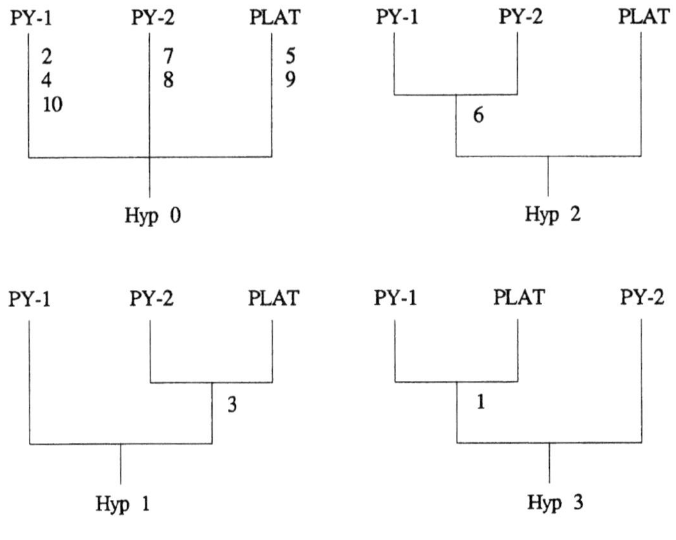

Figure 11.2 — Three taxon statements of the main groups in the Platycerioideae. For legend see Figure 11.1 and the text.

into two parts that are not closest related to each other, one part being closest to *Platycerium,* should be tested by the distribution of character states of characters not yet studied.

The ant-ferns

Ant-ferns are Polypodiaceae that occur in Malesia and America. There is much debate regarding the monophyly and the generic delimitation in this group. Some recent authors (e.g., Pichi Sermolli 1977) have placed all of the polypodiaceous ant-ferns together in one 'natural' group, whereas others held different opinions.

As regards the accommodation of the Malesian ant-ferns Holttum (1954) and others placed the species with a scaly rhizome and sori situated on the blade, together with other species with round scales, in *Phymatodes s.s.* (= *Phymatosorus*) which Holttum regarded as 'a very natural group'. The ant-ferns with naked rhizomes and with sori situated on reflexed fertile lobes with constricted bases were placed by him in *Lecanopteris s.s.* On the other hand Pichi Sermolli (1977) recognised within the group comprising the Malesian ant-ferns, apart from *Lecanopteris s.s.* with naked rhizomes and marginal sori, the genus *Myrmecopteris* (syn. *Myrmecophila,* nom. illeg.), thus accommodating part of *Phymatodes sensu* Holttum (1954). In the meantime it has been demonstrated that new species of *Lecanopteris* could be described with a naked rhizome and with sori as in *Phymatosorus* (Jermy & Walker 1979; Hennipman & Verduyn 1987; Hennipman 1990b). Thus the soral position does not warrant the recognition of two different genera. Detailed studies on the indument of living plants of five species of *Lecanopteris s.s.* held in cultivation at Utrecht further showed that at least the rhizome apices of all 'naked' *Lecanopteris* species are set with hairs apart from tiny scales similar to the smaller scales as found on the rhizomes of the *Myrmecophila* group. On the basis of present knowledge there is no support for recognition of two genera of Malesian ant-ferns.

The American ant-ferns (the potato-ferns) are generally referred to *Solanopteris,* a tropical genus of three small-sized epiphytic species forming ant-gardens (Gómez 1974). *Solanopteris* has unique character states like the presence of rhizome tubers apart from its paraphyses. The macro- and micromorphology of its species are quite different from those of the Malesian species. When including the micromorphological characters in our analyses the relationship of *Solanopteris* appears to be closest to *Microgramma* including *M. (Pleopeltis) percussa,* as they share the possession of uniquely shaped frond scales, which is an inferred synapomorphy (Baayen & Hennipman 1987). Whether or not *Solanopteris* should be united with *Microgramma* seems a matter of personal judgement.

In conclusion, the cladistic analysis of a datamatrix also including micromorphological characters show that the Asian and American ant-ferns belong to two quite different evolutionary lines within the subfamily Polypodioideae. The systematic position of *Solanopteris* is in the tribe Polypodieae, that of *Lecanopteris* is still obscure.

Character analyses

Character analyses on selected features, using representative samples of Polypodiaceae, and applying also new methods and techniques, were executed in order to trace possible new sets of characters and inferred apomorphic character states. The analyses included stomata (Sen & Hennipman 1981), paraphyses and sterile frond indument (Baayen & Hennipman 1987), venation patterns (Hetterscheid & Hennipman 1984), spores (Hennipman 1990a; van Uffelen & Hennipman 1985), sporogenesis (van Uffelen, in prep.) and rhizome anatomy. It was anticipated that such studies could be helpful in refining existing classifications, as well as in classifying groups of ferns of which the relationships were still unknown or disputed.

Characters from the microscopical or lower levels are often regarded by other authors as independent of selective pressure as they are less likely to be manipulated by the environment. Therefore, such characters might be of significant phylogenetic importance. The following illustrates a character analysis of the spores executed in cooperation with Wim Star, Tuhinsri Sen, Gerda van Uffelen and Betty Verduyn.

Diversity of polypodiaceous spores

An extensive study of the exospore and the perispore executed with the light microscope (LM), scanning electron microscope (SEM), combined with the transmission electron microscope (TEM), revealed the existence of a completely new set of characters, and the recognition of a number of quite distinct types of spores. The spores of Polypodiaceae are monolete, usually bean-shaped, with a linear aperture. The mature wall consists of two essentially different layers, the exospore and the perispore. In quite a few species the perispore is indistinct. This may account for the general idea that a perispore might be absent in Polypodiaceae (Alston 1956), or that its presence in some species is at least questionable (Lloyd 1981). However, all polypodiaceous spores do have a perispore and quite a few distinct perispore types can be recognised (Hennipman 1990a; Lloyd 1981; van Uffelen & Hennipman 1985; Roos 1985).

One of the types of spores that can be recognised is the spore type with a smooth outer surface of the exospore – the plesiomorphous state – and a relatively thick perispore. The outer surface ornamentation of the latter can be more or less smooth, tuberculate or verrucate. The perispore is often provided with unique baculae or spines; it is always very electron dense when seen with TEM. This type of perispore is an inferred apomorphy. It describes a group of ill-defined genera, the systematic position of which is much debated. This group of palaeotropical and mainly Asian ferns includes *Selliguea* (oldest generic name), *Crypsinus, Crypsinopsis, Grammatopteridium, Holcosorus, Oleandropsis, Phymatopteris,* and *Pycnoloma.* Three more genera may together with *Selliguea s.l.* constitute a monophyletic group: *Polypodiopteris, Arthromeris,* and *Christopteris* (*'Christiopteris'*).

Testing the relationship of *Polypodiopteris* with *Selliguea* as proposed earlier by Christensen (1937) and Copeland (1947), based on the characteristics of the spores, could start only recently when mature spores of the two species comprising this genus from Borneo became available. They appeared to be similar to those found in *Selliguea s.l. Polypodiopteris* deviates from typical *Selliguea* in having pectinate fronds. However, such fronds are found in other groups of Polypodiaceae as well and are likely to represent a parallel development.

Arthromeris differs from *Selliguea* in having articulated pinnae and a perispore with hollow instead of solid tubercles set with baculae. Its generic identity remains to be demonstrated in view of the variation of fronds found in *Selliguea s.l.*

The systematic position of *Christopteris,* a genus comprising two small-sized epiphytes from tropical Asia, is still ambiguous. It seems to belong to either this alliance or to the microsoroids. This genus has received much attention from many pteridologists in the past. It has, for instance, been regarded a primitive genus by Bower (1928), and as an isolated one within the Polypodiaceae by Pichi Sermolli (1977). The genus *sensu* Hennipman and Hetterscheid (1984) consists of two species, *C. tricuspis* from mainland Asia and *C. sagitta* from the Philippines. When found in the wild these two species look rather similar to species of *Selliguea* apart from being acrostichoid. The relationship of this genus to *Selliguea s.l.* could not be tested up till now as mature spores are not available from either two species. The living material collected during 1986 from Mt Apo, Mindanao, produces in Utrecht abundant sterile fronds only.

The crestate type of perispore; parallelism at the submicroscopical and ultrastructural level

One of the most difficult problems with which the fern phylogenist has to deal is homoplasy. Repetition of character states takes place in what are assumed to be entirely different lines of evolution. Parallel developments are not restricted to macromorphological characters but can also express themselves at submicroscopical and ultrastructural levels as is shown by the following example.

One of the perispore types that was recognised (Hennipman 1990a) is the winged or crestate perispore in which the perispore expresses itself in solid crests or folds which are usually reticulately connected. In some cases, however, the crests of the perispore run parallel to the aperture. The latter condition is found in one of the subgroups of *Pyrrosia,* the *P. angustata* group, Platycerioideae (van Uffelen & Hennipman 1985; Hovenkamp 1986), and in *Goniophlebium,* Polypodioideae (Hennipman 1990a; Rödl-Linder 1990).

The spores of *Pyrrosia* species show a surprising variation of different simple and complex morphological types using especially characters from the perispore (van Uffelen & Hennipman 1985). *Pyrrosia* and *Goniophlebium* are not closest relatives as each of them shares far more similarities with other groups. *Pyrrosia* is part of the subfamily Platycerioideae of which the monophyly is presently undisputed. *Goniophlebium* is part of the subfamily Polypodioideae.

The morphological similarities as seen with the SEM were further studied with the TEM. It was found that at high magnifications the crests of both *Pyrrosia angustata* and *Goniophlebium demersum* were built of fibrillae running parallel to each other. The similarities in micromorphological and ultrastructural characters as developed in *Pyrrosia angustata* and *Goniophlebium* certainly result from parallel evolution. An almost similar diversity and plasticity was found in the detailed structure of the exospore.

In conclusion, the general idea that spore characters are conservative in nature could not be supported. Characters of both the exospore and the perispore show great diversification and are most liable to change. Unique spore types found in Polypodiaceae can act as markers for tracing inferred monophyletic groups. This is especially significant in groups of Polypodiaceae of which the relationships used to be most ambiguous. On the other hand character states of spores – and even surprisingly complex ones – may express themselves in submicroscopical and ultrastructural characteristics in a similar way in quite different evolutionary lines.

Acknowledgements

I wish to express my gratitude to a great number of talented individuals who have added significantly to the progress of work on Polypodiaceae in the Netherlands and abroad. I am especially thankful to Prof. Dr. C.H. Haufler, Dr. M.C. Roos, Mrs. G.A. van Uffelen and Mrs. G.P. Verduyn for help in various ways. Mr. R.V. Hensen critically read the manuscript.

References

Alston, A.H.G. 1956. The subdivision of the Polypodiaceae. Taxon 5: 23–25.

Baayen, R.P., & E. Hennipman. 1987. The paraphyses of the Polypodiaceae (Filicales). Beitr. Biol. Pflanzen 62: 251–347.

Barrington, D.S., C.H. Haufler & Ch.R. Werth. 1989. Hybridization, reticulation, and species concepts in the Ferns. Amer. Fern J. 79: 55–64.

Barrington, D.S., C.A. Paris & T.A. Ranker. 1986. Systematic inferences from spore and stomate size in the ferns. Amer. Fern J. 76: 149–159.

Bosman, M.T.M. 1986. Preliminary results of monographic studies in Microsorum. Acta Bot. Neerl. 35: 117.

Bower, F.O. 1928. The ferns. Vol. 3. The leptosporangiate ferns. Cambridge.

Ching, R.C. 1940. On natural classification of the family 'Polypodiaceae'. Sunyatsenia 5: 201–268.

Christensen, C. 1937. Taxonomic Fern-Studies IV. Revision of the Bornean and New Guinean ferns collected by O. Beccari and described by V. Cesati and J.G. Baker. Dansk Bot. Ark. 9: 33–52.

Copeland, E.B. 1947. Genera Filicum. Chronica Botanica Comp. Waltham, Mass.

Gómez, L.D. 1974. Biology of the potato-fern, Solanopteris brunei. Brenesia 4: 37–61.

Haufler, C.H. 1987. Electrophoresis is modifying our concepts of evolution in homosporous pteridophytes. Amer. J. Bot. 74: 953–966.

Haufler, C.H. 1989a. Species concepts in pteridophytes: Introduction. Amer. Fern J. 79: 33–35.

Haufler, C.H. 1989b. Species concepts in pteridophytes: Summary and synthesis. Amer. Fern J. 79: 90–93.

Haufler, C.H., & D.E. Soltis. 1986. Genetic evidence indicates that homosporous ferns with high chromosome numbers may be diploid. Proc. Natl. Acad. U.S.A. 83: 4389–4393.

Haufler, C.H., & J.S. Sweeney. 1989. Electrophoretic evidence that a reciprocal gene silencing mechanism can promote genetic diversity in polyploids. Amer. J. Bot. 76, Suppl.: 203.

Haufler, C.H., & M.D. Windham. 1989. The inscrutable Polypodium vulgare complex: insights from Asia. Amer. J. Bot. 76, Suppl.: 202–204.

Hennipman, E. 1977. A monograph of the fern genus Bolbitis (Lomariopsidaceae). Leiden Bot. Ser. 2: 1–331.

Hennipman, E. 1990a. The significance of the SEM for character analysis of spores of Polypodiaceae (Filicales). The Systematic Association, Special Volume 41: 23–44. Clarendon Press, Oxford.

Hennipman, E. 1990b. The systematics and phytogeography of the Malesian ant-fern genus Lecanopteris (Filicales, Polypodiaceae). Proc. Intern. Symp. Systematic Pteridology Beijing 1988: 39–47.

Hennipman, E. & W.L.A. Hetterscheid. 1984. The emendation of the fern genus Christiopteris, including the transference of two taxa to the microsorioid Polypodiaceae. Bot. Jahrb. Syst. 105: 1–10.

Hennipman E., K.U. Kramer & P. Veldhoen. 1990. Polypodiaceae. In: K.U. Kramer & P.S. Green (eds.), The Families and Genera of Vascular Plants. Pteridophytes and Gymnosperms. Springer. In press.

Hennipman, E., & M.C. Roos. 1982. A monograph of the fern genus Platycerium (Polypodiaceae). Verh. Kon. Ned. Akad. Wet., Natuurk., Ser. 2, 80: 1–126.

Hennipman, E., & G.P. Verduyn. 1987. A taxonomic revision of the genus Lecanopteris (Polypodiaceae) in Sulawesi, Indonesia. Blumea 32: 313–319.

Hetterscheid, W.L.A., & E. Hennipman. 1984. Venation patterns, soral characteristics, and shape of the fronds of the microsorioid Polypodiaceae. Bot. Jahrb. Syst. 105: 11–47.

Hickey, R.J., W.C. Taylor & N.T. Luebke. 1989. The species concept in Pteridophytes with special reference to Isoëtes. Amer. Fern J. 79: 78–89.

Holttum, R.E. 1947. A revised classification of Leptosporangiate ferns. J. Linn. Soc. Bot. 53: 123–186.

Holttum, R.E. 1954. A revised Flora of Malaya. 2. Ferns of Malaya.

Holttum, R.E. 1972. Posing the problems. In: A.C. Jermy, J.A. Crabbe & B.A. Thomas, The phylogeny and classification of the Ferns: 1–10. Acad. Press, London.

Hovenkamp, P.H. 1986. A monograph of the fern genus Pyrrosia (Polypodiaceae). Leiden Bot. Ser. 9: 1–280.

Jarrett, F.M. 1980. Studies in the classification of the leptosporangiate ferns: 1. The affinities of the Polypodiaceae sensu stricto and the Grammitidaceae. Kew Bull. 34: 825–833.

Jermy, A.C., & T.G. Walker. 1975. Lecanopteris spinosa – a new ant-fern from Indonesia. Fern Gaz. 11: 223–256.

Levin, D.A. 1970. Hybridization and evolution – a discussion. Taxon 19: 167–171.

Lloyd, R.M. 1981. The perispore in Polypodium and related genera (Polypodiaceae). Canad. J. Bot. 59: 175–189.

Lovis, J.D. 1977. Evolutionary patterns and processes in ferns. Adv. Bot. Res. 4: 229–415.

Manton, I. 1954. Problems of cytology and evolution in the Pteridophyta. Cambridge Univ. Press, London and New York.

Pichi Sermolli, R.E.G. 1977. Tentamen Pteridophytorum genera in taxonomicum ordinam redigendi. Webbia 31: 313–512.

Paris, C.A., F.S. Wagner & W.H. Wagner Jr. 1989. Cryptic species, species delimitation, and taxonomic practice in the homosporous ferns. Amer. Fern J. 79: 46–54.

Ravensberg, W.J., & E. Hennipman. 1986. The species of Pyrrosia formerly referred to Drymoglossum and Saxiglossum (Filicales, Polypodiaceae). Leiden Bot. Ser. 9: 281–310.

Rödl-Linder, G.A 1990. A monograph of the fern genus Goniophlebium (Polypodiaceae). Blumea 34: 277–423.

Roos, M.C. 1985. Phylogenetic systematics of the Drynarioideae (Polypodiaceae). Thesis, Utrecht.

Roos, M.C. 1986. Ibid. Verh. Kon. Ned. Akad. Wet., Natuurk., Ser. 2, 85: 1–317.

Sen, U., & E. Hennipman. 1981. The stomates of the Polypodiaceae (Filicales). Blumea 27: 175–201.

Shivas, M.G. 1961. Contributions to the cytology and taxonomy of species of Polypodium in Europe and America. I. Cytology. II. Taxonomy. J. Linn. Soc. (Bot.) 58, 370: 13–25; 27–38.

Strazewski, H. Ritter von. 1915. Die Farngattung Platycerium. Flora (Jena) 108: 271–310.

Uffelen, G.A. van, & E. Hennipman. 1985. The spores of Pyrrosia Mirbel (Polypodiaceae), a SEM study. Pollen et Spores 27: 155–197.

Wagner Jr, W.H. 1951. Cytotaxonomic analysis of evolution in Pteridophyta. Evolution 5: 177–181.

Wagner Jr, W.H. 1954. Reticulate evolution in the Appalachian aspleniums. Evolution 8: 103–118.

Wagner Jr, W.H. 1962. Irregular morphological development in hybrid ferns. Phytomorphology 12: 87–100.

Wagner Jr, W.H. 1970. Biosystematics and evolutionary noise. Taxon 19: 146–151.

Walker, T.G. 1985. Cytotaxonomic studies of the ferns of Trinidad 2. The cytology and taxonomic implications. Bull. Br. Mus. Nat. Hist. (Bot.) 13: 149–249.

Yatskievych, G., & R.C. Moran. 1989. Primary divergence and species concepts in ferns. Amer. Fern J. 79: 36–45.

12 A review of natural vegetation studies in Malesia, with special reference to Indonesia

KUSWATA KARTAWINATA

Summary

Many of the recent studies on vegetation and ecology in the Flora Malesiana region have been carried out in Malaysia and Papua New Guinea. Forests have received greater attention than other ecosystems. The ecology of various types of rain forest ecosystems has been summarised by Whitmore (1984). In Indonesia the mangrove ecosystem has been studied in some detail, while other ecosystems hardly have been investigated. Van Steenis (1935) gave a comprehensive qualitative description of various vegetation types of Indonesia, which, although somewhat outdated, is still the most complete. Several general vegetation maps of Malesia are available. Detailed large-scale vegetation maps of Sumatra have recently been produced. Synthesis of available ecological data and information on Sumatra and Sulawesi has also been attempted recently. A great deal of intensive floristic and ecological study is still urgently needed, before the natural vegetation in the region disappears because of the current rate of utilisation, conversion and destruction.

Introduction

Available references on Malesian vegetation and ecology reveal that most investigations have been carried out in Malaysia. The studies have been largely concentrated on forests, particularly the lowland dipterocarp forest. Much of the information has been synthesised by Whitmore (1984a) in his Tropical Rain Forest of the Far East. For Papua New Guinea, Paijmans (1976), Gressitt (1982), Johns (1985, 1987a & b), Brouns (1987) and Grubb and Stevens (1985) provide additional information on vegetation, biogeography and ecology. Beyond the accounts referred to by Whitmore (1984a), little is known about the Philippines. The information generated there from the Forest Resources Inventory programme, launched since 1983, is concerned only with dominant timber trees and is of little value for floristic interpretation (Madulid, pers. comm.).

The following brief account illustrates the progress of vegetation and ecological research in Indonesia during the last three decades, but in no way the data are exhaustive or critically reviewed. Only natural communities are dealt with.

P. Baas et al. (eds.), The Plant Diversity of Malesia, 121–132.

Vegetation and plant ecological studies in Indonesia

Botanical studies in Indonesia date back to the pre-Linnean time. Vegetation studies and plant ecological investigations, however, have barely started and only during the last two decades the work on the subject by resident and visiting specialists has increased substantially.

Vegetation mapping

Published vegetation maps covering Indonesia include the 'Atlas van tropisch Nederland' (van Steenis 1938), 'Vegetation Map of Malaysia' (van Steenis 1958), and 'Map of Forest of Indonesia' (Hannibal 1950, cited by Dilmy & Kostermans 1958). The latter was based on van Steenis' map and the results of the field strip surveys made by the Forest Service. Some of the surveying strips were maintained as permanent plots until the early 1970s when the logging activities encroached the area.

As a byproduct of the intensified forest inventory of commercial timber trees to meet the need of increasing logging activities, complemented with aerial photographs, the Direktorat Bina Program (1980) published a 'Map of Forest Stands of Indonesia' (scale 1 : 2,750,000). BIOTROP and the 'Institute de la Carte International du Tapis Végétal' published vegetation and bioclimatic maps of Sumatra (scale 1 : 1,000,000), based on satellite images and field survey data (Lamounier *et al.* 1983, 1986, 1987). Other maps published recently include the 'Land Use Map' (scale 1 : 2,500,000) (Tata Guna Tanah 1980) and the 'Original Vegetation Types of the Sundaic and Wallacean Subregions' (MacKinnon & MacKinnon 1986). The later was used by Whitmore (1984b) as one of the bases for his 'Vegetation Map of Malesia'.

Large-scale vegetation maps often complement the management plans of various national parks, nature reserves and other protected areas, prepared by PHPA (the Directorate-General of Nature Conservation and Forest Protection) in cooperation with WWF and/or FAO. The floristic composition of various vegetation types, however, is often lacking, except for the list of common species. The most detailed large-scale vegetation map is that of the Ujung Kulon National Park (scale 1 : 75,000) and each vegetation type is complemented with a floristic enumeration and a list of character species (Hommel 1987).

The RePProT (Regional Planning Programme for Transmigration) produced accurate land use, land suitability and land capability maps (with scale 1 : 250,000). The land use maps indicate the vegetation types well, but floristic indicators are lacking. The maps could constitute a good base for further detailed vegetation studies. Vegetation maps on wetlands are contained in survey reports prepared by the Asian Wetland Bureau and PHPA (Silvius 1987).

The recent vegetation maps become quickly obsolete due to the rapid destruction of natural vegetation, especially lowland forest throughout the country.

This is attributed to logging, development projects (e.g. tree and crop plantations, and transmigration), smallholder conversion including shifting cultivation, and loss due to fire. The current best estimate of the annual primary forest destruction rate is about 900,000 hectares, of which almost half is attributed to smallholder conversion (unpublished World Bank Report).

Floristic composition and ecological studies

Lowland vegetation

The work by van Steenis (1935) on 'Maleische vegetatieschetsen' is by far the most complete, qualitative, and detailed description of the vegetation of Indonesia. It includes also various ecological factors responsible for the patterning of the vegetation, correlating the climate, soil, geology, and water availa-

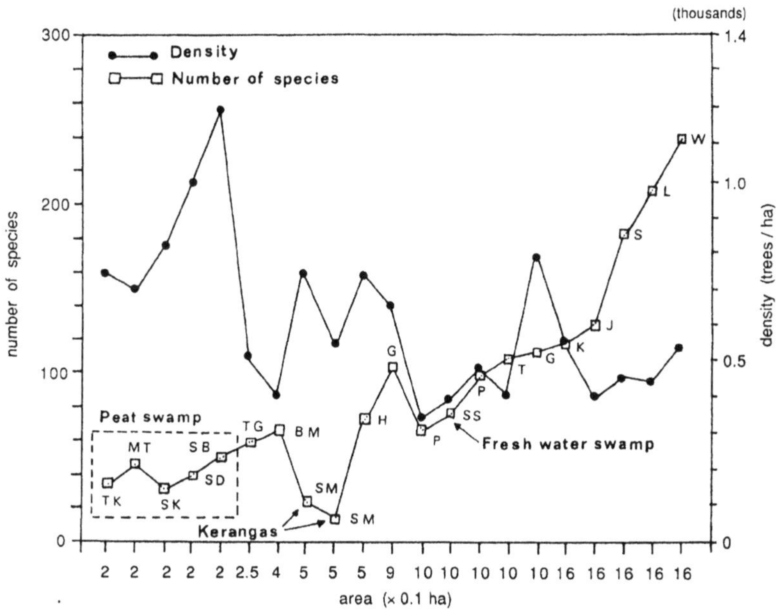

Figure 12.1 — Number of species and density of trees (diam. > 10 cm) in plots of different size in lowland forests of Indonesia. BM: Biak Mentelang, G: Gunung Gadut, H: Halmahera, J: Jaro, K: Ketambe, L: Lempake, MT: Muara Tolam, P: Pleihari, S: Sekundur, SB: Sungai Sebangau, SD: Sungai Durian, SK: Sungai Siak Kecil, SM: Semboja, SS: Sungai Sesayap, T: Toraut, TG: Tanah Grogot, TK: Teluk Kiambang, W: Wanariset. [Sources: Abdulhadi *et al.* 1989; Anderson 1976; Jafarsidik 1980; Kartawinata *et al.* 1981; Kohyama *et al.* 1989; Mogea 1980; 1984; Riswan 1987a & b; Suhardjono & Wiriadinata 1984; Whitmore & Sidiyasa 1986; Whitmore *et al.* 1987.]

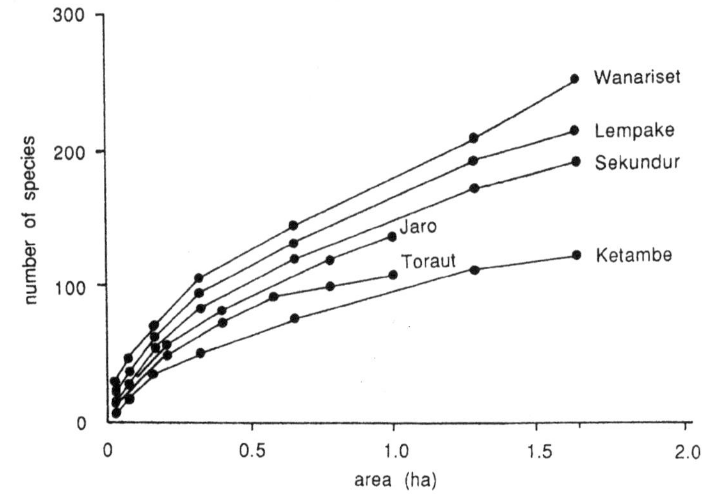

Figure 12.2 — Species-area curve for trees with dbh > 10 cm in some samples of lowland forests in North Sumatra (Ketambe and Sekundur), East Kalimantan (Lempake, Jaro and Wanariset), and North Sulawesi (Toraut). (After Abdulhadi *et al.* 1989; Whitmore & Sidiyasa 1986.)

bility with vegetation types. Van Steenis (1957) used habitat factors as the basis for constructing the outline of vegetation of Malesia and this approach was also adopted by Whitmore (1984a) in his tropical forest classification in the Far East. The outline has been further refined and applied to Indonesia (Kartawinata 1978, 1981) and the latest refinement is presented elsewhere (Kartawinata 1989).

During the last two decades various studies and inventories were undertaken mostly in Kalimantan and Sumatra by the National Institute of Biology (now Center for Research and Development in Biology) and the Forest Research Institute (now Forest Research and Development Center). These include botanical exploration, enumeration of tree species and ecology of the effects of human activities on the lowland forests. Tree species enumeration within plots (including permanent ones) is aimed at gathering information on floristic composition of vegetation types in the area explored and species diversity. In Figure 12.1 examples are summarised of the results obtained, illustrating species richness in various forests.

In East Kalimantan dipterocarp species are prevalent in the lowland forests, except in freshwater swamp forests. In a permanent plot of 10.5 hectare lowland dipterocarp forest at Wanariset in East Kalimantan, 406 species of trees (dbh > 10 cm), 178 genera and 57 families were recorded, with a density of 534 trees/ha, a mean basal area of 29 sq.m/ha (Partomihardjo *et al.* 1984).

The dipterocarp species are the largest trees and dominate the forest and the number of species amounts to 30 (7.4%) with a total basal area of 79.5 sq.m (26%). This is the richest forest in the country, where in a 1.6-hectare plot 239 species are recorded (Figure 12.2) (Kartawinata *et al.* 1981). The Lempake forest differs in tree composition and structural characteristics from that of Wanariset. It contains less dipterocarps, amounting to only 13 species in a plot of 1.6 ha with a density of 27 trees per hectare and a basal area of 10 sq.m per hectare (Riswan 1987a).

In 25 plots of primary forest at the Gunung Leuser National Park, North Sumatra, 332 tree species, 179 genera, and 68 families were recorded (Abdulhadi *et al.* 1984). The forest is less diverse than those at Lempake and Wanariset. Dipterocarps are less prominent and only 12 species of the family with a density of 23 trees/ha are recorded. Meliaceae, however, are more abundant, having 22 species with a mean density of 74 trees/ha, second only to Euphorbiaceae. The forest has a low stature with a mean height of 15–23 m, depending on the topography.

The species richness varies a great deal, as indicated also by MacKinnon (1982) for various forest types. In some instances the low number of species may not indicate low species richness, but may be attributed to inadequacy of field identification. In some cases plants are identified by their vernacular names only without a complete set of voucher specimens. The results are definitely unreliable, since only few species have consistent vernacular names. Collection of voucher specimens presents problems also as good (fertile or sterile) herbarium specimens are sometimes difficult to collect, the researcher thus relying only on fallen leaves. This is one of the constraints in securing good data on tree enumeration, and has been a nagging problem as has been pointed out by Dilmy and Kostermans (1958).

Mangroves are the best studied vegetation type. Their floristic composition and structure are well known although they vary from one place to another contingent upon habitat conditions. Data on mangrove ecology can be found in several proceedings of national seminars (Soemodihardjo *et al.* 1979, 1984; Soerianegara *et al.* 1986, 1987), in a review (Soemodihardjo 1985), a research report (Ogino & Chihara 1988), and other separate papers.

Montane vegetations

Floristic enumeration of montane vegetation has received little attention. In Java, van Steenis (1972) synthesised available information on vegetation and ecology. Quantitative assessment was carried out at the Gunung Gede-Pangrango by Meijer (1959), who recorded 355 species in a one-hectare plot, and Yamada (1975, 1976a & b, 1977), who provided data on floristic composition along the altitudinal gradient.

Outside Java, quantitative floristic assessment has been limited to East Kalimantan (Bratawinata 1986) and Irian Jaya (Hope *et al.* 1976). Bratawinata shows the altitudinal distinction of primary forest in the Apo Kayan area. The number of tree species does not always decrease with increasing elevation, but species composition, prevalent species and structural characteristics do change. The only comprehensive study of high altitude vegetation was carried out by Hope *et al.* (1976) in the Sudirman Range, particularly around the permanently snow-capped Mt Jaya. Not less than 24 community types occurring in varied habitats could be identified, ranging from upper montane forest to alpine tundra. They represent typical vegetations in the highlands of Irian Jaya and perhaps the whole of New Guinea.

Dynamics of vegetation

The dynamics of vegetation received the greatest emphasis in ecological studies. Successions after volcanic eruption on the Krakatau Islands and vicinity have been the focus of several studies (e.g. Whittaker *et al.* 1984, 1989; Richards & Wiriadinata 1985; Tagawa 1984; Tagawa *et al.* 1985; Soeriaatmadja 1985; Kartawinata *et al.* 1985). The forests thus far developed during the last 100 years are still in a seral stage. The return to conditions similar to a primary forest will take a very long time since sources of propagules in Java and Sumatra are declining.

Studies on successions in naturally formed gaps have been just initiated in lowland and montane forests (Partomihardjo *et al.* 1987; Sriyanto 1987; Mackie *et al.* 1987). The gaps in primary forest constitute 16–28%, with a size mostly of less than 500 sq.m and regeneration within them consists mostly of primary forest species, while in secondary forest the gaps constitute 13% with a size range of 18–186 sq.m. Gaps created by traditional timber extraction by the natives are comparable to natural treefall gaps in size and in density of seedlings they contain. Gaps resulting from logging activities, however, are much bigger, amounting to more than 30%, and a gap can be as big as one hectare. In such a big gap regeneration consists mostly of secondary forest species.

Studies on regeneration after human disturbance deal with the effects of logging and shifting cultivation. Substantial data have been collected from studies on logged-over forests conducted by the Forest Research Institute. The studies are concerned primarily with the regeneration of dipterocarps, and the regeneration is identified by commercial species group and not by species; hence it is difficult to make a floristic interpretation. Environmental effects of mechanised logging in Indonesian forests have been summarised by Kartawinata (1981). It is shown that on average 50% of the residual trees are damaged, regeneration is dominated by secondary forest species, and there is a sign of genetic erosion due to the creaming or extraction of the best trees (dipterocarps particularly) as well as loss of species.

Practices of traditional shifting cultivation by Kenyah Dayak do not lead to permanent forest destruction (Mackie *et al.* 1987). If left fallow for more than 100 years the old field communities resulted therefrom would acquire structure and composition of primary forest. However, the rate of return to primary forest composition is dependent upon the proximity of seed sources. The recovery is very slow and estimated to take 148–478 years for a young secondary forest to reach maturity (Riswan 1982; Riswan & Kartawinata 1988a; Riswan *et al.* 1985).

Experimental studies of succession after clear cutting and burning of lowland dipterocarp forest and kerangas forest in East Kalimantan (Riswan 1982; Riswan & Kartawinata 1988b, 1989) show that seedlings were the main component in the vegetation recovery process in dipterocarp forest, in contrast to vegetative resprouts in kerangas. Different regimes of disturbances lead to differences in species composition, but in kerangas forest composition is similar. The pattern of secondary succession is critically determined by the successional process during the first 6-month period. Burning altered the soil properties in dipterocarp forest but soil fertility returned to the original level within 1.5 years, but in kerangas forest the situation is different.

Other ecological aspects

'The ecology of Sumatra' (Whitten *et al.* 1984) and 'The ecology of Sulawesi' (Whitten *et al.* 1987) have been recently published; 'The ecology of Kalimantan' is now in preparation. These studies are the synthesis of available information and contain good accounts on plant ecology. The book on the ecology of Sulawesi includes also keys to plant species of mangroves and estuaries, aquatic macrophytes and tree ferns.

Plant-animal interactions received great interests of various specialists (e.g. Rijksen 1978; Whitten 1980). Although the studies emphasised the ecology of primates, they contain good accounts on structure and composition of vegetation, phenology and seed dispersal. Current information on production ecology is very limited and studies have been carried out mainly in lowland forests of Sumatra and Kalimantan (e.g. Schaik & Mirmanto. 1985; Yamakura *et al.* 1986) and montane forest of Java (Yamada 1976). Litter production of the lowland forest of Sumatra is estimated to be 7.6–11.5 tons/ha/year and that of Kalimantan is 7.2 tons/ha/year while that of the montane forest is 5.9 tons/ha/year. The above ground biomass of a lowland forest of East Kalimantan is estimated to be 509 ton/ha. Little is known about nutrient cycling in various ecosystems. The studies by Bruijnzeel (1983, 1984, 1985) and Riswan (1982) deal with some aspects of nutrient cycling.

Long-term ecological studies of peat swamp, fresh water, kerangas lowland and montane forests, are currently being carried out at the Gunung Palung Nature Reserve, West Kalimantan by Dr. Mark Leighton of Harvard University. The studies cover such aspects as vegetation structure and composition, regeneration, phenology, and plant-animal interaction as the bases for forest management. Much of the palaeo-ecological studies have been carried out by Flenley and his students and the results are incorporated into his book 'The equatorial rain forest: a geological history' (Flenley 1979).

Conclusion

Research has been focused mainly on vegetation and dynamics of the lowland rain forests of Sumatra and Kalimantan. Little information has been collected on fresh water and peat swamp forests, montane forests outside Java, seasonal forests, savannas and grasslands. Not much work has been done on such aspects as phenology, autecology, plant-soil relationships, reproductive ecology, and production ecology. Lack of trained personnel is one of the factors in the slow progress of studies on vegetation and ecology in Indonesia.

Selected references

Abdulhadi, R., K. Kartawinata & R. Yusuf. 1984. Pola hutan di Ketambe, Taman Nasional G. Leuser, Aceh. In: S. Wirjoatmodjo (ed.), Laporan Teknik 1982–1983: 207–214. LBN-LIPI, Bogor.

Abdulhadi, R., E. Mirmanto & R. Yusuf. 1989. Struktur dan komposisi petak hutan Dipterocarpaceae di Ketambe, Taman Nasional G. Leuser, Aceh. Ekologi Indonesia 1: 29–36.

Anderson, J.A.R. 1976. Observations on the ecology of five peat swamp forests in Sumatra and Kalimantan. Soil Research Institute, Bogor, Bulletin 3: 45–55.

Bratawinata, A.A. 1986. Bestandesgliederung eines Bergenwaldes in Ostkalimantan/Indonesien nach floristischen und strukturellen Merkmalen. Ph.D. Thesis, Göttingen.

Brouns, J.J.W. 1987. Quantitative and dynamic aspects of a mixed seagrass meadow in Papua New Guinea. Aquatic Botany 29: 433–473.

Bruijnzeel, L.A. 1983. Hydrological and biogeochemical aspects of man-made forests in South-Central Java, Indonesia. Thesis, Amsterdam.

Bruijnzeel, L.A. 1984. Elemental content of litterfall in a lower montane rain forest in Central Java, Indonesia. Malayan Nature J. 37: 199–208.

Bruijnzeel, L.A. 1985. Nutrient content of litterfall in coniferous forest plantations in Central Java, Indonesia. J. Trop. Ecol. 1: 353–372.

Dilmy, A., & A.J.G.H. Kostermans. 1958. Research on the vegetation of Indonesia. In: Proceedings of the Kandy Symposium, Unesco: 28–31. Paris.

Direktorat Bina Program. 1980. Peta tegakan hutan Indonesia. Direktorat Bina Program, Direktorat Jendral Kehutanan, Bogor.

Flenley, J.R. 1979. The equatorial rain forest: a geological history. Butterworths, London.

Gressitt, J.L. (ed.). 1982. Biogeography and ecology of New Guinea. Junk Publ., The Hague.

Grubb, P.J., & P.F. Stevens. 1985. The forests of the Fatima Basin and Mt. Kerigoma, Papua New Guinea, with a review on montane and submontane rain forests in Papuasia. Publication BG/5, Research School of Pacific Studies, A.N.U., Canberra.

Hommel, P.W.F.M. 1987. Landscape ecology of Ujung Kulon (West Java, Indonesia). Soil Research Institute, Wageningen.

Hope, G.S., J.A. Peterson, U. Radok & I. Alison (eds.). 1976. The equatorial glaciers of New Guinea, Balkema, Rotterdam.

Jafarsidik, Y.S. 1980. Floristic inventory of Pleihari Game Reserve, S. Kalimantan. Lembaga Penelitian Hutan, Bogor, Laporan no. 344.

Johns, R.J. 1985. The vegetation and flora of the South Naru area (Madang Province), Papua New Guinea. Klinkii 3: 70–83.

Johns, R.J. 1987a. A provisional classification of the dipterocarp forests of Papua New Guinea. In: A.J.G.H. Kostermans (ed.), Proceedings of the Third Round Table Conference on Dipterocarps: 175–211. Unesco, Jakarta.

Johns, R.J. 1987b. The natural regeneration of Anisoptera and Hopea in Papua New Guinea. In: A.J.G.H. Kostermans (ed.), Proceedings of the Third Round Table Conference on Dipterocarps: 213–233. Unesco, Jakarta.

Kartawinata, K. 1978. Ecological zones of Indonesia. In: Papers presented at the 13th Pacific Science Congress, Vancouver, Canada: 51–58. LIPI, Jakarta.

Kartawinata, K. 1981. The classification and utilization of forests in Indonesia. In: R.A. Carpenter (ed.), Assessing tropical forest lands: Their suitability for sustainable uses: 163–174. Tycooly International, Dublin.

Kartawinata, K. 1989. Progress on vegetation and ecological studies in Malesia with special reference to Indonesia. MAB Regional Seminar on Methods of Biological Inventory and Cartography for Ecological management. Tokyo, October 1989.

Kartawinata, K., R. Abdulhadi & T. Partomihardjo. 1981 Composition and structure of a lowland dipterocarp forest at Wanariset, East Kalimantan, Indonesia. Malay. For. 44: 397–418.

Kartawinata, K., A. Apandi & T.B. Suselo. 1985. The forest of Peucang Island, Ujung Kulon National Park, West Java. In: D.S. Sastrapradja and eight others (eds.), Proc. Symp. on 100 years development of Krakatau and its surroundings I: 448–452. LIPI, Jakarta.

Kohyama, T., M. Hotta, K. Ogino, Syahbuddin & E. Mukhtar. 1989. Structure and dynamics of forest stands in Gunung Gadut, West Sumatra. In: M. Hotta (ed.), Diversity and plant-animal interactions in equatorial rain forests, Report of the 1987–1988 Sumatra Research, Sumatra Nature Study (Botany), Kagoshima University.

Lamounier, Y., A. Gadrinab & Purnajaya. 1983. International map of vegetation and environmental conditions: Southern Sumatra, scale 1 : 1000,000. Institut de la Carte Internationale du Tapis Végétal and SEAMEO-BIOTROP, Toulouse and Bogor.

Lamounier, Y., Purnajaya & Setiabudi. 1986. International map of vegetation and environmental conditions: Central Sumatra, scale 1 : 1000,000. Institut de la Carte Internationale du Tapis Végetal and SEAMEO-BIOTROP, Toulouse and Bogor.

Lamounier, Y., Purnajaya & Setiabudi. 1987. International map of vegetation and environmental conditions: Northern Sumatra, scale 1 : 1000,000. Institut de la Carte Internationale du Tapis Végétal and SEAMEO-BIOTROP, Toulouse and Bogor.

Mackie, C., T.C. Jessup, A.P. Vayda & K. Kartawinata. 1987. Shifting cultivation and patch dynamics in upland forest in East Kalimantan, Indonesia. In: Y. Hadi (ed.), Workshop on Impact of Man's Activities on Tropical Upland Forest Ecosystems: 465–518. Faculty of Forestry, University Pertanian Malaysia, Serdang.

MacKinnon, J.M. 1982. National conservation plan for Indonesia, vol 1. Introduction, evaluation methods and overview of national nature richness. FAO Field Report 34, FO/INS/78/061, Bogor.

MacKinnon, J., & K. MacKinnon. 1986. Review of the protected area system in the Indo-Malayan realm. IUCN, Gland.

Meijer, W. 1959. Plant sociological analysis of montane rain forest near Tjibodas, West Java. Acta Bot. Neerl. 8: 277–291.

Mogea, J. 1980. Komposisi flora pohon hutan primer di Biak Mentelang, Kutacane, Aceh Tenggara. In: A. Budiman & K. Kartawinata (ed.), Laporan Teknik 1979–1980: 137–139. LBN-LIPI, Bogor.

Mogea, J. 1984. Struktur dan komposisi hutan primer dan sekunder di Tanah Grogot, Kalimantan Timur. In: S. Wirjoatmodjo (ed.), Laporan Teknik 1982–1983. LBN-LIPI, Bogor.

Ogino, K., & M. Chihara (eds.). 1988. Biological system of mangroves: A report of East Indonesian mangrove expedition. Ehime University, Ehime.

Paijmans, K. (ed.). 1976. New Guinea vegetation. Elsevier, Amsterdam.

Partomihardjo, T., R. Abdulhadi, K. Kartawinata & T. Uji. 1984. Struktur dan komposisi hutan Dipterocarpaceae tanah rendah Wanariset, Kalimantan Timur. In: S. Wirjoatmodjo (ed.), Laporan Teknik 1982–1983: 195–200. LBN-LIPI, Bogor.

Partomihardjo, T., R. Yusuf, S. Sunarti, Purwaningsih, R. Abdulhadi & K. Kartawinata. 1987. A preliminary note on gaps in a lowland dipterocarp forest in Wanariset, East Kalimantan. In: A.J.G.H. Kostermans (ed.), Proceedings of the Third Round Table Conference on Dipterocarps: 241–253. Unesco, Jakarta.

Richards, K., & H. Wiriadinata. 1985. Krakatau and biogeographical theory. In: D.S. Sastrapradja and eight others (eds.) Proceedings of Symposium on 100 years development of Krakatau and its surroundings: 358–363. LIPI, Jakarta.

Rijksen, H.D. 1978. A field study on Sumatran Orang Utans (Pongo pygmaeus abellii Lesson 1827). Ecology, behaviour and conservation. Meded. Landbouwhogeschool Wageningen 78-2: 1–420.

Riswan, S. 1982. Ecological studies on primary, secondary and experimentally cleared mixed dipterocarp forest and kerangas forest in East Kalimantan, Indonesia. Ph.D. Thesis, University of Aberdeen.

Riswan, S. 1987a. Structure and floristic composition of a mixed dipterocarp forest at Lempake, East Kalimantan. In: A.J.G.H. Kostermans (ed.), Proceedings of the Third Round Table Conference on Dipterocarps: 435–457. Unesco, Jakarta.

Riswan, S. 1987b. Kerangas forest at Gunung Pasir, Sambodja, East Kalimantan. Its structural and floristic composition. In: A.J.G.H. Kostermans (ed.), Proceedings of the Third Round Table Conference on Dipterocarps: 471–494. Unesco, Jakarta.

Riswan, S., & K. Kartawinata. 1988a. A lowland dipterocarp forest 35 years after pepper plantation in East Kalimantan, Indonesia. In: S. Soemodihardjo (ed.), Some ecological aspects of tropical forest of East Kalimantan: A collection of research reports. MAB Indonesia Contribution No. 48: 1–40.

Riswan, S., & K. Kartawinata. 1988b. Regeneration after disturbance in kerangas (heath) forest in East Kalimantan, Indonesia. In: S. Soemodihardjo (ed.), Some ecological aspects of tropical forest of East Kalimantan: A collection of research reports. MAB Indonesia Contribution No. 48: 61–85.

Riswan, S., & K. Kartawinata. 1989. Regeneration after disturbance in lowland dipterocarp forest in East Kalimantan, Indonesia. Ekologi Indonesia 1: 1–8.

Riswan, S., J.B. Kentworthy & K. Kartawinata. 1985. The estimation of temporal process in tropical rain forest: a study of primary mixed dipterocarp forest in Indonesia. J. Trop. Ecol. 1: 171–182.

Schaik, C.P., & E. Mirmanto. 1985. Spatial variation in the structure and litterfall of Sumatran rain forest. Biotropica 17: 196–205.
Silvius, M.J. (ed.). 1987. The Indonesian wetlands inventory. A preliminary compilation of existing information on wetlands of Indonesia. Vol. 1. AWB/Interwader/PHPA, Bogor.
Soemodihardjo, S. 1985. Current state of knowledge on Indonesian mangrove. Panitia Program MAB Indonesia-LIPI, Jakarta.
Soemodihardjo, S., A. Nontji & A. Djumali. 1979. Prosiding Seminar Ekosistem Hutan Mangrove. Panitia Program MAB Indonesia-LIPI, Jakarta.
Soemodihardjo, S. and six others (eds.). 1984. Prosiding Seminar II Ekosistem Mangrove. Panitia Program MAB Indonesia-LIPI, Jakarta.
Soeriaatmadja, R.E. 1985. Vegetation analysis of plant communities at the Rakata Kecil Island. In: D.S. Sastrapradja and eight others (eds.), Proceedings of Symposium on 100 years development of Krakatau and its surroundings: 437–455. LIPI, Jakarta.
Soerianegara, I. and four others (eds.). 1986. Diskusi panel daya guna dan batas lebar jalur hijau mangrove. Panitia Program MAB Indonesia-LIPI, Jakarta.
Soerianegara, I. and five others (eds.). 1987. Prosiding ekosistem mangrove. Panitia Program MAB Indonesia-LIPI, Jakarta.
Sriyanto, A. 1987. Pengaruh rumpang terhadap permudaan alam dan struktur tegakan di hutan hujan tropika pegunungan. M.Sc. Thesis, Fakultas Pasca Sardjana, IPB, Bogor.
Steenis, C.G.G.J. van. 1935. Maleische vegetatieschetsen. Tijdschr. Kon. Nederl. Aardrijksk. Genoot. 52: 25–67, 171–203, 363–398.
Steenis, C.G.G.J. van. 1938. Plantengeographie. In: Atlas van Tropisch Nederland 7 & 7a. Kon. Nederl. Aardrijksk. Genoot. & Topografischen Dienst in Nederl.-Indië, Batavia.
Steenis, C.G.G.J. van. 1957. Outline of vegetation types in Indonesia and some adjacent regions. In: Proceedings of the 8th Pacific Science Congress 4: 61–97.
Steenis, C.G.G.J. van. 1958. The vegetation map of Malaysia. Unesco, Paris.
Steenis, C.G.G.J. van. 1972. The mountain flora of Java. E.J. Brill, Leiden.
Suhardjono, & H. Wiriadinata. 1984. Komposisi pohon hutan rawa di Sesayap, Kalimantan Timur. In: S. Wirjoatmodjo (ed.), Laporan Teknik 1982–1983: 185–187. LBN-LIPI, Bogor.
Tagawa, H. (ed.). 1984. Interim report of Grant-in-aid for overseas research in 1982 and 1983. Researches on the ecological succession and the formation process of volcanic ash soils on the Krakatau Islands, Kagoshima University, Kagoshima.
Tagawa, H., E. Suzuki, T. Partomihardjo & A. Suriadarma. 1985. Vegetation and succession on the Krakatau Islands, Indonesia. Vegetatio 60: 131–145.
Tata Guna Tanah. 1980. Peta penggunaan tanah. Direktorat tata Guna Tanah, Direktorat-Jendral Agraria, Jakarta.
Whitmore, T.C. 1984a. Tropical rain forest of the Far East, Oxford University Press, Oxford.
Whitmore, T.C. 1984b. A vegetation map of Malesia. J. Biogeogr. 11: 461–471.
Whitmore, T.C., & K. Sidiyasa. 1986. Composition and structure of a lowland rain forest at Toraut, Northern Sulawesi. Kew Bull. 41: 747–756.
Whitmore, T.C., K. Sidiyasa & T.J. Whitmore. 1987. Tree species enumeration of 0.5 hectare on Halmahera. Gard. Bull., Singapore 40: 31–34.
Whittaker R.J., M.B. Bush & K. Richards. 1989. Plant recolonization and vegetation succession on the Krakatau Islands, Indonesia. Ecol. Monogr. 59: 59–123.
Whittaker, R.J., K. Richards, H. Wiriadinata & J.R. Flenley. 1984. Krakatau 1883 to 1983: a biogeographical assessment. Progress in Physical Geography 8: 61–81.
Whitten, A.J. 1980. The kloss gibbon in Siberut rain forest. Ph.D. Thesis, Cambridge.

Whitten, A.J., S.J. Damanik, J. Anwar & N. Hisyam. 1984. The ecology of Sumatra. Gadjah Mada University Press, Yogyakarta.

Whitten, A.J., M. Mustafa & G.S. Henderson. 1987. The ecology of Sulawesi. Gadjah Mada University Press, Yogyakarta.

Yamada, I. 1975. Forest ecological studies of the montane forest of Mt. Pangrango, West Java. I. Stratification and floristic composition of the montane rain forest near Cibodas. The Southeast Asian Studies 13: 402–426.

Yamada, I. 1976a. Forest ecological studies of the montane forest of Mt. Pangrango, West Java. II. Stratification and floristic composition of the forest vegetation of the higher part of Mt. Pangrango. The Southeast Asian Studies 13: 513–534.

Yamada, I. 1976b. Forest ecological studies of the montane forest of Mt. Pangrango, West Java. III. Litter fall of the tropical montane forest near Cibodas. The Southeast Asian Studies 14:194–229.

Yamada, I. 1977. Forest ecological studies of the montane forest of Mt. Pangrango, West Java. IV. Floristic composition along the altitude. The Southeast Asian Studies 15: 226–254.

Yamakura, T., A. Hagihara, S. Sukardjo & H. Ogawa. 1986. Aboveground biomass of tropical rain forest stand in Indonesian Borneo. Vegetatio 68: 71–82.

13 The illusionary concept of the climax

R. J. JOHNS

Summary

The widespread view of tropical rain forest as a climatic climax is questioned. An understanding of the dynamics of rain forest ecosystems can best be gained by viewing the rain forest as a dynamic and unstable ecosystem. This approach will have important consequences on the interpretation of speciation in tropical areas. The existence of very diverse populations of many species in areas of inherent instability possibly strengthens the argument for punctuated equilibrium. In areas such as New Guinea where the ecosystems are very unstable, management practices for rain forest areas must be adapted to allow for the expected differences in regeneration strategies shown by the major commercial species.

Introduction

Our understanding of the dynamics of the rain forest vegetation of tropical regions has been hampered by the general acceptance of the climax theory and its uncritical application in tropical rain forest ecology. In 1952 Richards published his book 'The Tropical Rain Forest'. However, it is unfortunate that Richards interpreted rain forest ecology following the classical Clementsian approach and that his cautionary notes on the variations expected in tropical ecosystems are seldom read. Despite these questions Richards still viewed most areas of tropical rain forest as a climatic climax which is "generally regarded as a position of stability in the development of vegetation." It should be noted that Richards rejected the monoclimax view, the climatic climax, but proposed the recognition of azonal climaxes.

The general acceptance of climax theories has had the result that subsequent studies of tropical rain forest ecology have been restricted and insufficient attention has been paid to the dynamics of rain forest ecosystems. This is the result of the traditional interpretation of lowland tropical rain forest as a climax vegetation type "unchanged through eons of time" (Richards 1952). It should however be noted that Richards recognised that "many ecological concepts and theories will be greatly modified through the study of tropical rain forest communities." Since the publication of Richards the three most widely used texts for teaching tropical rain forest ecology are those of Flenley (1979), Mabberley (1983) and Whitmore (1984). All introduce concepts which question, though not directly, the concept of climax tropical rain forest.

P. Baas et al. (eds.), The Plant Diversity of Malesia, 133–146.

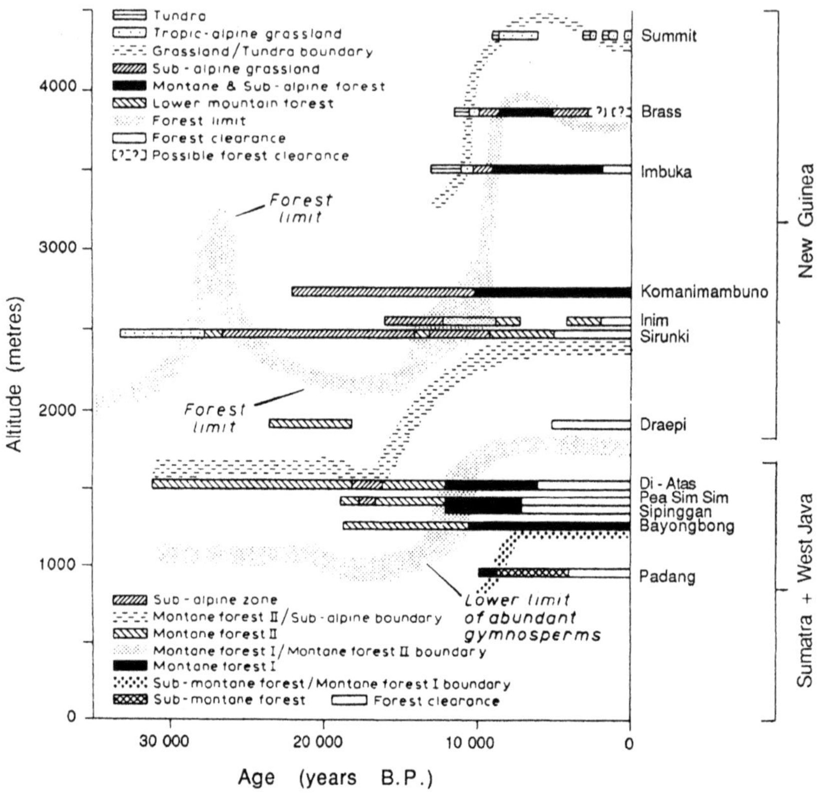

Figure 13.1 — Altitudinal variation in vegetation zones in Pleistocene. (From Flenley 1979.)

Whitmore (1984), through his detailed study of the impact of cyclones on the vegetation of Kolombangara (Whitmore 1969), illustrated one important factor leading to ecosystem instability in some parts of the tropics. This study was paralleled by the work of Webb (1958) in North Queensland. Flenley in 1979 (see Flenley 1984) describes the massive altitudinal fluctuations of vegetation that occurred in the montane tropics during Pleistocene times. Although the effects of the Pleistocene glaciations were seen particularly at higher altitudes (Figure 13.1), there was apparently a major contraction of rain forest in the lowlands as a result of reduced rainfall (Walker 1972). The changes of tropical vegetation at high altitudes during the Pleistocene are paralleled by ecological instability in many areas of lowland tropical rain forest. There appears to be a continuous cycle of rain forest disturbance especially in the Papuasian region (Johns 1986).

The climax - Clements reviewed

Clements in 1914, 1916, and 1936 (see Clements 1936) developed his concepts of plant succession and the climax theory from an extensive review (over 300 papers) of the 'available' literature. Most articles, however, were based on the research conducted in temperate vegetation communities, notably that written on North America and Eurasia. Clements assumed that all successions in an area converged towards a single, stable, mature plant community, termed the climax. Richards himself pointed out that little reference was made by Clements, or other proponents of the climax theory, to studies of tropical rain forest communities. As van Steenis emphasised several times, the richest type of vegetation in number of species, volume and density, is found in the tropics, and it is not the impoverished anthropogenic vegetation of Europe (or North America) which should be the starting point of one's investigations. It is doubtful if many of the biological theories developed in the temperate zone can be applied unquestionably to tropical vegetation.

If the current literature on the dynamics of the rain forest ecosystem had been available to Richards when he wrote his text it is doubtful that the three major sections in his book would be titled: Floristic Composition of Climax Communities, Primary Successions, and Tropical Rain Forest under Limiting Conditions. Despite the fact that Richards himself questioned the reality of applying the concepts developed by Clements to studies in tropical areas, research workers have subsequently tended to ignore his cautionary comments and have applied the climax theories with little question. This is partly due to the acceptance of the argument that diverse communities were of necessity old and stable communities (i.e. climax communities). Where convenient the fact that the communities were climax was accepted "because they were very diverse in species." This diversity argument was in vogue for several decades following the publication of Richards' book.

The description of the diverse climax communities of the tropical rain forest became, unfortunately, a very sterile study. A profile diagram, a list of the number of tree species with a d.b.h. over 10 cm, a list of the number (and species) of epiphytes, climbers, shrubs, seedlings and herbs: this became the aim of tropical vegetation studies. It robbed the science of the stimulation arising from studying the dynamics of the processes operating in the ecosystem: converting a potentially dynamic science into static descriptions of climax rain forest.

The assault on the universal applicability of the climax theory has arisen from several sources. The question of ecosystem instability has placed in question one of the basic tenets needed to apply climax theories throughout the regions dominated by tropical rain forest. The apparent instability of the rain forest environment during the Pleistocene glaciations first illustrated that large areas of diverse, mature, rain forest occupies sites which were recently under quite different types of vegetation (Figure 13.1). Initially studies were made in mountain areas in the tropics. Of particular importance are the studies reported in Flenley (1979, 1984) which include the research of G. Hope, D. Walker and others in Papua New Guinea (also see summary by Morley & Flenley 1987). These studies showed that montane forests in the tropics have undergone massive altitudinal movements, with the major vegetation zones varying over 1000 m during the Pleistocene as the glacial advances in the tropical mountains waxed and waned. Moreover, during the periods of major glacial advance, large areas now covered with luxuriant rain forest, supported a dry, parched, savannah vegetation (Nix & Kalma in Walker 1972). Extensive areas of lowland rain forest were recently (at least geologically) covered by a completely different vegetation type.

Studies in geologically unstable areas such as North-West Borneo (Beaman & Beaman 1990, this volume), New Guinea (Johns 1986), Central America and also the mountain regions of South America built up an impressive chronicle of ecological instability. Some of the most diverse regions of tropical rain forest support forest dominated by mature, diverse rain forest species that are functionally components of a secondary, not a climax, forest. Van Steenis (1958), in his paper on the biological nomad theory, provided an important key to understanding these concepts. Recognition of the group of long-lived nomad species, those species which cannot regenerate under a closed canopy but can attain a considerable age and dominate areas of rain forest, has stimulated the study of regeneration strategies of tropical trees. In Papua New Guinea the list of long-lived nomads is impressive and includes many important commercial species. The common tree genera *Anisoptera, Hopea, Vatica, Calophyllum, Diospyros, Intsia, Albizia, Elaeocarpus, Terminalia,* and at higher altitudes *Castanopsis, Lithocarpus* (some species), *Eucalyptopsis, Araucaria,* and *Agathis* are all examples of long-lived nomad species in Papuasia. It must be noted that some of these genera do include both nomad and dryad species. *Eucalyptus deglupta,* the only true rain forest eucalypt, is also secondary. It grows along river banks, particularly on volcanic material.

Another important characteristic of many tropical species is their ability to adopt different ecological roles at different altitudes. High altitude forests dominated by *Nothofagus grandis,* a common species in montane New Guinea, have saplings commonly in the subcanopy at higher altitudes. In contrast the extensive, almost pure stands, of *Nothofagus grandis* at lower altitudes have little *Nothofagus* regeneration in the subcanopy. Dense regeneration of shade-tolerant tree genera in the Lauraceae, Myristicaceae etc. occurs throughout the forests. Even following logging regeneration of *Nothofagus* in open areas seems to be by root suckers, not from seed. The canopy dominated by *Nothofagus* is apparently being replaced by a variety of shade-tolerant species which regenerate freely in the low light intensity conditions in the subcanopy of the forests.

Womersley (see Womersley & McAdam 1957) explained the decline of *Araucaria* in montane forests as a consequence of the replacement of forests dominated by Gondwanic elements by a recent invasion of Malesian tropical rain forest elements. The apparent decline of *Araucaria,* however, is a simple successional phenomenon. Following extensive rain forest fires some 300 years ago (Johns 1986), *Araucaria* became dominant in the secondary forests. However, *Araucaria* cannot regenerate successfully under a closed canopy and consequently the *Araucaria* forest is replaced by a broadleaf forest dominated by shade-tolerant species, which now form a closed subcanopy in undisturbed forest areas (Havel 1971; Enright in Gressitt 1982). In the Jimi Valley similar processes can be seen with the invasion of *Araucaria* into old garden sites following the depopulation of the valley as a result of malaria.

Of particular importance in many areas is the frequency of disturbance. Studies in Papuasia indicate that disturbances occur at regular intervals. Consequently, unless a climax community can become rapidly established, it will occur rarely in tropical rain forest areas due to instability factors. Taylor (1957) estimated that for a mature, 'climax' community to develop following volcanic eruptions, a period of stability of some 700 years is required. Studies in Papua New Guinea would indicate that, even discounting the influence of man on tropical communities, ecosystems are seldom stable long enough for a climax to develop. Even in areas of comparative geological stability environmental factors are not stable over extended periods. See section on 'Components of ecosystem instability' for further examples of disturbance.

Diversity of tropical rain forest communities

If speciation can be explained by the theory of punctuated equilibrium (van Welzen & Vermeulen 1989) then the diversity of rain forest communities in unstable areas can be better understood. It is interesting that many tropical genera are their most diverse in regions where the ecosystem is apparently unstable. In contrast Connell (1971) has argued that species diversity will be low in disturbed areas because the components of the mature phase cannot survive. This proposal is not supported in Papuasia where despite destructive phenomena the elements of the mature forest are quite diverse though the forests are seldom stable enough for the development of a climax community. For examples of taxa which show their greatest diversity in regions of disturbance, see the section on 'Speciation in tropical mountains'.

What is a climax community?

Whitaker (1974) defined the climax community as a "self-maintaining, steady-state condition in relation to a particular habitat." He concluded that a climax is interpretative; it cannot be directly proved from nature. It is essential that the community is self-maintaining. Whitaker (l.c.) provides a detailed summary of the concepts of the climax.

Components of ecosystem instability

In a recent paper on the instability of the tropical rain forest ecosystem in Papuasia (Johns 1986) a list was made of the most important factors leading to continual destruction. Several of these factors are limited to areas of crustal instability, Papuasia, Java, Sumatra, North-West Borneo, the Andes and parts of Central America. Earthquakes, which can cause massive destruction of rain forest, and volcanoes, destructive both through blast effects and the influence of volcanic ash, are both largely 'confined' to the margins of plates.

Other factors seem quite independent of geological instability. Cyclones (tornadoes) are major causes of rain forest instability in the Philippines, Solomon Islands (Whitmore 1969), Vanuatu (Gillison, pers. comm.), Papua New Guinea (Johns 1986), Fiji, and Queensland (Webb 1958). Browne (1949) and Wyatt-Smith (1955) reported that several hundred square kilometres of forests in Keantan (Malay Peninsula) were devastated by a storm (cyclone?). Wood (1970) reported that over half the potential commercial forests in Samoa were destroyed by cyclones in 1961, 1966 and 1968. In Central America cyclonic winds are common and similarly destructive in their effects on tropical rain forest. Localised cyclonic winds are common in areas of Papua New Guinea (Johns 1986) and extensive areas of 'secondary lower montane forests', particularly forests dominated by *Castanopsis acuminatissima* originate due to these local disturbances. Individual wind storms effect up to 40 plus hectares.

Another major climatic factor destroying, on a local scale, forest vegetation is lightning strikes. Anderson (1966) made observations on the effects of lightning on dipterocarp forests dominated by *Shorea* sp. in Borneo. Detailed studies by Johns in 1980, published in 1982, showed extensive destruction of circular patches in mangrove forest in stands dominated by *Rhizophora apiculata* in the Labu Lakes area, south of Lae (Papua New Guinea). It was observed in several patches that trees towards the margin of the patches had the sides of trees facing towards the supposed strike dead or dying. Arentz (1988) proposed that drought could be a contributing factor, caused presumably by increased salt concentrations, as a result of decreased fresh water inputs into the mangrove ecosystem from the neighbouring river systems. Similar circular patches occur throughout mangrove communities in Malesia and also in North-East Queensland where they have been associated with infection by *Phytophthora* (Pegg *et al.* 1980). *Phytophthora* has been isolated from both healthy and diseased patches of mangroves in Papua New Guinea (Arentz & Simpson in Johns 1982). There was no evidence that the patches were a result of the presence of these pathogens.

The El Niño, the Southern Oscillation, appears to have had a major impact on ecosystem stability, at least in Malesian rain forest communities. Although many reports were listed of extensive rain forest droughts and fires in the literature from Papuasia (reviewed in Johns 1986) the significance of the El Niño Southern Oscillation was not realised until the 1982 droughts and fires which decimated extensive areas of rain forest in Kalimantan and also in Sabah (Woods 1987). The destruction by fire of some 2 million hectares of 'primary' lowland dipterocarp forest emphasises the potential impact of drought on rain forest communities. Few ecologists will believe that rain forests can burn. Van Steenis told the author that there were reports of extensive rain forest fires in Borneo in the 1880's–1890's. 1886 was another period of major El Niño related droughts. Is all the dipterocarp forest of lowland Borneo as old and stable as some authors would suggest? A summary of the major El Niño related droughts is published in Johns (1989).

Speciation in tropical mountains

The consequences of the Pleistocene glacial advances upon the zonation of vegetation types on tropical mountains have been summarised for Papuasia by Morley & Flenley (1987) and Flenley (1984). Lowering of the tree line by over 1200 m in the mountain areas of Papuasia caused, upon recolonisation, a large number of isolated mountain peaks (equivalent to islands) on which local speciation has occurred. The study by Vink (1970), of the genus *Drimys* (Winteraceae), gives a classic example of a genus in the process of active speciation. Survival of a multitude of distinct (at least locally), self perpetuating entities of *Drimys* on each major mountain region, suggests that the comparatively recent instability resulted in widespread speciation: punctuated equilibrium.

As a result of their more distinctive flowers the equivalent events, recolonisation of tropical mountains, have resulted in recognition of a large number of species in the genus *Rhododendron* from the different mountain regions. There is no question about the ability of the 'species' of *Rhododendron* to hybridise, a feature regularly observed particularly along roadsides, where distinct species freely interbreed. Argent *et al.* (1988) include a typical illustration of the probable effects of post-Pleistocene speciation on the occurrence of *Rhododendron* species in the Kinabalu area (Table 13.1 and Figure 13.2). Field studies (Johns, unpub.) showed that within 15 years of the Lamington eruption in 1952, *Rhododendron* species were common on the upper volcanic post-eruptive debris on the summit of the mountain, indicating their ability to recolonise vacant slopes. *Rhododendrons,* as with other Ericaceae like *Vaccinium* and *Dimorphanthera,* are common secondary species at higher altitudes in Malesia.

Several other taxonomic studies in Papuasia have shown similar patterns of speciation. Detailed studies of the two fern genera, *Grammitis* (Parris 1986) and

Table 13.1 — Distribution of *Rhododendron* in Western Sabah. (From Argent *et al.* 1988.)

I - Long Pasia

bagobonum
borneense
crassifolium
durionifolium
  subsp. *subahense*
himantodes
javanicum
  subsp. *cockburnii*
  subsp. *moultonii*
lanceolatum
micromalayanum
nieuwenhuisii
orbiculatum
pneumonanthum
stenophyllum
yongii

II - G. Lumarku

borneense
crassifolium
cuneifolium
durionifolium
  subsp. *subahense*
exuberans

(II continued)

longiflorum
micromalayanum
orbiculatum
stenophyllum
yongii

III - Kimanis

bagobonum
borneense
crassifolium
fallacinum
javanicum
  subsp. *brookeanum*
    var. *kinabaluense*
  subsp. *moultonii*
longiflorum
orbiculatum
praetervisum
stapfianum
suaveolens

IV - G. Trus Madi

borneense
crassifolium

(IV continued)

cuneifolium
fallacinum
himantodes
javanicum
  subsp. *brookeanum*
    var. *kinabaluense*
lowii
polyanthemum
rugosum
stenophyllum
variolosum

V - G. Alab

bagobonum
borneense
crassifolium
cuneifolium
durionifolium
fallacinum
himantodes
javanicum
  subsp. *brookeanum*
    var. *kinabaluense*
  subsp. *moultonii*

(V continued)

longiflorum
orbiculatum
praetervisum
retivenium
rugosum
stapfianum
stenophyllum
suaveolens
yongii

VI - G. Kinabalu

abietifolium
acuminatum
bagobonum
borneense
buxifolium
crassifolium
cuneifolium
ericoides
exuberans
fallacinum
himantodes

(VI continued)

javanicum
  subsp. *brookeanum*
    var. *kinabaluense*
  subsp. *gracile*
lowii
maxwellii
nervulosum
orbiculatum
polyanthemum
praetervisum
retivenium
rugosum
stapfianum
stenophyllum
suaveolens
variolosum

VII - G. Tambuyukon

baconii
crassifolium
cuneifolium
longiflorum
meijeri
stapfianum
suaveolens

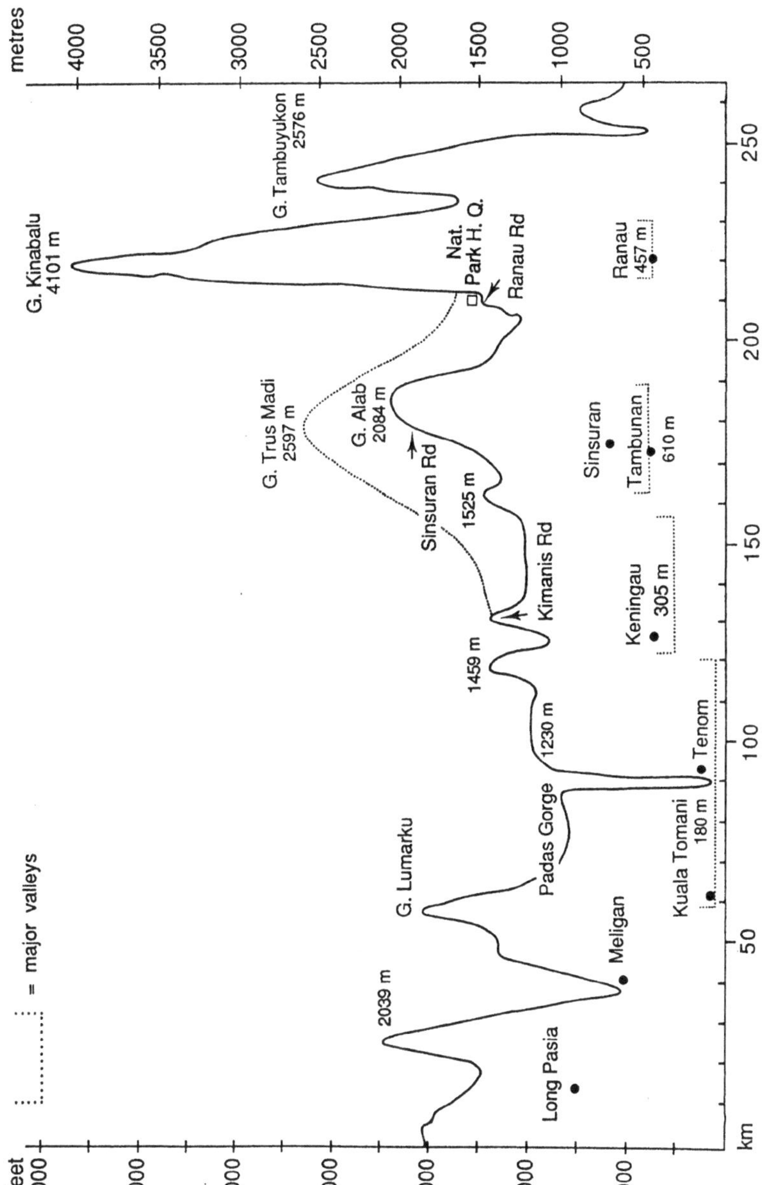

Figure 13.2 — Cross section of the major collecting sites for *Rhododendron* in Western Sabah. (From Argent *et al.* 1988.)

Cyathea (Holttum 1963), show patterns similar to those reported in *Rhododendron* from Kinabalu by Argent *et al.* in 1988. Lowland genera also illustrate similar patterns of speciation. In the Myrsinaceae *Rapanea* has undergone marked speciation in Papuasia (Sleumer 1986) and similar patterns are also found in *Schefflera* (Frodin, pers. comm.). *Terminalia* (Coode 1978) and *Elaeocarpus* (Coode 1981) have also undergone marked species radiations in Papuasia as have *Sericolea* (Coode 1981; van Balgooy 1986) and *Macaranga* (Whitmore in Airy Shaw 1980). Van Welzen & Vermeulen (1989) and van Welzen (1989) have used punctuated equilibrium to explain variability in complex species in areas showing marked ecosystem instability. Detailed studies of *Guoia* and *Bulbophyllum* provided the major examples on which they based these ideas. Other examples by these authors were *Drimys, Impatiens, Cupaniopsis* and *Epiblastus* from New Guinea and *Xanthophyllum, Nephelium* and the Dipterocarpaceae from Borneo.

Many 'secondary' genera show particular diversity in tropical areas such as Papuasia. Epiphytic genera are in essence secondary, the great diversity of epiphytic orchid genera probably being a direct result of the speciation occurring in these disturbed environments. Diversity in tropical genera in Papuasia occurs in many cases in herbs (*Begonia, Impatiens*), epiphytes (*Piper* and many orchid genera), shrubs and tree species (including *Myristica* and *Horsfieldia*) which are functionally secondary in nature. Certain genera such as *Garcinia, Saurauia, Calophyllum, Ficus* and *Elaeocarpus* also illustrate speciation in genera which inhabit secondary communities.

Other genera in Papuasia show very strong evidence for speciation but lack of isolation, or perhaps survival of the 'complete' progenitors, makes species recognition, although philosophically justified, not possible in terms of current taxonomic practice. *Drimys* is a good example of these problems (Vink 1970). 'Species' are quite distinct on individual mountains but material from adjacent mountains intergrades. Hence the conflict between the ecologist who can recognise 'locally distinct, ecological species' occupying different sites, and the taxonomist who can 'find' intermediate specimens often from regions quite remote from the area where the ecologist is collecting.

Does an 'intermediate' specimen collected from an area remote from areas of divergence necessarily negate local speciation events? Are such specimens merely 'noise', survivals by chance of local speciation which should have little significance in our interpretation of species evolution? A botanical key for local regions might be at least as good a base for species identification as a key for a large biogeographical region such as those for Flora Malesiana. Does a speciation event producing entity B which occurs several thousand, perhaps only several hundred kilometres, from a speciation event resulting in entities A and C, but producing a form intermediate between A and C, decrease the biological significance of the second occurrence, or does it mean that these 'local' species should no longer be recognised? Interbreeding does not occur, species isolation is complete, and the local species breed true. Without the interference of man, these distinct populations (entities) are isolated ecologically and genetically and should be treated as distinct species. Even on a local scale where two populations of a bird-pollinated species are isolated by landslides (or another disturbance factor) the populations can be completely isolated as rain forest birds do not cross large (over 20–30 m) clearings. Thus local speciation can occur. Are insect pollinators similarly restricted in their flight?

Of equal interest, and also unexplained, is the occurrence of non variability in species occurring in areas of ecological instability. What are the factors that restrict variability in widespread plants such as *Castanopsis acuminatissima* and *Octomeles sumatrana*? These species are quite distinct from plants such as *Anisoptera thurifera* (Ashton 1982; Johns 1987), *A. costata* (Ashton 1982) and *Drimys piperita* (Vink 1970) which show very complex patterns of variability. They appear very set from the genetic viewpoint but despite their ability to invade a wide variety of ecological sites as aggressive secondary species, they show very little variability in natural populations. What is the nature of the genetic base which restricted variability in the secondary populations in Papuasia?

Significance of instability to rain forest management

Silvicultural systems have been developed to enable the sustained production of forest resources in tropical regions. Inherent in these systems is the recognition of different ecological requirements for the major groups of rain forest seedlings (de Vogel 1980); generally grouped very broadly in forestry literature into shade tolerant and light demanding species. Rain forest management systems were developed first in Malaya, with the Malayan Uniform System, specifically designed to facilitate the re-establishment of forests dominated by dipterocarp species with shade-tolerant seedlings. Studies showed that the dipterocarp species had the following characteristics: seedlings which were shade-tolerant and could grow slowly under the parent canopy and species with intermittant 'mast flowering' (presumably to reduce seedling predation). Such features are 'expected' of components of climax rain forest communities: the dryad species of van Steenis.

Management of 'secondary' rain forests was modelled as the Uniform Shelterwood System, which was developed mainly in tropical East Africa. The dominant tree species in forests managed by this system is composed of species with light demanding seedlings. More detailed studies such as those conducted by de Vogel (1980) are required to see how typical these characters are of the seedlings of many rain forest species. Many destructive phenomena do not result in a complete destruction of all canopy trees or the existing regeneration within the rain forest.

In Papua New Guinea selective rain forest logging is quite varied in its effects on the vegetation. Unlike those areas dominated by dipterocarps in Borneo where volume per hectare are in excess of 100 cubic metres, most forests in Papuasia have very low volumes of extractable timber, usually less than 35 m^3 per ha. Combined with the small diameter of the trees, most individuals being less than 50 cm d.b.h., and the great diversity of most forest types in terms of floristic composition the current exploitation of the forests of Papuasia does not pose a serious threat to the preservation of the biodiversity of the forest communities or species conservation. Exceptions do occur where chip logging is practised, or where there is a follow-up development of agricultural crops such as cocoa or oil palm (both are of marginal economic value). Indeed some studies (Caroline Sargent, pers. comm.) indicate that sustained yield management of rain forest in Papuasia could offer a better, sustainable return than the replacement of the resource by agricultural crops. Because of the dominance of the forest canopy by long-lived nomad species (the major source of commercial timber) and their abundant regeneration in gaps created by selective logging of the rain forest, the long term management of these ecosystems seems a practical possibility.

Conclusions

The concept of the climax has been uncritically applied for the interpretation of the biodiversity of tropical vegetation, mainly as a result of the widespread use of the 'classical' texts on tropical rain forest ecology. This has resulted in the view of rain forest as stable, niche filled, diverse communities whose variation and biodiversity is consequent upon a long term, almost geological stability. In reality the rain forest, particularly in areas of geological instability, is part of a dynamic ecosystem which is continually changing and evolving, seldom (if ever) sufficiently stable to reach the stage of self perpetuation: the climax. This long term ecological instability probably acts as a major evolutionary pump (punctuated equilibrium). Diversity and instability are correlated in many tropical rain forest ecosystems.

References

Airy Shaw, H.K. 1980. The Euphorbiaceae of New Guinea. Kew Bull. Additional Series. 8. Macaranga by T.C. Whitmore pp. 123–162.

Anderson, J.A.R. 1966. A note on two tree fires caused by lightning in Sarawak. Malay. Forester 29: 18–20.

Arentz, F. 1988. Stand-level dieback etiology and its consequences in the forests of Papua New Guinea. Geo Journal 17 (2): 209–215.

Argent, G., A. Lamb, A. Phillips & S. Collenette. 1988. Rhododendrons of Sabah. Sabah Parks Publication No. 8. 146 pp.

Ashton, P.S. 1982. Dipterocarpaceae. Flora Malesiana. I, 9^2: 237–552.
Balgooy, M.M.J. van. 1986. A revision of Sericolea (Elaeocarpaceae). Blumea 28: 103–141.
Beaman, J.H., & R.S. Beaman. 1990. Diversity and distribution patterns in the flora of Mount Kinabalu. In: P. Baas, C. Kalkman & R. Geesink (eds.), The plant diversity of Malesia: 147–160. Kluwer Acad. Publ., Dordrecht, Boston, London.
Browne, F.G. 1949. Storm forest in Kelantan. Malay. Forester 12: 28–33.
Clements, F.E. 1936. Nature and structure of the climax. J. Ecol. 24: 252–284.
Connell, J.H. 1971. On the role of natural enemies in preventing competitive exclusion in some animals and rainforest trees. In: J.P. Boer & G.R. Gradwell (eds.), Dynamics of numbers in populations: 298–312. Advanced Study Institute, Wageningen.
Coode, M.J.E. 1978. Combretaceae. Handbooks of the Flora of Papua New Guinea. Vol. 1: 43–110.
Coode, M.J.E. 1981. Elaeocarpaceae. Handbooks of the Flora of Papua New Guinea. Vol. 2: 38–185.
Flenley, J.R. 1979. The equatorial rainforest: a geological history. London. 162 pp.
Flenley, J.R. 1984. Late Quaternary changes of vegetation and climate in the Malesian mountains. Erdwissenschaftliche Forschung 18: 261–267.
Gressitt, L.J. (ed.). 1982. Biogeography and ecology of New Guinea. Junk, The Hague.
Grey-Wilson, C. 1980. Impatiens in Papuasia. Studies in Balsaminaceae I. Kew Bull. 34: 661–688.
Havel, J.J. 1971. The Araucaria forests of New Guinea and their regenerative capacity. Journ. Ecol. 59: 203–214.
Holttum, R.E. 1963. Cyatheaceae. Flora Malesiana II, 1^2: 65–176.
Johns, R.J. 1982. A report on the mangrove forests of the Labu Lakes area, Morobe Province. Forestry Department. P.N.G. University of Technology, Lae, Papua New Guinea.
Johns, R.J. 1986. The instability of the tropical ecosystem in New Guinea. Blumea 34: 21–60.
Johns, R.J. 1987. The natural regeneration of the dipterocarp forests of Papua New Guinea. In: A.J.G.H. Kostermans (ed.), Proceedings of the Third Round Table Conference on Dipterocarps: 213–233. MAB/UNESCO.
Johns, R.J. 1989. The influence of drought on tropical rain forest vegetation in Papua New Guinea. Mountain Research and Development 9 (3): 248–251.
Mabberley, D.J. 1983. Tropical rain forest ecology. Tertiary Level Biology XI. 156 pp.
Morley, R.J., & J.R. Flenley. 1987. Late Cainozoic vegetational and environmental changes in the Malay Archipelago. In: T.C. Whitmore (ed.), Biogeographical evolution of the Malay Archipelago. Oxford Monographs of Biogeography. Oxford Science Publications.
Parris, B.S. 1986. A taxonomic revision of the genus Grammitis in New Guinea. Blumea 29: 13–222.
Pegg, K.G., N.C. Gillespie & L.L. Fosberg. 1980. Phytophthora sp. associated with mangrove death in coastal Queensland. Australian Plant Pathology 9: 6–7.
Richards, P.W. 1952. The tropical rain forest: an ecological study. Cambridge.
Sleumer, H. 1986. A revision of the genus Rapanea in New Guinea. Blumea 31: 245–269.
Steenis, C.G.G.J. van. 1958. Rejuvenation as a factor for judging the status of vegetation types. The biological nomad theory. In: Proc. Symp. on Humid Tropics Vegetation, Kandy. UNESCO, Paris.
Taylor, B.W. 1957. Plant succession on recent volcanoes in Papua. Journ. Ecol. 45: 233–243.
Vink, W. 1970. The Winteraceae of the Old World. I. Pseudowintera and Drimys - Morphology and taxonomy. Blumea 33: 411–421.

Vogel, E.F. de. 1980. Seedlings of dicotyledons. Pudoc, Wageningen. 465 pp.
Walker, D. 1972. Bridge and Barrier: the natural and cultural history of the Torres Strait. RSPS, Publication BG/3, Canberra.
Webb, L.J. 1958. Cyclones as an ecological factor in tropical lowland rain forest, North Queensland. Austr. J. Bot. 6: 220–228.
Welzen, P.C. van. 1989. Guioa Cav. (Sapindaceae): Taxonomy, phylogeny and historical biogeography. Leiden Botanical Series Vol. 12. Rijksherbarium/Hortus Botanicus, Leiden.
Welzen, P.C. van, & J.J. Vermeulen. 1989. Punctuated equilibrium: complex species explained. Programme and Abstracts, Flora Malesiana Symposium, August 1989. Leiden.
Whitaker, R.H. 1974. Climax concepts and recognition. In Handbook of Vegetation Science 8: 137–154.
Whitmore, T.C. 1969. The vegetation of the Solomon Islands. Phil. Trans. Roy. Soc. B 225: 259–270.
Whitmore, T.C. (ed.). 1981. Wallace's Line and Plate Tectonics. Clarendon Press, Oxford.
Whitmore, T.C. 1984. Tropical rain forests of the Far East. 2nd Ed. Clarendon Press, Oxford.
Womersley, J.S., & J.B. McAdam. 1957. The forests and forest conditions in the territories of Papua and New Guinea. Govt Printer, Port Moresby.
Wood, T.W.W. 1970. Wind damage in the forest of Western Samoa. Malay. Forester 33: 92–99.
Woods, P. 1987. Drought and fire in tropical rain forests in Sabah: an analysis of rainfall pattern and some ecological effects. In: A.J.G.H. Kostermans (ed.), Proceedings of the Third Round Table Conference on Dipterocarps: 367–387. MAB/UNESCO.
Wyatt-Smith, J. 1955. Storm forest in Kelantan. Malay. Forester 17: 5–11.

14 Diversity and distribution patterns in the flora of Mount Kinabalu

JOHN H. BEAMAN and REED S. BEAMAN

Summary

Mount Kinabalu (4,101 m) in northern Borneo, encompassing an area of about 700 km^2, has one of the richest floras in the world. The flora includes c. 4,000 species of vascular plants, about one-third of which are known from a single collection. A high percentage of the species have extremely restricted distributions, frequently associated with ultramafic outcrops. Ideal conditions for a diverse flora and rapid evolutionary rates apparently result from several factors including a vast range of climatic conditions, numerous geologically recent habitats on a diversity of substrates (particularly ultramafic outcrops), regularly recurring El Niño droughts that may drive catastrophic selection, precipitous topography resulting in strong reproductive isolation over short distances, and small population size of many species which may be susceptible to genetic drift. Numerous apparent neo-endemics in many genera and families suggest that frequent speciation events in the recent past may have contributed significantly to the great diversity of the flora. Short- and long-range dispersal of some plants pre-adapted to montane environments also may have contributed to the high floristic diversity. Some species may be relictual, but the relict nature of the flora is not considered to be as significant as previously thought. Because 40 percent of the flora is known from a single locality and most species are very restricted in occurrence, the flora is highly endangered by shifting and permanent agriculture, various development projects, and mining activities, notwithstanding the park status of a portion of the mountain.

Introduction

Mount Kinabalu (4,101 m), centrally located in the Flora Malesiana region, is the highest mountain in Borneo and between the Himalayas and New Guinea. In 1963 van Steenis presented an important paper on the phytogeography of the mountain flora of Kinabalu for a symposium concerning the 1961 Royal Society Expedition (van Steenis 1964). We are pleased to be able to participate in the present symposium commemorating Professor van Steenis and to have the opportunity to consider some aspects of a flora that was of paramount interest to him and to which he contributed in very many ways without ever having set foot on the mountain.

In the 1963 symposium van Steenis stated that his contribution could only be tentative because there was no complete plant list for Mount Kinabalu. We are now attempting to remedy this situation by developing a computerised inventory

P. Baas et al. (eds.), The Plant Diversity of Malesia, 147–160.

of the entire vascular plant flora. He also indicated that the massif is so large, its flora so rich, and the occurrence of species sometimes so local, that further intensive search would be highly promising, notably on the almost unexplored north and east slopes. In the 27 years since van Steenis spoke those words there has been a second Royal Society Expedition and extensive collecting has been done by the Sabah Forest Department. Many other botanists individually have conducted field work there, including several members of the Rijksherbarium staff. Notwithstanding the additional quarter century of field work, however, what van Steenis said in 1963 about floristic richness, extreme localisation of species, and the need for additional field work still pertains. Another towering figure in the study of the Kinabalu flora, Professor E.J.H. Corner (1978), thought it to be the richest and most remarkable assemblage of plants in the world. This paper will consider some aspects of the high diversity of Kinabalu plant species and their extreme localisation.

Floristic research on Mount Kinabalu

The first detailed account of the flora of Mount Kinabalu was a masterful enumeration by Stapf (1894), who recognised 364 species, 140 of them published for the first time. The next major paper on this flora was by Gibbs (1914), who listed 203 species (excluding non-Kinabalu localities), of which 38 were described as new. Important as these publications are, they include less than ten percent of the presently known flora. Meijer (1965) prepared an account of the Kinabalu flora to serve as a guide for participants in the excursion of the Humid Tropics Symposium of UNESCO in Southeast Asia (June-July, 1963). His guide is incomplete, but it represents the only account published in recent times which attempts to list a significant segment of the flora.

Stapf's (1894) account was based largely on specimens collected in 1892 by G.D. Haviland. Gibbs, the first western woman to climb the mountain, added to this list her own collections made in 1910. Subsequent to the work of Stapf and Gibbs, Joseph and Mary Strong Clemens spent two months collecting on Kinabalu in 1915; their collections were worked up by E.D. Merrill at Manila (orchids by Ames and Schweinfurth). The Clemenses returned to Kinabalu in August, 1931, and were there until December 31, 1933, except for about five months in the second half of 1932 when they traveled to Bogor for pre-identification and sorting of their collections. These pre-identifications, incidentally, were made largely by van Steenis. During the approximately two years the Clemenses were in residence on Mount Kinabalu, living in a 'rest house' at Dallas, a 'lodge' built for them at Tenompok, and in a tent on the Penibukan ridge, they amassed an enormous collection of about 9,000 numbers, now widely distributed among herbaria of the world, but with many specimens still not thoroughly studied.

In 1961 and 1964 Professor Corner led two Royal Society expeditions to Mount Kinabalu, each of about four months duration, during which approximately 3,600 collections were made on the eastern shoulder and side. The Royal Society specimens were determined chiefly at Kew, where the largest set is deposited. A general report and some special reports were published on the 1961 expedition (Corner 1964). A report on the 1964 expedition has not been published.

Since 1960 the Sabah Forest Department (formerly British North Borneo Forest Department) has had a number of plant collectors active on Mount Kinabalu. Dr. Willem Meijer, who was Forest Botanist from 1959 to 1968, made several collecting trips to the mountain and also stimulated work by other Forest Department personnel. We have seen about 2,200 Forest Department specimens. Our own collection from Kinabalu amounts to about 2,300 numbers, which were obtained from July 1983 to August 1984. Much of our field time was spent on the eastern and northern slopes (the east side because the forest was being cut down for a golf course and other developments on the Pinosuk Plateau, where we could collect efficiently from felled trees; the north side because of lack of previous work there).

Together with a number of collaborators we are presently preparing an inventory of this flora which will be published in book form and additionally will be available as a computer database (Beaman and Regalado, 1989). The database, which presently includes about 18,000 specimen records, is still in preliminary form with considerable work remaining to be done in revising determinations and standardising localities. The data in this paper concerning numbers of taxa and other parameters that may seem precise must therefore be taken as approximate. Nonetheless, we believe the data are sufficiently complete to permit generalised conclusions about diversity and distribution of the Kinabalu flora.

Species diversity

Nearly 4,000 species of vascular plants have been collected within the c. 700 km^2 area of Mount Kinabalu. Over 180 families and 950 genera occur in the flora. The high species diversity apparently results from a combination of factors, among which the most important are: 1) great altitudinal and climatic range from tropical rain forests near sea level to freezing alpine conditions at the summit; 2) precipitous topography causing effective geographic and reproductive isolation of species over short distances; 3) geological history of the Malay Archipelago; 4) a diverse geology with many localised edaphic conditions, particularly the serpentine or ultramafic substrates; and 5) frequent climatic oscillations influenced by El Niño events. These factors will be discussed in greater detail below.

Table 14.1 — Speciose genera including 20 or more taxa in the flora of Mount Kinabalu.

Major taxon	Family	Genus	Number of taxa
Pteridophytes	Aspleniaceae	*Asplenium*	32
	Athyriaceae	*Diplazium*	33
	Cyatheaceae	*Cyathea*	22
	Grammitidaceae	*Grammitis*	28
Monocotyledons	Orchidaceae	*Bulbophyllum*	57
		Coelogyne	31
		Dendrobium	30
		Dendrochilum	28
		Eria	40
Dicotyledons	Actinidiaceae	*Saurauia*	23
	Araliaceae	*Schefflera*	22
	Begoniaceae	*Begonia*	30
	Clusiaceae	*Garcinia*	22
	Dipterocarpaceae	*Shorea*	28
	Elaeocarpaceae	*Elaeocarpus*	42
	Ericaceae	*Diplycosia*	29
		Rhododendron	32
	Fagaceae	*Lithocarpus*	46
	Gesneriaceae	*Cyrtandra*	29
	Lauraceae	*Litsea*	53
	Meliaceae	*Aglaia*	21
	Moraceae	*Ficus*	98
	Myrsinaceae	*Ardisia*	33
	Myrtaceae	*Syzygium*	66
	Piperaceae	*Piper*	29
	Rubiaceae	*Lasianthus*	21
		Psychotria	29
		Urophyllum	20
	Symplocaceae	*Symplocos*	24
	Urticaceae	*Elatostema*	36

Stapf (1894) and van Steenis (1964) explained the origin of the Kinabalu mountain flora on the basis of former ranges that connected Borneo with both the Asian mainland and New Guinea. Stapf wrote, "If Kinabalu once was in immediate connection with the highland of New Guinea, or what was then its equivalent, the presence of many of the Austral-Antarctic types in the Kinabalu flora would become intelligible ... In fact, the assumption of such highlands seems to me the *conditio sine qua non* for understanding the history of the flora of Kinabalu, and of the highland flora of Malaya generally." In agreement with these assumptions, van Steenis (1964) wrote, "Concluding, I feel that the arguments for a former mountain connexion running from Asia towards the Austral region cannot be denied and also that Kinabalu or other high mountains in Borneo made part of this transtropical bridge for temperate genera."

Both Stapf and van Steenis were writing at a time before plate tectonic theory and its impact on plant geography were developed (cf. Raven & Axelrod 1974; Whitmore 1981, 1987). The flora of Kinabalu has not received any phytogeographic attention subsequent to elaboration of this theory over the past 20 years. A new analysis is badly needed, but it is doubtful that fundamental improvement in understanding of phytogeographic relationships would be possible using the same data Stapf and van Steenis worked over, even in light of tectonic theory. Some of the conclusions undoubtedly would be different, but these authors were working with a seriously inadequate data set. Van Steenis (1964) emphasised that "the interest, and *uniqueness of Kinabalu lies largely with its mountain flora* [italics his]." Later in the paper he stated, "Plant geographically the foothill zone, though extremely rich in species, yields few problems innate to Kinabalu."

Lacking an extensive listing of the species, the earlier authors were not able to visualise what now seem some truly exciting evolutionary and phytogeographic aspects of the flora of Mount Kinabalu. These involve the numerous genera with great numbers of species; 73 genera have 10 or more species and 30 genera have 20 or more. The two largest genera are *Ficus* with 98 taxa and *Syzygium* with 66. Table 14.1 lists speciose genera which include 20 or more species. Important questions concerning phytogeography, evolution, and speciation of the flora now need to be addressed that have been overlooked by Stapf, van Steenis, and other authors because of insufficient data. It may be especially instructive to examine relationships of potential neo-endemic species in speciose genera.

Another kind of data unavailable to Stapf and van Steenis concerns the nature of the substrate on which individual species occur. From our field work and from herbarium specimen locality data, we have been greatly impressed with the importance of areas with ultramafic geology that harbour rare or endemic species, and the frequency of species in genera with numerous species that occur principally on ultramafic substrates. Even the rare epiphytes are often restricted to ultramafic areas. Unfortunately, many of the early collection records do not indicate the nature of the geology, but this sometimes can be established on the basis of the locality at which a particular specimen was collected. The scleromorphic form of plants from ultramafic outcrops in many cases also seems indicative.

Mount Kinabalu is one of the geologically most recent major massifs in the world. It is an adamellite pluton, 150 km^2 in area, which radiometric age determinations have shown to range between 9.0 and 4.9 million years (Jacobson 1978). The mountain has been uplifted diapyrically in the last 1.5 million years, and may still be rising at a rate of about 0.3 cm/year (Myers 1978). During the Pleistocene the summit supported an ice cap 5 km^2 in extent. Deglaciation of the summit occurred about 9,200 years ago (Jacobson 1978). Moraines have been observed as low as about 3,000 m (Myers 1978).

The Kinabalu climate is another factor that must be included in evolutionary and phytogeographic considerations. From base to summit the climatic conditions range from humid tropical to alpine. During the Pleistocene the vegetational zones must have been compressed, with alpine and subalpine conditions occurring at considerably lower elevations than at present. Perhaps of as great significance as the different climatic zones is the fact that the climate has undergone striking variations in recent time and probably during the Pleistocene as well. The El Niño/Southern Oscillation phenomenon, resulting in droughts in Borneo, occurs on the average of every four years (Cane 1983). The 1983 drought was a particularly severe one, of once in 100 years magnitude (Beaman *et al.* 1986) and resulted *inter alia* in the death of a number of *Leptospermum* trees on Kinabalu. Plant mortality also has been recorded after other recent droughts (Lee & Lowry 1980; Lowry *et al.* 1973; Smith 1979).

It seems possible that the model of edaphic endemism and catastrophic selection in speciation, as outlined by Lewis (1962) and Raven (1964), could be relevant to the flora of Mount Kinabalu. Raven noted that this mode of origin of new species most likely has been of greatest importance in areas characterised by extreme climatic fluctuation. Catastrophic selection on Kinabalu may have been driven by the frequent El Niño droughts. Numerous ultramafic outcrops offer the unusual edaphic conditions this hypothesis requires. Raven and Axelrod (1978) suggest that the process may be most important in annuals of mediterranean climates. We believe the concept merits examination with respect to the woody species of Mount Kinabalu.

Mount Kinabalu is an ideal evolutionary laboratory for other reasons as well. In the short time during which it has been elevated, numerous new habitats have been created that could be occupied by newly evolved plants. It is noteworthy that some of the rarest and most interesting species occur on old landslides, i.e., geologically recent habitats. The upper slopes, far distant from other similar habitats, simulate a new oceanic island system such as Hawaii. The terrain

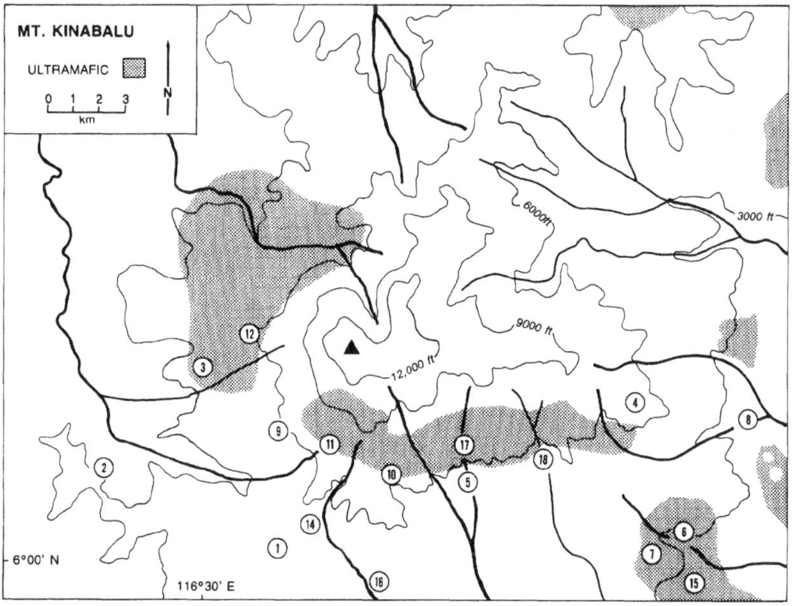

Figure 14.1 — Situation map of Mount Kinabalu showing occurrence of ultramafic outcrops and positions of the 18 most important localities from which rare species have been collected. Locality numbers on the map correspond to locality numbers in Tables 14.3 and 14.4.

presents many situations where immigrant populations can be effectively isolated from donor populations. Rapid adaptive radiation could be facilitated under these conditions.

Environmental conditions on Kinabalu are so diverse and so localised as to virtually enforce occurrence of highly localised populations, particularly on the many small and widely scattered ultramafic outcrops at different elevations (Figure 14.1). The physiognomy of numerous sharp ridges and profound valleys results in circumstances that severely limit movement of both pollinators and seed-dispersal agents. Kiester *et al.* (1984) have made a case for the diversification of *Ficus* being promoted by the drift of population subunits, temporally isolated by differences in flowering time. The likelihood of a role for genetic drift in the small populations on Mount Kinabalu is intriguing, as is the possibility of catastrophic selection. Both forces may be acting jointly, a possibility suggested by Grant (1971: 116).

Van Steenis (1964) considered the array of mountain species on Kinabalu to be distinctly relictual. In discussing the age of the flora, he concluded that there was no evidence that the flora is young (van Steenis 1967). The ideas of both Stapf (1894) and van Steenis (1964, 1969) on land bridges and an ancient flora populated from former ranges now eroded away no longer seem tenable. Some taxa may be relict (e.g., *Trigonobalanus*), but the extraordinary richness of the flora now seems better explained mainly on the basis of rapid adaptive radiation, catastrophic selection and drift, and long-distance (or short-distance) dispersal of pre-adapted plants from distant and neighbouring mountain systems. Lee and Lowry (1980), for example, have presented cogent arguments for the derivation of *Leptospermum recurvum* from *L. flavescens* (= *L. javanicum*) on Mount Kinabalu after the last Pleistocene glaciation.

Gentry (1982), Gentry and Dodson (1987), and Gentry (1989) have discussed 'explosive' speciation in northern Andean cloud forests, suggesting that this phenomenon has been extremely important in generating the high overall species richness in the Neotropics. Luteyn (1989) notes explosive radiation in Andean Ericaceae associated with an unstable montane environment. Van Welzen (1989) relates complex species and sibling species to rapid evolutionary change in geologically unstable areas and associates this mode of speciation with the model of punctuated equilibria (Eldredge & Gould 1972). The patterns of diversity we observe in the rich Kinabalu flora are in accord with these views on rapid speciation. The Kinabalu example indicates the likelihood that rapid speciation may be occurring in orogenetically active montane environments throughout the tropics, not just the Neotropics as emphasised by Gentry (1989).

A likely scenario for evolution and diversification of the Kinabalu flora was anticipated by Corner (1964) in discussing fig diversity when he stated, "The evolution of subg. *Ficus* seems to have occurred where there has been relatively recent mountain-making, and to have employed regionally the section or series which happened to be on the spot." It may be possible to generalise this hypothesis to much of the Kinabalu flora, but careful analysis of taxonomic characters and distribution of species in representative genera will be required, along the lines of the research by Lee and Lowry (1980) with *Leptospermum*, which will challenge botanists for many years to come.

Species distributions

That many species on Mount Kinabalu are of very local occurrence, as suggested by van Steenis (1964), is now well documented by the data in Table 14.2. This table shows that fully 40 percent of the flora is known from just a single locality. Not shown in the table is the fact that 1,336 species have been collected only once. Many of these have not been rediscovered since they were collected by the Clemenses nearly 60 years ago. Species of frequent occurrence are the exception, since only 104 species (2.6 percent of the flora) have been collected in more than ten localities. Except for the nearly insignificant gymnosperms, all the major groups (i.e., pteridophytes, monocots, dicots) show close parallels to the overall figures for frequency of occurrence (Table 14.2).

Table 14.2 — Number of localities from which species are known on Mount Kinabalu.

Number of localities	Number of species: pterido-phytes	gymno-sperms	mono-cots	dicots	total	percent of total
1	211	3	345	1046	1605	40.3
2	126	7	183	502	818	20.5
3	85	2	93	273	453	11.4
4	64	2	69	187	322	8.1
5	44	0	34	130	208	5.2
6	34	5	20	111	170	4.3
7	25	0	23	67	115	2.9
8	17	0	12	56	85	2.1
9	7	0	8	43	58	1.5
10	5	1	9	29	44	1.1
11	4	0	4	21	29	0.7
12	1	0	6	15	22	0.5
13	0	0	2	10	12	0.3
14	0	0	2	8	10	0.2
15	1	0	0	12	13	0.3
>15	0	0	6	12	18	0.4

Since there is such an enormous number of local species, we have searched the database to find the localities where rare species are concentrated. These data are shown in Table 14.3, where the 18 most important localities are listed, together with the numbers of species found only there and at one to four additional localities. The location of these areas is indicated in Figure 14.1. From this figure it is evident that almost all of the important localities are on the south side of the mountain, a bias resulting from collector activity. The top three localities (Tenompok, Dallas, and Penibukan) are places where the Clemenses lived and collected extensively for extended periods in 1931–33. The locality fourth in importance (Eastern Shoulder) was the subject of the first Royal Society Expedition. Even though collector bias strongly affects the data, it is obvious that if the flora had not been diverse in these areas, the collectors would not have spent so much time and effort there. Furthermore, since each of the important localities has been relatively intensively collected, and still large numbers of species are known only from single localities, additional credence is given to the assertion of extreme species localisation.

A fact that must be emphasised about the important localities is that the majority of them (11 of 18; Table 14.3) have partly or entirely ultramafic substrates. Since ultramafic outcrops frequently are of limited size, it follows that species

Table 14.3 — Most important localities on Mount Kinabalu, based on occurrence of rare species known only from one to five localities.

	Locality	n_0	n_1	n_2	n_3	n_4
1.	Tenompok	131	150	127	106	88
2.	Dallas	164	136	99	81	46
3.	Penibukan*	110	118	88	79	49
4.	Eastern Shoulder	55	57	40	54	41
5.	Mesilau River	28	44	33	64	46
6.	Lohan River*	81	52	38	19	15
7.	Hampuan Hill*	67	44	22	7	8
8.	Poring Hot Springs	50	43	25	17	12
9.	Gurulau Spur*	14	27	28	41	34
10.	Liwagu/Mesilau Rivers	26	35	23	35	24
11.	Summit Trail*	22	26	32	30	22
12.	Marai Parai*	23	35	26	18	22
13.	Penataran River*	22	26	24	24	18
14.	Park Headquarters	22	31	22	25	13
15.	Kulung Hill*	42	31	23	6	8
16.	Sosopodon*	27	23	14	25	20
17.	Mesilau Cave*	16	30	28	14	17
18.	Bembangan River*	14	19	19	23	23

Columns labeled n_0–n_4 indicate the number of species occurring at a particular locality that also occur at 0 (n_0) to 4 (n_4) other localities.

* Locality partly or entirely on ultramafic substrates.

Table 14.4 — Most important localities on Mount Kinabalu, number of species at each locality, percentage of total number of species occurring at each locality, and percentage of species occurring at each localityy that also occur at zero to four other localities.

	Locality	no. of spp.	% of total spp.	p_0	p_1	p_2	p_3	p_4
1.	Tenompok	963	24.2	13.6	15.6	13.2	11.0	9.1
2.	Dallas	712	17.9	23.0	19.1	13.9	11.4	6.5
3.	Penibukan	696	17.5	15.8	17.0	12.6	11.4	7.0
4.	Eastern Shoulder	459	11.5	12.0	12.4	8.7	11.8	8.9
5.	Mesilau River	418	10.5	6.7	10.5	7.9	15.3	11.0
6.	Lohan River	275	6.9	29.5	18.9	13.8	6.9	5.5
7.	Hampuan Hill	195	4.9	34.4	22.6	11.3	3.6	4.1
8.	Poring Hot Springs	186	4.7	26.9	23.1	13.4	9.1	6.5
9.	Gurulau Spur	320	8.0	4.4	8.4	8.8	12.8	10.6
10.	Liwagu/Mesilau Rivers	233	5.8	11.2	15.0	9.9	15.0	10.3
11.	Summit Trail	227	5.7	9.7	11.5	14.1	13.2	9.7
12.	Marai Parai	270	6.8	8.5	13.0	9.6	6.7	8.1
13.	Penataran River	196	4.9	11.2	13.3	12.2	12.2	9.2
14.	Park Headquarters	205	5.2	10.7	15.1	10.7	12.2	6.3
15.	Kulung Hill	141	3.5	29.8	22.0	16.3	4.3	5.7
16.	Sosopodon	201	5.0	13.4	11.4	7.0	12.4	10.0
17.	Mesilau Cave	238	6.0	6.7	12.6	11.8	5.9	7.1
18.	Bembangan River	212	5.3	6.6	9.0	9.0	10.8	10.8

Columns labeled p_0–p_4 indicate the number of species occurring at a particular locality that also occur at 0 (p_0) to 4 (p_4) other localities.

which occur on them also have small population sizes. Three localities (Lohan River, Hampuan Hill, and Kulung Hill) are particularly noteworthy for having 30 percent or more of their species known only from those areas (Table 14.4). The total number of species found at each of the important localities (Table 14.4), as might be expected, corresponds closely with the number of highly localised species found at those localities.

Conservation issues

The procedures for recognising the most important localities provide an objective means for identifying areas on Mount Kinabalu that should have highest priority for conservation. Some of the localities, however, are already essentially devastated. The Tenompok area is largely converted to temperate vegetable gardens or the forest otherwise felled. Dallas is in an area of extensive shifting agriculture. At least the upper slopes of the Penibukan area are still in pristine condition, as is the Eastern Shoulder. The Lohan River, Hampuan Hill, and Kulung Hill localities were degazetted from Kinabalu Park in 1984, but possibly they may again be brought under jurisdiction of the Forest Department or Sabah Parks. Unfortunately, the forests of Hampuan Hill and Kulung Hill were partially destroyed in 1987 or 1988 by illegal and unsuccessful shifting agriculture (W. Meijer, pers. comm.). Some species of rare orchids in this area also were recently stripped (P. Cribb, pers. comm.). Poring Hot Springs and Gurulau Spur are mostly within the Park boundaries, as is the Summit Trail, Marai Parai, the Penataran River and Park Headquarters, and thus theoretically protected.

The future of the Mesilau River area, Sosopodon, Mesilau Cave, and the Bembangan River, all on or near the Pinosuk Plateau, is bleak. Government development projects that include a dairy farm, golf course, expensive house sites, and experimental plantations and gardens already have nearly or completely destroyed the forests in this area. Corner (1964) characterised the Pinosuk Plateau as " ... the least known region of the mountain, richest in life, and most deserving of fuller study." A large degazettement in 1984 legitimised further impact on this area, an assault that began in the 1970s when a 10 mi^2 area for the Mamut Copper Mine was removed from Kinabalu Park less than ten years after its founding. Considering the rapid and apparently inexorable destruction of forests on the lower slopes of Mount Kinabalu in the past 25 years, much of which formerly constituted portions of a national park, it is questionable as to whether much, if any, of the original vegetation below the 2000-meter contour will remain for long. The extreme fragility and endangered status of this phenomenal resource are emphasised in that 40 percent of the species are known from a single locality and 30 percent have been collected only once. We hope the data presented here will help call to the attention of both the scientific community and responsible government agencies the unique and irreplaceable natural biological features of Mount Kinabalu.

Acknowledgements

Recent research upon which this paper is based has been supported by the U.S. National Science Foundation Grants BSR-8507843 and 8822696. Figure 14.1 was prepared by Kimberly Medley. Jacinto C. Regalado, Jr. participated extensively in obtaining the herbarium specimen data.

References

Beaman, J.H., & J.C. Regalado, Jr. 1989. Development and management of a microcomputer specimen-oriented database for the flora of Mount Kinabalu. Taxon 38: 27–42.

Beaman, R.S., J.H. Beaman, C.W.Marsh & P.V. Woods. 1986 [1985]. Drought and forest fires in Sabah in 1983. Sabah Soc. J. 8: 10–30.

Cane, M.A. 1983. Oceanographic events during El Niño. Science 222: 1189–1195.

Corner, E.J.H. 1964. Royal Society Expedition to North Borneo, 1961: Reports. Proc. Linnean Soc. London 175: 9–56, 21 pl.

Corner, E.J.H. 1978. Plant Life, Chapter 6. In: M. Luping, Chin Wen & E.R. Dingley (eds.), Kinabalu: Summit of Borneo: 112–178. Sabah Society Monograph 1978. Sabah Society, Kota Kinabalu, Sabah, Malaysia.

Eldredge, N., & S.J. Gould. 1972. Punctuated equilibria: an alternative to phyletic gradualism. In: T.J.M. Schopf (ed.), Models in Paleobiology: 82–115. San Francisco.

Gentry, A.S. 1983 ['1982']. Neotropical floristic diversity: phytogeographical connections between Central and South America, Pleistocene climatic fluctuations, or an accident of the Andean orogeny? Ann. Missouri Bot. Gard. 69: 557–593.

Gentry, A.S. 1989. Speciation in tropical forests. In: L.B. Holm-Nielsen, I.C. Nielsen & H. Balslev (eds.), Tropical forests: Botanical dynamics, speciation and diversity: 113–134. Academic Press, London, etc.

Gentry, A.S., & C.H. Dodson. 1987. Diversity and phytogeography of Neotropical vascular epiphytes. Ann. Missouri Bot. Gard. 74: 205–233.

Gibbs, L.S. 1914. A contribution to the flora and plant formations of Mount Kinabalu and the highlands of British North Borneo. J. Linn. Soc., Bot. 42: 1–240, 8 pl.

Grant, V. 1971. Plant speciation. Columbia Univ. Press, New York, London.

Jacobson, G. 1978. Geology, Chapter 5. In: M. Luping, Chin Wen & E.R. Dingley (eds.), Kinabalu: Summit of Borneo: 101-110. Sabah Society Monograph 1978. Sabah Society, Kota Kinabalu, Sabah, Malaysia.

Kiester, A.R., R. Lande & D.W. Schemske. 1984. Models of coevolution and speciation in plants and their pollinators. Amer. Naturalist 124: 220–243.

Lee, D.W., & J.B. Lowry. 1980. Plant speciation on tropical mountains: Leptospermum (Myrtaceae) on Mount Kinabalu, Borneo. Bot. J. Linnean Soc. 80: 223–242.

Lewis, H. 1962. Catastrophic selection as a factor in speciation. Evolution 16: 257–271.

Lowry, J.B., D.W. Lee & B.C. Stone. 1973. Effect of drought on Mount Kinabalu. Malayan Nature J. 26: 178–179.

Luteyn, J.L. 1989. Speciation and diversity of Ericaceae in Neotropical montane vegetation. In: L.B. Holm-Nielsen, I.C. Nielsen & H. Balslev (eds.), Tropical forests: Botanical dynamics, speciation and diversity: 297–310. Academic Press, London, etc.

Meijer, W. 1965. A botanical guide to the flora of Mount Kinabalu. Symposium on ecological research on humid tropics vegetation: 325–364. Government of Sarawak, and UNESCO.

Myers, L.C. 1978. Geomorphology, Chapter 4. In: M. Luping, Chin Wen & E.R. Dingley (eds.), Kinabalu: Summit of Borneo: 91–94. Sabah Society Monograph 1978. Sabah Society, Kota Kinabalu, Sabah, Malaysia.

Raven, P.H. 1964. Catastrophic selection and edaphic endemism. Evolution 18: 336–338.

Raven, P.H., & D. I. Axelrod. 1974. Angiosperm biogeography and past continental movements. Ann. Missouri Bot. Gard. 61: 539–673.

Raven, P.H., & D. I. Axelrod. 1978. Origin and relationships of the California flora. Univ. Calif. Publ. Bot. 72: 1–134.

Smith, J.M.B. 1979. Vegetation recovery from drought on Mt. Kinabalu. Malayan Nature J. 32: 341–342.

Stapf, O. 1894. On the flora of Mount Kinabalu, in North Borneo. Trans. Linn. Soc. London, Bot. 4: 69–263, pl. 11–20.

Steenis, C.G.G.J. van. 1964. Plant geography of the mountain flora of Mt. Kinabalu. In: E.J.H. Corner (ed.), A discussion on the results of the Royal Society Expedition to North Borneo, 1961. Proc. Roy. Soc. Ser. B, 161: 7–38.

Steenis, C.G.G.J. van. 1967. The age of the Kinabalu flora. Malay. Nat. J. 20: 39–43.

Steenis, C.G.G.J. van. 1969. Plant speciation in Malesia, with special reference to the theory of non-adaptive saltatory evolution. Biol. J. Linnean Soc. 1: 97–133.

Welzen, P.C. van. 1989. Guioa Cav. (Sapindaceae): Taxonomy, phylogeny, and historical biogeography. Leiden Bot. Series 12: 1–315.

Whitmore, T.C. (ed.). 1981. Wallace's line and plate tectonics. Clarendon Press, Oxford.

Whitmore, T.C. (ed.). 1987. Biogeographical evolution of the Malay Archipelago. Clarendon Press, Oxford.

15 Altitudinal zonation of the rain forests in the Manusela National Park, Seram, Maluku, Indonesia

I.D. EDWARDS, R.W. PAYTON, J. PROCTOR and S. RISWAN

Summary

Nine 0.25 ha permanent sample plots have been established at a range of altitudes from sea level to 2500 m on the southern sides of Gunung Kobipoto and Gunung Binaia, on the island of Seram. All trees over 10 cm diameter were assessed and detailed analyses were made of the soils in each plot. Soil pH decreases with greater altitude; organic carbon increases, humus forms change progressively from mull to mor and finally hydromor or peat; and rooting depth decreases. However, the influence of contrasting parent materials on drainage, base saturation and soil depth are apparent on both mountains. Both lowland and montane forests on Seram display a low tree species diversity compared with many rain forests in the Far East. Tree species richness declines with altitude above 600 m but is greater on Gunung Binaia between 1000 m and 1500 m than at corresponding altitudes on Gunung Kobipoto. The montane forest at 2400 m on Gunung Binaia shows the effects of fire.

The vegetation and soils on the Manusela mountains can be divided into five zones; lowland, lower montane, montane, subalpine and alpine. On Gunung Binaia the lower montane zone can be further divided into a lower Fagaceous forest and an upper facies dominated by Myrtaceae and Lauraceae. On the smaller mountain the montane zone occurs at a lower altitude; this demonstrates the *Massenerhebung* effect but may also reflect differences in parent material and soil.

Introduction

Seram is one of the largest islands (17,500 km^2) of the Moluccas, lying to the west of New Guinea and forming part of the Maluku Province of Indonesia. The majority of the island is mountainous, rising to 3027 m above sea level in the peak of Gunung Binaia, which forms the highest point on the central Merkele ridge and much of central Seram remains covered by primary rain forest. Very little modern ecological research has been completed in Seram mainly due to its remoteness and the inaccessibility of the island's interior.

The uniqueness of the Seram environment has recently been recognised by the Indonesian Government through the creation of the Manusela National Park which covers a large area in the centre of the island, stretching from north to south coast and including the peaks of Gunung Kobipoto (1577 m) and Gunung Binaia (3027 m) (Figure 15.1). The National Park encompasses a wide range of

P. Baas et al. (eds.), The Plant Diversity of Malesia, 161–175.

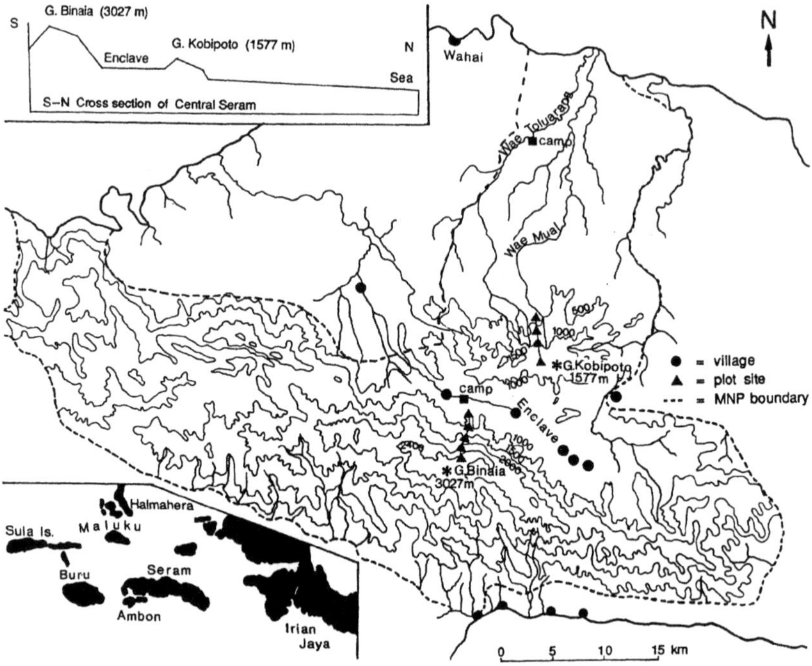

Figure 15.1 — Map of Central Seram showing Manusela National Park.

habitats from lowland rain forest, through different types of montane forest, to shrubby *Rhododendron* and high altitude tree-fern grassland. Until now little was known of the vegetation ecology and virtually nothing of the soils along the altitudinal and environmental gradient.

Elsewhere there have been several studies on the altitudinal zonation of forests on tropical mountains (Richards 1952; Grubb 1977), and those from the Far East were reviewed by Whitmore (1984). Detailed studies on the altitudinal changes in soil development from locations in Malaysia and Indonesia include those of Askew (1964), Reynders (1964) and Burnham (1975), and in addition to climatic changes with altitude, soil conditions have often been suggested as causes of forest zonation (Grubb 1977). There has, however, been relatively little quantitative integrated soil and ecological research relating to the structure and species composition of montane forests (Tanner 1977; Edwards & Grubb 1977), and until the recent work of Proctor *et al.* (1988) and Heaney & Proctor (1990) few such studies have investigated the unbroken altitudinal sequences of forest and soil from near sea level to the tree line. The Manusela National Park is particularly well suited to this type of investigation as the national park status means that the forest is unlikely to be affected by intensive felling and therefore permanent study plots could be established to monitor both medium and long term changes. The presence of high altitude limestones provided an additional interest to the location.

Methods

During July to November 1987, nine permanent sample plots, 50 × 50 m, were laid out within Manusela National Park. These formed two altitudinal sequences, the first over non-calcareous rocks on the north side of Gunung Kobipoto with plots at 370, 600, and 1000 m and, near the summit, at 1470 m; the second on predominantly calcareous rocks on the north side of Gunung Binaia with plots at 1060, 1200, 1640, and 2100 m and, near the tree line, at 2400 m.

During the selection of forest plots the uniformity of aspect, slope and soil type were first established by auger surveys and measurements. A soil profile pit was then excavated to represent the modal soil type of each study plot selected. This was described and the principal soil horizons were sampled for chemical, physical and mineralogical analyses using the methods described in the FAO (1977) and in Hodgson (1976). Orientated Kubiena tin samples (8 × 5 × 3 cm) were collected for soil thin section preparation and soil microscopy. Soil variation within the study plots was established by augering points determined by a stratified random sampling strategy. Ten soil samples from each plot were examined by this method, and samples of 0–15 cm of topsoil were collected for chemical analysis.

Soil change with altitude was further investigated at 100 m intervals on north-facing mountain slopes. Soils were classified provisionally using the FAO-UNESCO (1973) system (Table 15.1), although the classes allocated will probably need revision once soil analytical data become available. Soil pH was measured in water (1 : 2.5 ratio) and organic carbon determined by oxidation with dichromate according to methods in Avery & Bascomb (1974).

In each permanent sample plot, all trees > 10 cm diameter at breast height (dbh) were marked with a numbered aluminium tag, the dbh measured and the tree co-ordinates recorded. The average height of the canopy and emergent trees (if any) were estimated.

Voucher specimens were collected within each plot for all species of tree (> 10 cm dbh). Few specimens collected were in a fertile condition. Plant specimens were first pressed in newspaper, soaked in 75% methylated spirit and then stored in sealed polythene bags until they reached herbaria where they could be dried. The first set of each voucher specimen was identified at the Herbarium Bogoriense (BO) and duplicate sets were sent to the Royal Botanic Garden Edinburgh (E), and distributed to other herbaria, including Kew (K) and Leiden (L).

Table 15.1 — Summary of selected soil data for plots.

Alt. (m)	Soil depth (cm)		Soil* texture	Drainage class	Soil pH		% Org. C	Humus form	Rooting depth	Provisional classification (FAO/UNESCO 1973)
	Topsoil	To rock			0–10 cm	50 cm				
Gunug Kobipoto										
370	15	100	SL	well	6.1	6.2	3.6	mull	80	Chromic Cambisol
600	10	110	SL	well	4.8	5.1	5.2	mull	55	Dystric Cambisol
1000	11	84	L	well	4.5	4.9	25	mor	42	Dystric Cambisol
1470	20	74	P/C	very poor	3.4	4.5	39	peat	20	Humic Gleysol
Gunug Binaia										
1060	8	120	L/CL	moderate	5.6	4.9	4.6	mull	80	Orthic Luvisol
1200	24	24	SL	well	NA	NA	NA	mull	24	Ranker
1640	10	250+	SL/CL	well	5.3	6.7	4.2	mull	65	Orthic Luvisol
2100	12	150+	L/CL	well	3.2	4.6	37	mor	34	Humic Luvisol
2400	22	22	SP	well	3.7	–	36	hydromor	22	Humic Ranker

* USDA soil texture classes (SL = sandy loam, L = loam, CL = clay loam, C = clay, P = peat, SP = sandy peat; / denotes 'over').
NA = not available.

Physical environment and soils

Climate

The climate of central Seram is hot and permanently humid in the lowlands. Temperature decreases with altitude and rainfall increases up to the level of persistent cloud cover but the pattern of precipitation is different on the northern and southern sides of the central Gunung Binaia-Merkele mountain ridge. Generally rain occurs throughout the year and no part of the Park regularly receives less than 60 mm of rain during any month, although Wahai in the northern part of the Park has an annual average rainfall of 2100 mm compared with over twice this (4522 mm) in Tohuru on the south coast. Typically the north side of the island is wettest during June to August, with a peak of over 1000 mm average rainfall in July, but during July 1987 the season was uncharacteristically dry. Normally the summits of Gunung Binaia and Gunung Kobipoto are cloud-covered, clearing only for brief periods in the morning.

Altitudinal sequence of soils on Gunung Kobipoto

The Kobipoto massif consists largely of high grade metamorphic rocks including gneisses, pelitic schists and quartzites. However, these are often overlain by steeply faulted unmetamorphosed laminated siltstones, micaceous and quartzose sandstones, and conglomerates. The mountain is steeply dissected by narrow valleys or ravines with intervening steeply sloping ridges, the flanks of which are subject to soil creep, slumping and landslides. This complex geological outcrop gives rise to an altitudinal 'banding' of potential soil parent materials but their effects on soil development are often masked by the occurrence of widespread colluvial deposits containing a mixture of rocks including gneiss, schist, siltstone and sandstone.

The soils encountered from the footslopes to the summit (Table 15.1) are mainly varieties of leached brown soils (Cambisols) and poorly drained hydromorphic soils (Gleysols). Cambisols occupy the steep upper and midslopes of the ridge flanks from 300 m to the summit. They range from sandy loams and loams over parent material derived from sandstone or gneiss, to clay loams and silty clay loams over those derived from siltstones, mudstones or pelitic schists. Soils increase in their acidity from slightly acid (pH 5.5–6.5) Chromic Cambisols below 500 m to moderately acid (pH 4.5–5.5) Dystric Cambisols with mull humus between 500 and 1000 m. Above 1000 m, strongly acid (pH < 4.5) Dystric Cambisols with mor humus forms occur and tree roots tend to be concentrated in the deep organic surface layers of the soil. Soils below this altitude tend to be deep rooted with a moderate reserve of weatherable minerals and plant nutrients in their subsoils.

The occurrence of mudstones and siltstones near the summit of Gunung Kobipoto appears to result in permanently waterlogged clayey Humic Gleysols with very acid, wet peaty topsoils over slowly permeable gleyed subsoils on ridge flanks with slopes of less than about 20°. Well-drained Dystric Cambisols still occur on ridge crests.

Table 15.2 — Summary of the plot data: floristics and forest structure.

Alt.	N	S	c	d	BA	Dominant species	%BA
Gunung Kobipoto							
370	135	26	0.08	2.24	32.9	*Drypetes longifolia*	16
						Planchonella nitida	13
						Astronia macrophylla	13
600	94	29	0.09	2.99	37.7	*Lithocarpus* sp.	32
						Litsea robusta	20
						Shorea sp.	16
1000	173	25	0.13	1.90	50.1	*Lithocarpus* sp.	43
						Weinmannia	11
1470	155	18	0.16	1.44	38.1	*Phyllocladus hypophyllus*	37
						Myrtaceae	18
						Trimenia papuana	13
Gunung Binaia							
1060	139	44	0.05	3.73	40.7*	*Aglaia ganggo*	28*
						Syzygium sp.	21*
1200	140	29	0.10	2.45	36.4	*Castanopsis buruana*	35
						Lithocarpus sp.	29
						Syzygium sp.	11
1640	179	28	0.130	2.09	54.6	*Syzygium* sp.	60
2100	98	17	0.133	1.72	41.0	*Engelhardia serrata*	27
						Dacrycarpus imbricatus	22
						Saurauia sp.	17
2400	338	13	0.169	0.71	40.4	*Quintinia* sp.	31
						Phyllocladus hypophyllus	28
						Myrica javanica	12
						Vaccinium sp.	11

* BA & dominance based on 0.13 ha section of plot

N = number of trees; S = number of species; c = index of dominance = $\Sigma\, ni/N^2$ (ni = number of individuals); d = index of species richness = $S/\sqrt{N}$; BA = basal area of trees ≥ 10 cm dbh m^2/ha; %BA = species basal area / total basal area.

Altitudinal sequence of soils on Gunung Binaia

Gunung Binaia and the Merkele ridge consist largely of hard, grey, sometimes crystalline, oolitic and bioclastic Asinepe limestone (Audley-Charles *et al.* 1979), but a zone of more argillaceous rocks appears to occur between 1750 and 1900 m. Argillaceous limestones, marls and carbonate mudstones of the Saman Saman limestone group and the less resistant Nief Beds were recorded by Audley-Charles on the lower slopes descending into the intermontane valley known here as the 'enclave'.

The moderate lower slopes of Gunung Binaia, between 700 m and about 1000 m altitude, are mainly over mudstones and marls mantled by non-calcareous residuum, or colluvial deposits containing chert, mudstone and some sandstone. The soils were provisionally classified as varieties of leached, loamy Orthic Luvisols with mull humus forms (Table 15.1). Steeper dissected terrain over limestone interbedded with clastic sedimentary rocks forms a succession of steeply sloping ridges and intervening valleys or ravines between about 1000 and 1500 m altitude. A particularly steep section occurs above 1250 m. Soils are dominantly coarse loamy Eutric and Dystric cambisols with mull humus forms or shallow Rankers. The more base-rich soils occur on the steeper slopes which directly overlie limestone.

Slopes are rather less steep between 1500 and 1750 m and subdued karst features, including sculpted rocks, limestone pinnacles and sink holes are common. Within this zone leached loamy Cambisols over limestone are still dominant, with shallow Rankers and Rendzinas on upper ridge flanks, but poorly-drained Gleysols with slowly permeable clayey subsoils in topographic hollows. Deeper, moderately acid Orthic Luvisols with more base-rich clayey subsoils also continue to occur on less steeply shelving slopes over colluvial deposits, as found on the plot at 1640 m.

Soils in the montane zone above 1900 m have very acid mor humus forms, but vary from shallow Humic Rankers on slope deposits over siliceous chert, to deeper soils (Humic Cambisols or Humic Luvisols) with well-drained, acid, clayey subsoils below thin-layered colluvial deposits. The former soils dominate the very steep mountain slopes above 2400 m where thick siliceous chert screes mantle the underlying limestone.

Floristics and forest structure

Staff at Herbarium Bogoriense, Rijksherbarium in Leiden, Royal Botanic Gardens Kew and Royal Botanic Garden Edinburgh examined over 200 voucher specimens from the nine plots. These were divided into 163 taxa, of which 78 were identified to specific level, 49 to genus, 11 to family and the remaining 25 have still to be determined. Over 100 species were recorded which

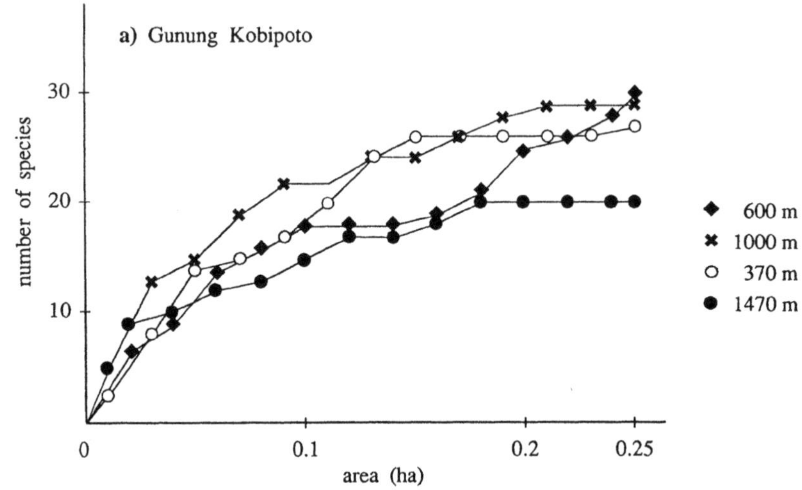

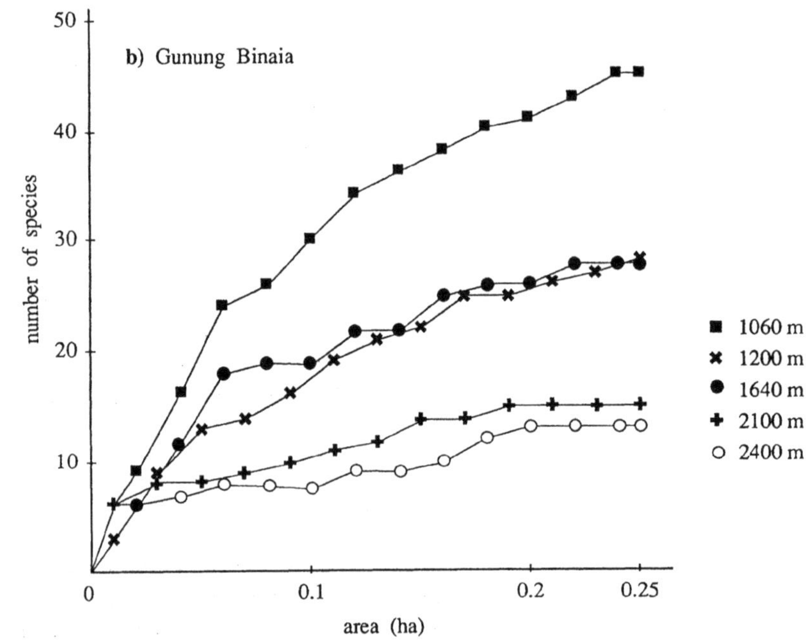

Figure 15.2 — Species-area curves.

are not on the published list of trees in the Manusela National Park. Management Plan (Smiet & Siallagan 1981).

Table 15.2 shows the number of trees, ≥ 10 cm dbh (N), number of tree species (S), total basal area and dominant species (with a basal area ≥ 10% of the total basal area) in each of the nine plots. The table also shows the indices of species richness ($d = S/\sqrt{N}$; Odum 1971) and dominance ($c = \Sigma(ni/N)^2$; Odum 1971). Basal area is greatest at 1000 m on Gunung Kobipoto and 1640 m on Gunung Binaia and declines in the plots above and below these altitudes. Figure 15.2a & b shows the relationship between the number of species and the size of plot from 0.01 to 0.25 ha. The species-area curves generally flatten out at 0.25 ha, suggesting the number of species would not have been greatly increased by enlarging the plot size. Only in the plot at 1060 m on Gunung Binaia does it appear that increasing plot size might have given a higher estimate of the tree species diversity.

Table 15.3 shows the distribution of the species which occurred in two or more plots. Table 15.4 gives a tentative altitudinal zonation based on structure, floristics, soils and other characteristics.

Discussion

The forest plots on Gunung Kobipoto and Gunung Binaia show low tree species diversity compared with many rain forests in the Far East (Whitmore 1984), including Borneo (Proctor *et al.* 1983) and New Guinea (Paijmans 1970). Relatively low tree species richness within small plots was also found in two other investigations of the Moluccan tree flora. An enumeration carried out by Whitmore *et al.* (1987) in rain forest containing *Lithocarpus celebicus, Castanopsis buruana* and *Agathis dammara,* at an altitude of 630 m on the island of Halmahera, about 300 km north of Seram, recorded 76 species of trees > 10 cm dbh, from 31 families within a plot 0.5 ha. This gives a species richness index of 3.94, which is somewhat higher than any of the plots in the present study. The only other enumeration carried out on Seram, by Sidiyasa & Tantra (1984) in the northern lowland forest of Manusela National Park, found only 54 species in 27 families, within four 20 × 1000 m transects, total area 8 ha, giving a species richness index of 0.97, lower than all the plots in the present study except that at 2400 m on Gunung Binaia. The low tree species diversity on Seram, which appears to occur over a range of soil types, is probably related to the relatively recent emergence of the island which is on the edge of the Australasian continental land mass (Audley-Charles 1981, 1987; George 1981) and to the varying levels of isolation experienced over the past 5 million years.

The data presented in Tables 15.2 and 15.3 show how the generally low tree diversity is affected by disturbance and altitude. The highest species diversity was recorded in the plot at 1060 m on Gunung Binaia, a large part of which has

Table 15.3 — Altitudinal distribution of tree species on Gunung Kobipoto and Gunung Binaia.

	Gunung Kobipoto				Gunung Binaia				
Species / altitude (× 10^{-1})	37	60	100	147	106	120	164	210	240
Chisocheton sandoricarpus	* —	*			* - - -	- - -	*		
Astronia macrophylla	* —	*							
Planchonella nitida	* —	*							
Aporusa sp.	* —	*					*		
Glochidion arborescens	*							*	
Eugenia stipularis	* —	*			* —	*			
Microcos sp.	* —	*							
Lithocarpus sp.		* —	*		* —	*			
Eugenia sp. 1		*	?		*				
Planchonella firma		* —	*		* —	* —	*		
Garcinia sp.		* —	*						
Cryptocarya paniculata		* —	*						
Saurauia sp.			*		* - - -	- - -	* —	*	
Prunus arborea			* —	*	* - - -	- - -	* —	* —	*
Magnolia sp.			*				*		
Strombosia ceylanica			*		*				
?Coelostegia griffithii			* —	*					
Platea excelsa			*		* - - -	- - -	*		
Fagraea blumei			* —	*		*			
Phyllocladus hypophyllus			* —	*				* —	*
Polyosma ilicifolia			*					*	
Zygogynum sp.			*				* —	*	
Elaeocarpus sp.			* —	*					
Ilex sp.				*					*
Dacrycarpus imbricatus				*				* —	*
Trimenia papuana				*				*	
Xanthomyrtus sp.				*					*
Canarium sp.					* —	*			
Ostodes paniculata					* - - -	- - -	*		
Tabernaemontana sp.					* - - -	- - -	*		
Rapanea sp.					* - - -	- - -	- - -	* —	*
Weinmannia blumei					* - - -	- - -	- - -	*	
Engelhardia serrata					* - - -	- - -	* —	* —	*
Macaranga hispida					* —	*			
Homalanthus populneus					* - - -	- - -	*		
Casearia sp.					* —	* —	*		
Adinandra sp.							* —	*	
Levieria sp.							* —	*	
Symplocos sp.							* —	*	

received relatively recent disturbance. Similar disturbance is widespread at this altitude and is probably due to logging by people from villages in the enclave below. The felling of large trees and resulting destruction of small patches of forest enabled secondary species like *Trevesia sundaica, Macaranga hispida, Artocarpus elasticus* and *Bombax* sp. to become established. The presence of these species in addition to tree species which are characteristic of primary rain forest such as *Prunus arborea, Platea excelsa* and *Chisocheton sandoricarpus* in undisturbed parts of the plot, probably accounts for this unusually high species diversity. Numerically, Fagaceous trees are not important in this disturbed forest (less than 4%) although *Lithocarpus* and *Castanopsis* are dominant a little higher up the mountain in undisturbed forest, representing 64% of the basal area of the plot at 1200 m altitude.

Table 15.2 shows that although the general trend is for species richness to de-crease with altitude, the plot at 600 m on Gunung Kobipoto has a higher tree diversity than the plot at 370 m on the steep slopes at the foot of the mountain. Soil conditions are unlikely to be responsible for this as the coarse loamy Dystric Cambisols at 600 m are of similar texture and depth, and are developed on similar parent material, to the redder Chromic Cambisols lower down the slope. It is more probable that high tree diversity is due to integration of lowland and montane tree species. The plot at 370 m appears to contain only lowland species, whereas the 600 m plot contained both lowland tree species (e. g. *Astronia macrophylla*) at their upper altitudinal limits and montane forest species (e.g. *Planchonella firma*) at their lower altitudinal limits. Proctor *et al.* (1988) found the highest diversity of tree species on Gunung Silam, Sabah also occurred between 600 and 700 m.

At altitudes between 1000 m and 1500 m, the forests on Gunung Binaia appear to be richer than those at corresponding altitudes on Gunung Kobipoto (Table 15.2) and some species which are relatively common on Gunung Binaia were not recorded on the smaller mountain (e. g., *Engelhardia serrata* and *Rapanea* sp.; see Table 15.3). This could be due to differences in the geology or soils between the two mountains. Although the limestone on Gunung Binaia was generally covered in a deep layer of decalcified residuum or colluvial deposits, the soils at 1640 m have a higher pH (5.3–6.7) and base content than any of the soils above 370 m on Gunung Kobipoto. Also differences in subsoil structure and drainage tend to result in deeper rooting on Gunung Binaia compared with Gunung Kobipoto. However, true calcareous soils (Rendzinas) were only recorded on the steepest, unforested and unstable slopes at altitudes above 2400 m and none of the tree species restricted to Gunung Binaia are known to be calcicoles. The herbaceous flora of Gunung Binaia includes calcicole species (e. g. Gesneriads and ferns) but these appear to be confined to shallow Rendzinas above the tree line or on exposed limestone present as subdued karst features below 1750 m (G. Argent & B. Parris, personal communication).

An alternative explanation for the greater tree species richness on Gunung Binaia could be the phenomena described by van Steenis (1972) whereby trees growing below their optimum altitudinal limits on large mountains rely on immigration of seed from higher altitudes. Consequently, these species are absent from forests at similar altitudes on smaller mountains.

The plot just below the tree line at 2400 m on Gunung Binaia has the lowest species richness recorded (Table 15.2). This plot contains a number of large conifer stumps (probably *Dacrycarpus imbricatus* or *Phyllocladus hypophyllus*), some in excess of 1 m diameter showing that at one time it contained much larger trees than those present today but that they were destroyed, probably by fire. This hypothesis is supported by the presence of charcoal identified microscopically in thin soil sections. However, broadleaved trees (e.g. *Vaccinium* sp., *Myrica javanica* and *Quintinia* sp.) appear to have survived the fire. The large number of multi-stemmed broadleaved trees in the plot suggests that many of these may have been burnt to the ground but later regenerated by producing coppice shoots. Conifers are unable to coppice but a number of small trees, presumably established as seedlings on the burnt ground, were recorded. Thus, the stand appears to be recovering from a catastrophic disturbance and as it matures the density of stems may be reduced by self-thinning. It seems unlikely, however, that many new species will be able to enter this closed forest and diversity will therefore remain low. Low species diversity and regular catastrophic fires are probably characteristics of montane forest at this altitude.

Several authors (e.g. Proctor *et al.* 1988; van Steenis 1984; Whitmore 1984) discuss the *Massenerhebung* effect, whereby vegetation zones are compressed on smaller mountains compared with larger ones, in relation to South-east Asian mountains. Van Steenis (1984) stated that it is the physiognomic zones only and not floristic composition that is telescoped on smaller mountains. However, on Seram at least, the distribution of the two conifers *Dacrycarpus imbricatus* and *Phyllocladus hypophyllus*, and montane broadleaved trees such as *Xanthomyrtus* sp., *Ilex* sp. and *Trimenia papuana* appears to demonstrate that the *Massenerhebung* effect applies to both floristics and forest structure. These montane species were not found below 1900 m on Gunung Binaia but they are all present in the mossy forest at 1460 m near the summit of Gunung Kobipoto, and *Phyllocladus hypophyllus* is present in the plot at 1000 m on Gunung Kobipoto. Other characteristics of montane forest including small leaf size, abundant bryophytes, low species diversity and very acid soils with mor humus forms, provide further evidence to show that this zone occurs about 400–500 m lower down on the smaller mountain. It should be noted, however, that the two mountains have different soils and the *Massenerhebung* effect is probably enhanced by the more base-rich, loamy Luvisols, with deep well-drained subsoils encountered on Gunung Binaia compared with very acid, poorly-drained, clayey Humic Gleysols at equivalent altitudes on Gunung Kobipoto. The former

Table 15.4 — Tentative altitudinal zonation of vegetation on Gunung Kobipoto and Gunung Binaia.

Zone	Altitude (Kobipoto)	Altitude (Binaia)	Characteristic families	General characters
Lowland	0 – 600	–	Guttiferae Melastomataceae Myristicaceae	Canopy > 35 m; buttresses frequent; pandans, palms, woody climbers, vascular epiphytes common; bryophytes few
Lower montane	500 – 1400	1000 – 1900	Fagaceae Myrtaceae Lauraceae	Canopy > 30 m; buttresses present; tree ferns, rattans, bamboo, non-woody climbers, epiphytes, bryophytes common
Upper montane	1400 – 1500	1900 – 2400	Ericaceae Myrtaceae Podocarpaceae	Canopy 30 m, emergent conifers ≥ 35 m; buttresses absent; tree ferns, vascular epiphytes common; bryophytes abundant
Subalpine	–	2400 – 2700	Cyatheaceae Ericaceae Rhamnaceae	Canopy < 10 m, stunted, multi-stemmed trees; bracken, tree ferns common; pendent lichens and bryophytes abundant
Alpine	–	2700 – 3000	Gramineae Ranunculaceae Compositae	Grasses and short herbs, scattered small shrubs and tree ferns; foliose and crustose lichens common

favour deep rooting and the persistence of mull humus at a greater altitude on Gunung Binaia and are likely to influence forest structure and floristic composition.

The vegetation on Gunung Binaia and Gunung Kobipoto can be divided into five zones based on the systems recognised by van Steenis (1984) and Whitmore (1984). Using the latter's nomenclature these are lowland rain forest, lower montane rain forest, upper montane rain forest, subalpine and alpine (see Table 15.4). The upper four zones are present on the north side of Gunung Binaia and the lower three zones on the north side of Gunung Kobipoto. Relatively sharp transitions occur between the upper zones but there appears to be a much more gradual change from lowland forest to lower montane forest on the bottom slopes of Gunung Kobipoto. Our data support dividing the lower montane zone on Gunung Binaia into two facies on the basis of species composition and soils: a lower Fagaceous forest (below 1600 m) dominated by *Lithocarpus* and *Castanopsis* and an upper belt (above 1600 m) dominated by large trees of *Syzygium* (which represented over 60% of the total basal area of the plot at 1640 m) and also characterised by the presence of bamboo. This division coincides with the replacement of soils with mull humus forms by soils with surface accumulation of very acid mor humus layers. It also corresponds with a tendency towards shallow rather than deep rooting by many of the tree species. These trends become much more marked in the montane zone above 1900 m.

Acknowledgements

This study received support from the British Ecological Society, the Natural Environment Research Council, Operation Raleigh Scientific Fund and the Royal Society of London. The authors are particularly grateful to Gill Fairweather for her invaluable assistance and to staff at Herbarium Bogoriense, Bogor; Pattimura University, Ambon; Soil Research Institute, Bogor; Rijksherbarium, Leiden (M.M.J. van Balgooy); the Royal Botanic Gardens, Kew (B.S. Parris); and the Royal Botanic Garden, Edinburgh.

References

Askew, G.P. 1964. The mountain soils of the east ridge of Kinabalu. Proc. Roy. Soc. Lond. B 161: 65–74.

Audley-Charles, M.G. 1981. Geological history of the region of Wallace's Line. In: T.C. Whitmore (ed.), Wallace's Line and Plate Tectonics. Clarendon Press, Oxford.

Audley-Charles, M.G. 1987. Dispersal of Gondwanaland: relevance to the evolution of angiosperms. In: T.C. Whitmore (ed.), Biogeographical evolution of the Malay Archipelago. Clarendon Press, Oxford.

Audley-Charles, M.G., D.J. Carter, A.J. Barber, M.S. Norvick & S. Tjokrosapoetro. 1979. Reinterpretation of the geology of Seram: implications for the Banda Arcs and Northern Australia. J. Geol. Soc. Lond. 136: 547–568.

Avery, B.W., & C.L. Bascomb. 1974. Soil survey laboratory methods. Soil Survey Technical Monograph no. 6. Harpenden.
Burnham, C.P. 1974. Altitudinal changes in soils on granite in Malaysia. Int. Congr. Soil Sci. 10 (6): 290–296.
Edwards, P.J., & P.J. Grubb. 1977. Studies of mineral cycling in a montane rainforest in New Guinea I. The distribution of organic matter in vegetation and soil. J. Ecol. 65: 943–969.
FAO. 1977. Guidelines for soil profile description. Food and Agricultural Organization of the United Nations, Rome.
FAO/UNESCO. 1973. Soil map of the World. Vol. 1. Legend FAO Paris.
George, W. 1981. Wallace and his line. In: T.C. Whitmore (ed.), Wallace's Line and Plate Tectonics. Clarendon Press, Oxford.
Grubb, P.J. 1977. Control of forest growth and distribution on wet tropical mountains. Ann. Rev. Ecol. Syst. 8: 83–107.
Heaney, A., & J. Proctor. 1990. Preliminary studies of rain forests on Volcán Barva, Costa Rica. J. Trop. Ecol. (in press).
Hodgson, J.M. 1976. Soil Survey Field Handbook. Soil Survey Technical Monograph. No. 5. Harpenden.
Odum, E.P. 1971. Fundamentals of Ecology. Saunders College Publishing, Philadelphia.
Paijmans, K. 1970. An analysis of four tropical rain forest sites in New Guinea. J. Ecol. 58: 77–101.
Proctor, J., J.M. Anderson, P. Chia & H.W. Vallack. 1983. Ecological studies in four contrasting rain forests in Gunung Mulu National Park, Sarawak. I. Forest Environment. J. Ecol. 71: 237–260.
Proctor, J., Y.F. Lee, A.M. Langley, W.R.C. Munro & T. Nelson. 1988. Ecological studies on Gunung Silam, a small ultra-basic mountain in Sabah, Malaysia. I. Environment, forest structure and floristics. J. Ecol. 76: 320–340.
Reynders, J.J. 1964. A pedo-ecological study of soil genesis from sea level to the eternal snow, Snow Mountains, central New Guinea. Brill, Leiden.
Richards, P.W. 1952. The Tropical Rainforest. Cambridge University Press.
Sidiyasa, K., & I.G.M. Tantra. 1984. Analisis flora pohon hutan dataran rendah Wae Mual, Taman Nasional, Seram-Maluku. Bulletin Penlitian Hutan 462: 19–34.
Smiet, F., & T. Siallagan. 1981. Proposed Manusela National Park, Central Seram, Management Plan 1982–1987. FAO Field Report 15. FAO, Rome.
Steenis, C.G.G.J. van. 1972. The Mountain Flora of Java. Brill, Leiden.
Steenis, C.G.G.J. van. 1984. Floristic altitudinal zones in Malesia. Bot. J. Linn. Soc. 89: 289–292.
Tanner, E.V.J. 1977. Four montane rain forests of Jamaica: A quantitative characterization of the floristics, the soils and foliar mineral levels and a discussion of the inter-relationships. J. Ecol. 65: 883–918.
Whitmore, T.C. 1984. Tropical rain forests of the Far East. 2nd. Ed. Clarendon Press, Oxford.
Whitmore, T.C., K. Sidayasa & T.J. Whitmore. 1987. Tree species enumeration of 0.5 hectare on Halmahera. Grdn. Bull. Sing. 40: 31–34.

16 The changing pattern of vertical stratification along an altitudinal gradient of the forests of Mt Pangrango, West Java

ISAMU YAMADA

Summary

The forest structure of the montane rain forest of Mt Pangrango, West Java, was examined by measuring tree height, height of the lowest living branch, and crown area. The crown mass of one plot was stratified into one-metre units and thenceforth the total crown area at each metre height level was calculated. Although the maximum total crown area was found at 25 m above the ground, a more or less similar crown area was found between the 5 and 35 m levels. As for species, *Schima wallichii* and *Castanopsis javanica* dominated the first layer with the largest crown area, while *Lithocarpus pseudomoluccus* and *Persea rimosa* occurred in lower frequency in the same layer. These four species were the main canopy species in the 1600 m plot. In the middle layer, *Polyosma ilicifolia* was dominant. In the lower layer, *Turpinia sphaerocarpa* and *Saurauia pendula* were the major species; the latter, especially, was abundant around 10 m above the ground with coverage similar to upper layer species. This stratification pattern changed with altitude and four groups of species were recognised, i.e., 1) upper layer species in montane forest, 2) species changing their habitat along a changing altitude, 3) subalpine species, and 4) lower layer species regardless of altitude.

Introduction

The vertical stratification of tropical rain forests has been studied from various points of view, such as the profile diagram approach (Davis & Richards 1933; Richards 1952; Rollet 1974, and many others), the production ecological approach (Monsi & Saeki 1953; Kira 1978), the crown depth diagram (Ogawa *et al.* 1965), the numerical approach (Hozumi *et al.* 1969; Yamakura 1987, 1988; Yamakura *et al.* 1989) and from many other viewpoints as reviewed in Bourgeron (1983) and Brünig (1983). Smith (1974) discussed stratification. Recently Pompa *et al.* (1988) studied the vertical structure of Mexican tropical rain forests, comparing several methods and discussing the existence of stratification.

There are many studies on montane zonation in the tropics and within Java; studies by van Steenis *et al.* (1972) and Seifriz (1923) are good examples. Ohsawa *et al.* (1985) studied Mt Kerinci, Sumatra, and compared its zonation with that of temperate regions.

P. Baas et al. (eds.), The Plant Diversity of Malesia, 177–191.

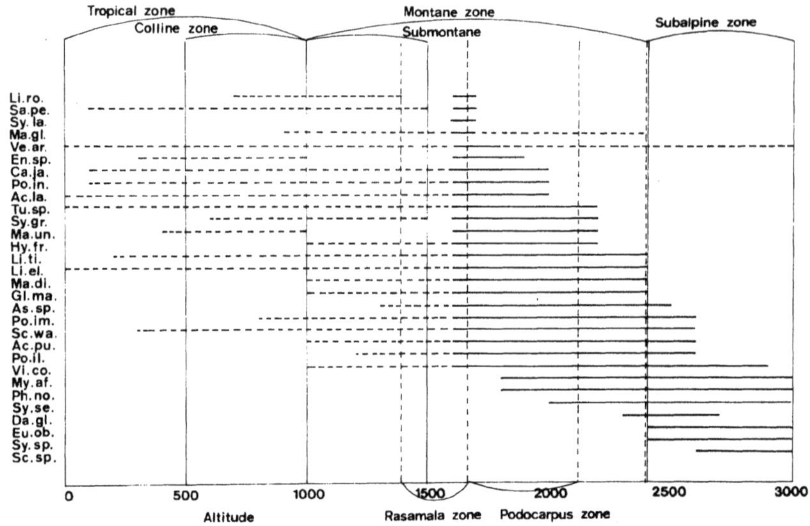

Figure 16.1 — The distribution of the main tree species on Mt Pangrango: the dotted line indicates the distribution range in Java, taken from the 'Flora of Java' by Backer; the solid line is the distribution determined in the present study. The zonation in the upper portion of the figure has been proposed by van Steenis *et al.* (1972) and the lower one by Seifriz (1923). Note that many of the species show overlapping altitudinal ranges and that no clear zonations are discernible. — *General legend for Figures 16.1–16.8:* abbreviations of species names:

Ac. la. : *Acer laurinum*
Ac. pu. : *Acronodia punctata*
Ar. ja. : *Ardisia javanica*
As. sp. : *Astronia spectabilis*
Ca. ja. : *Castanopsis javanica*
Da. gl. : *Daphniphyllum glaucescens*
De. fr. : *Decaspermum fruticosum* var. *polymorphum*
En. sp. : *Engelhardia spicata*
Eu. ob. : *Eurya obovata*
Fi. ri. : *Ficus ribes*
Fl. ru. : *Flacourtia rukam*
Gl. ma. : *Glochidion macrocarpum*
Hy.fr. : *Hypobathrum frutescens*
Le. fl. : *Leptospermum flavescens*
Li. el. : *Lithocarpus elegans*
Li. ps. : *Lithocarpus pseudomoluccus*
Li. ro. : *Lithocarpus rotundatus*
Li. ti. : *Lithocarpus tijsmannii*
Ma. di. : *Macropanax dispermus*
Ma. gl. : *Magnolia glauca*
Ma. un. : *Macropanax undulatus*
My. af. : *Myrsine affinis*
Ne. ja. : *Neolitsea javanica*
Pe. ri. : *Persea rimosa*
Ph. no. : *Photinia notoniana*
Po. il. : *Polyosma ilicifolia*
Po. im. : *Podocarpus imbricatus*
Po. in. : *Polyosma integrifolia*
Sa. pe. : *Saurauia pendula*
Sc. sp. : *Schefflera* sp.
Sc. wa. : *Schima wallichii*
Sy. fa. : *Symplocos fasciculata*
Sy. gr. : *Syzygium gracile*
Sy. la. : *Symplocos laurifolia*
Sy. se. : *Symplocos sessilifolia*
Sy. sp. : *Symplocos* sp.
Tu. sp. : *Turpinia sphaerocarpa*
Va. va. : *Vaccinium varingiaefolium*
Ve. ar. : *Vernonia arborea*
Vi. co. : *Viburnum coriaceum*

Although there are many separate articles about the above mentioned subjects, after the publication by Beard (1955) there are few papers showing changing patterns of vertical stratification along altitudinal gradients.

During my research on Mt Pangrango, West Java in 1969, I noticed a gradual change in the vertical stratification of the forest from 1600 m to the top. This paper deals with the changing pattern of vertical stratification of montane rain forests along altitudinal gradients.

Study sites and methods of study

The study site was chosen on Mt Pangrango (3019 m), West Java, where nine plots were set up from an altitude of 1600 m to the top of the mountain. Detailed reports of floristic composition and forest condition of the plots have already been published (Yamada 1975, 1976a, 1976b, 1977). The size of the plots varied from 1 ha at the lowest level to 400 m^2 at the summit. In each plot, dbh (diameter of breast height), tree height, and height to the lowest living branch of trees larger than 10 cm dbh were measured by diameter tape and hypsometer. Mapping of the location of each tree and drawing of the crown projection area was done at the 1600 m level plot.

Results and discussion

The distribution pattern of species on the mountain

The species found in the study plots were identified and the main species occurring are presented in Figure 16.1. This figure also includes the altitudinal distribution range in Java as a whole, as published in the 'Flora of Java'. The zonation classifications by van Steenis *et al.* (1972) and Seifriz (1923) are also indicated.

From this figure, we can see that the distribution range of many species found on this mountain extends to lower altitudes. The distribution pattern of most species covers various altitudinal zones and there is hardly any drastic change in species composition. This figure was prepared using data on species occurrence only; no quantitative data are included.

Figure 16.2 shows the distribution patterns of the main tree species based on basal area. While in figure 16.1 there was no clear indication of zonation, figure 16.2 shows distinctive trends by species.

Schima wallichii and *Podocarpus imbricatus* are the dominant species in the lower altitudes, whereas *Myrsine affinis* and *Eurya obovata* are dominant in the higher altitudes. When more species are added to the figure, the species distribution has a bi-modal trend which clearly indicates a depressed point at an altitude of 2300 m. Zonation according to this figure would be the montane zone

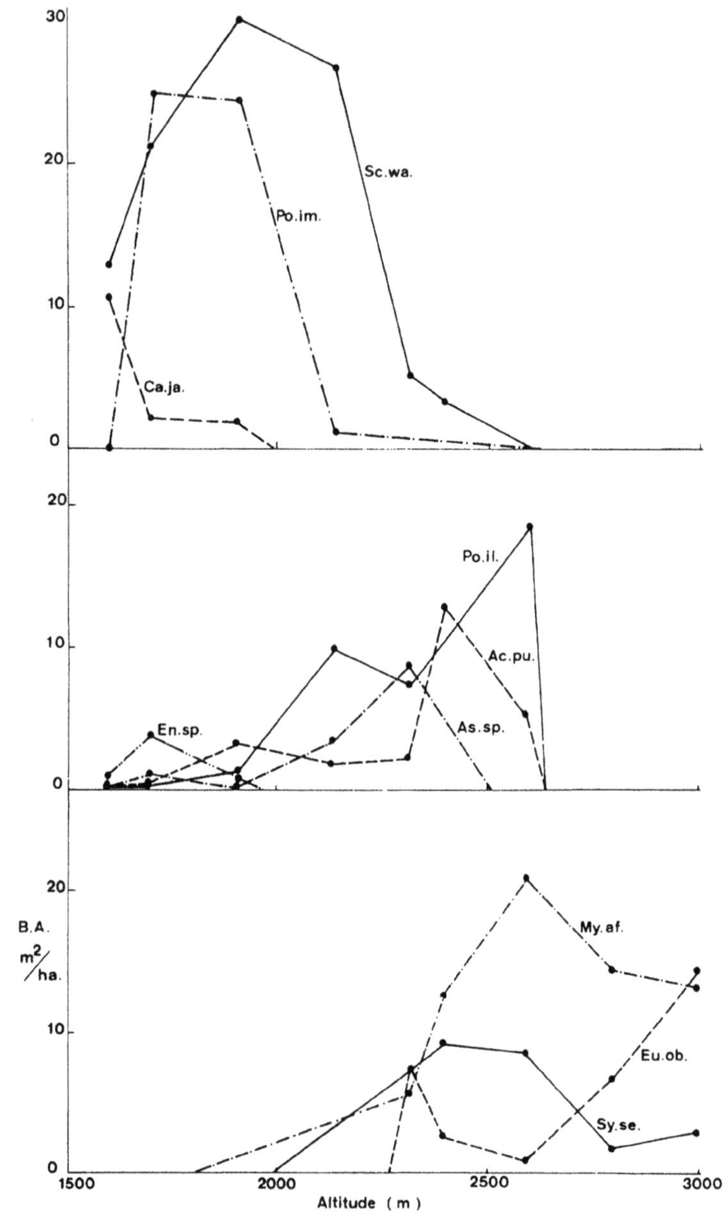

Figure 16.2 — The distribution patterns of the main tree species based on basal area. Dominant species in the upper layer in the montane zone (uppermost figure), species in the transitional zone (middle figure) and species in the subalpine zone (lowest figure) indicate a clear pattern in each zonation. For abbreviations see under Figure 16.1.

(the lower portion) and the subalpine zone (the higher portion). Between these two, there is a transitional zone around 2300 m above sea level. This zonation is basically identical to the one described by van Steenis *et al.* (1972).

Stratification of the montane forest

In order to find out the characteristics of species, we must look at the pattern of stratification. In the lowest plot, which was 1 ha in size, stratification was determined by measuring tree height, height of the lowest living branch, and crown area.

Table 16.1 presents several important statistics of stratification. Only trees larger than 10 cm dbh are included here. Four layers of trees can be distinguished by observing the canopies (Ogino 1974).

Table 16.1 — Statistics of the four layers in a one-hectare plot of the montane rain forest at 1600 m above sea level.

Layer	Maximum height	Average height	Number	Crown area	Average Crown area	Average Crown radius	Crown bulk	Average Crown bulk	Average Crown depth
	$\check{H}$ (m)	$\bar{H}$ (m)	N No/ha	CA (m^2)	$\overline{CA}$ (m^2)	$\overline{CR}$ (m)	CB (m^3)	$\overline{CB}$ (m^3)	$\overline{CD}$ (m)
Io	41	34.7	32 (8.1%)	3409.00 (20.9%)	106.5	5.8	44318.02 (28.0%)	1384.94	13.0
I	38	29.6	56 (14.2%)	4828.25 (29.6%)	86.2	5.2	54968.06 (34.7%)	981.57	11.4
II	36	19.9	134 (34.1%)	5025.90 (30.8%)	37.5	3.5	42524.45 (26.8%)	317.35	8.5
III	34.5	13.8	171 (43.5%)	3038.25 (18.6%)	17.8	1.8	16736.32 (10.6%)	97.87	5.5

Io layer: The entire crown surface is exposed to sunshine.

I layer: Most of the crown surface is exposed to sunshine but is partially covered by trees in the Io layer.

II layer: The crown surface is covered by trees in the upper Io and I layers, but still receives some sunshine.

III layer: The crown is completely covered by trees in the upper three layers. No direct sunshine reaches the crown.

The maximum height of this forest is 41 m. The average height of the uppermost layer is 34.7 m and the number of trees in this layer is 32, which is only 8% of the total. But the crown area covers 21%, crown bulk is 28% and the average crown depth is 13 m.

The average height of the I layer is 30 m, the number of trees is 56 (14.2%) and the crown area covers 30%. The crown bulk is the largest among the 4 layers at 34% and the average crown depth is 11.4 m. Although the number of trees belonging to the Io and I layers accounts for 22% of the total, the crown area covers 50%, and the crown bulk takes up 63%.

In the II layer, average height is 20 m and the number of trees is 134. The crown area covers 31% and the crown bulk is 27%. Average crown depth is 8.5 m. In the III layer, average height is 14 m and although the tree number is the largest, the crown area is as small as 19% and the average crown radius is 1.8 m. The crown bulk covers 10.6% and the average crown depth is 5.5 m.

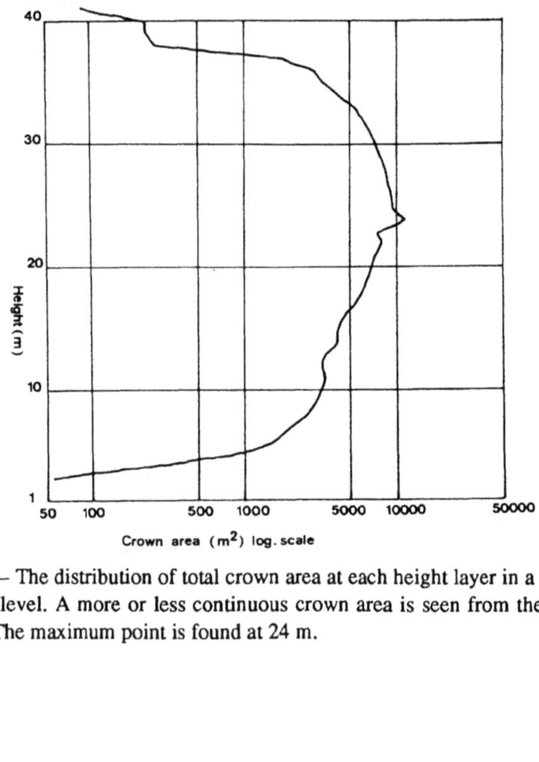

Figure 16.3 — The distribution of total crown area at each height layer in a 1 ha plot at 1600 m above sea level. A more or less continuous crown area is seen from the top layer to the lower layer. The maximum point is found at 24 m.

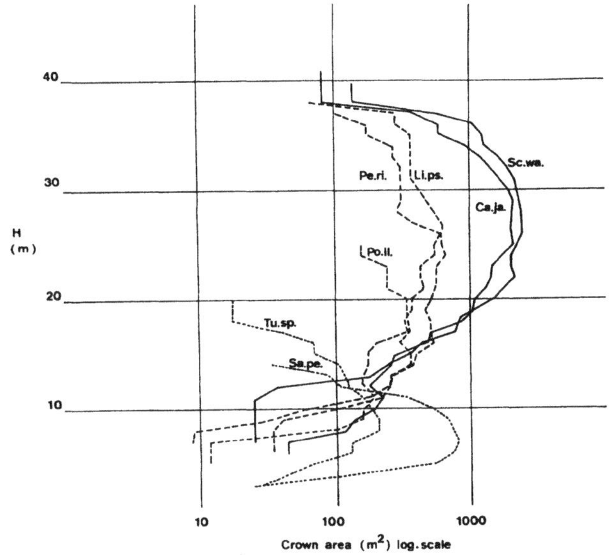

Figure 16.4 — The distribution patterns of the crown areas of the main tree species at each height layer in a 1 ha plot at 1600 m above sea level on Mt Pangrango, West Java. A very clear change in crown area is seen at the 12 m level. For abbreviations see Figure 16.1.

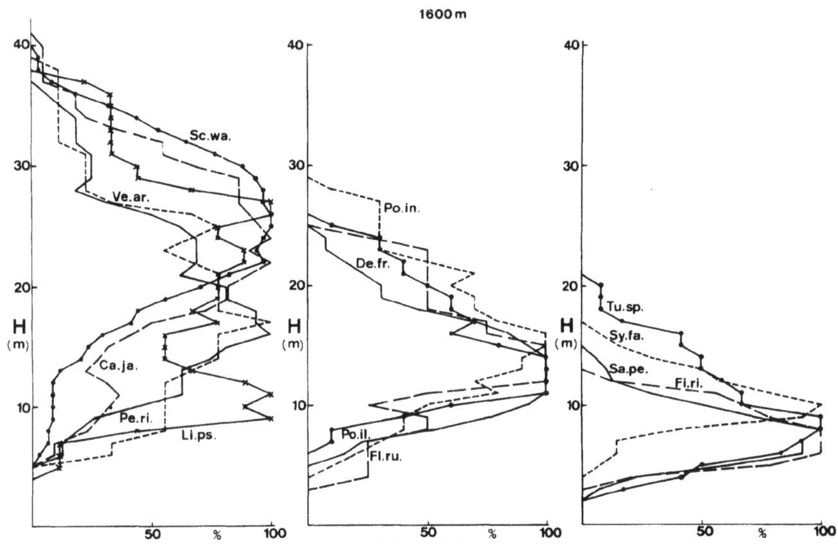

Figure 16.5 — Frequency of tree occurrence, by 1 m strata, expressed as a percentage of the maximum occurrence within the crown profile, at 1600 m above sea level on Mt Pangrango, West Java. Each species shows a clear crown distribution. For abbreviations see Figure 16.1.

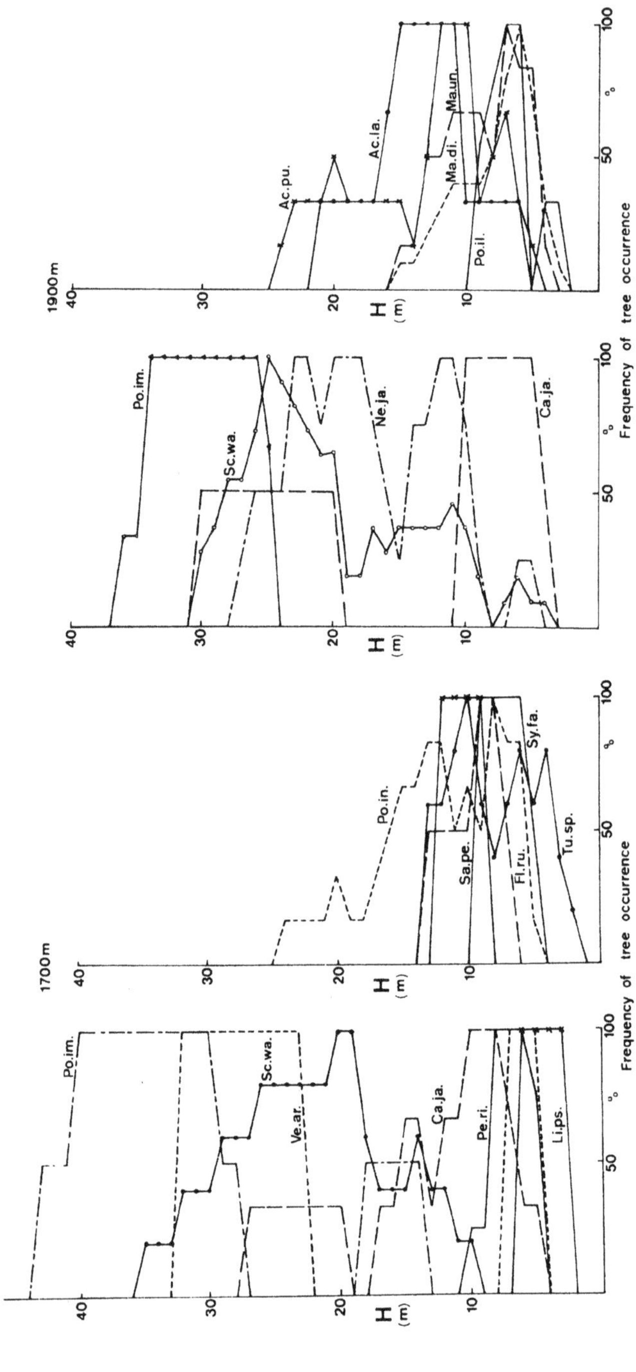

Figure 16.6 — For legend, see page 185.

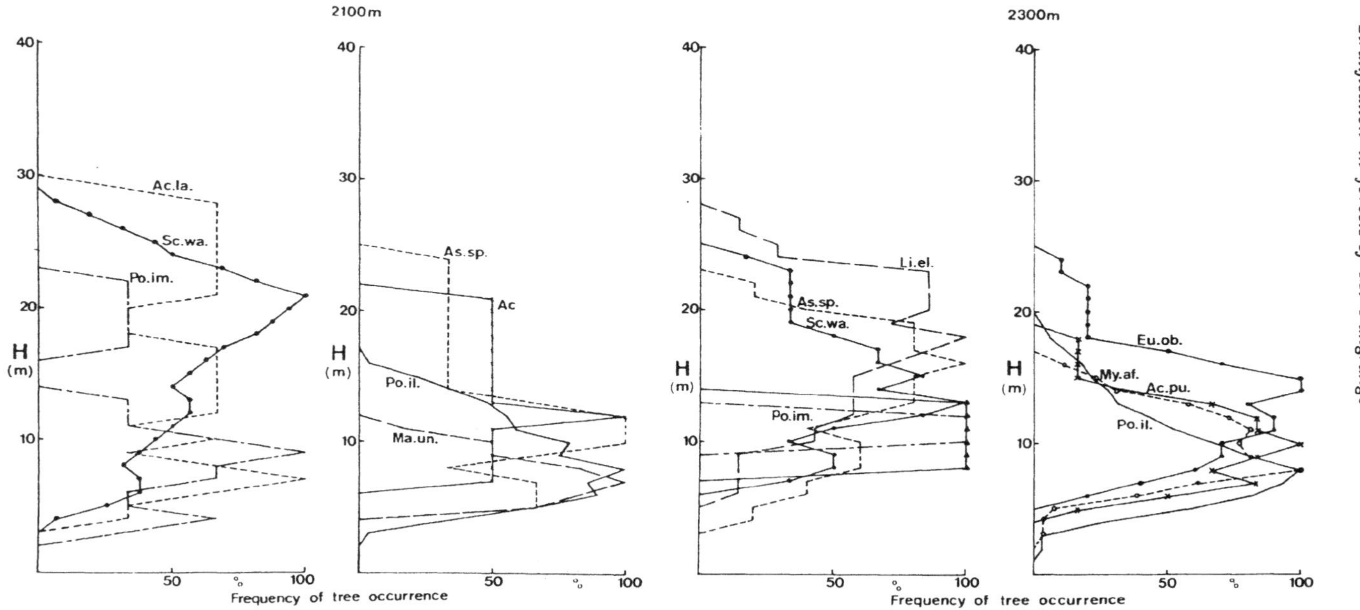

Figure 16.6 — Frequency of tree occurrence, by 1 m strata, expressed as a percentage of the maximum occurrence within the crown profile, at 1700 m to 2300 m above sea level on Mt Pangrango, West Java. Going up the mountain, tree height decreases, layers of stratification decrease in number and the species found in the lower altitudes are replaced by those found in the upper altitudes. For abbreviations see under Figure 16.1.

In Figure 16.3 the total crown area of the plot of each one-metre stratum is drawn on the X axis and height on the Y axis. Hence we can see proportionally how the crown area is distributed at different heights. The maximum crown area is found at 24 m.

Figure 16.4 shows the crown area distribution of all individuals of typical species in the plot. *Schima wallichii, Castanopsis javanica, Lithocarpus pseudomoluccus* and *Persea rimosa,* the main species in the uppermost and I layers, have some similarities. *Polyosma ilicifolia,* a II layer species, has a much lower crown area distribution. However, *Saurauia pendula* and *Turpinia sphaerocarpa,* typical III layer species, show a quite different pattern as seen here.

From this analysis of the crown area distribution, stratification is evident. In other words, this forest is basically composed of two definite layers, i.e., an upper stratum and a lower stratum bordering the 12 m level.

Change of stratification pattern along the altitudinal gradient

Since the crown area was measured in the lowest plot only, I designed another method to investigate stratification in other plots. The crown depth (total height minus height to the lowest living branch) was divided into one-metre stratum units. The number of trees of the same species appearing in each stratum was counted. Taking each species separately, the stratum with the highest number of trees was accorded 100% value and its tree number was used as a basis for the percentage calculation for the tree number in the other strata. Figure 16.5 shows the frequency of tree occurrence by 1 m stratum for all major species in the 1600 m plot. In this figure, we can see the distribution of individual trees at different heights, thus clearly indicating the dominant position of certain species in the stratification. Hence, in Figure 16.5, three groups can be distinguished: the trees predominating in the higher groups are *Schima wallichii, Vernonia arborea, Castanopsis javanica, Persea rimosa* and *Lithocarpus pseudomoluccus*; in the middle groups, *Polyosma integrifolia, P. ilicifolia, Decaspermum fruticosum* and *Flacourtia rukam*; and in the lowest groups, *Turpinia sphaerocarpa, Symplocos fasciculata, Saurauia pendula* and *Ficus ribes* are dominant.

Going up to the 1700 and 1900 m plots, *Podocarpus imbricatus* becomes the dominant species in the top layer, but at 2100–2300 m, the height of trees becomes much lower and the stratification pattern is not so clear (Figure 16.6). At the 2400 m and 2600 m plots, the composition changes to subalpine species and finally at the summit, the vegetation becomes mono-layered (Figure 16.7).

For an idea of the altitudinal distribution from the standpoint of stratification, the crown depths of each species in each plot were linked. Figure 16.8 shows these altitudinal distribution patterns for typical tree species viewed from the context of stratification.

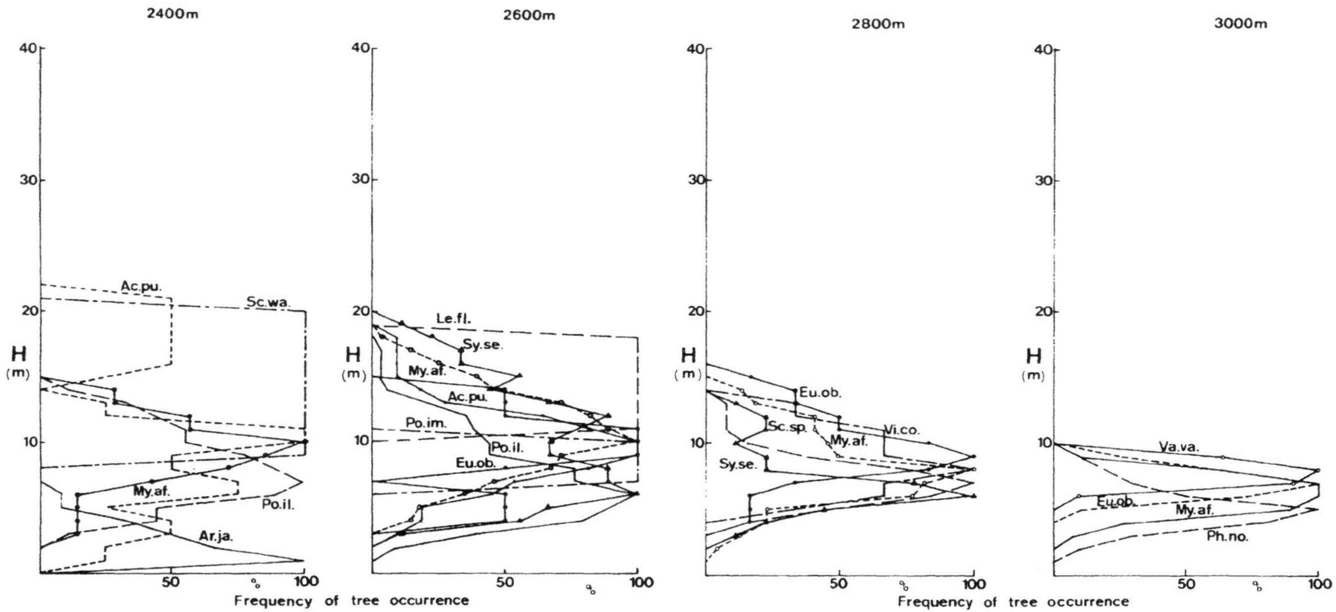

Figure 16.7 — Frequency of tree occurrence, by 1 m strata, expressed as a percentage of the maximum occurrence within the crown profile, at 2400 m to 3000 m above sea level on Mt Pangrango, West Java. In the subalpine zone, tree height becomes as low as 20 m at 2400 m above sea level and 10 m at the top of the mountain. The vegetation becomes mono-layered. For abbreviations see under Figure 16.1.

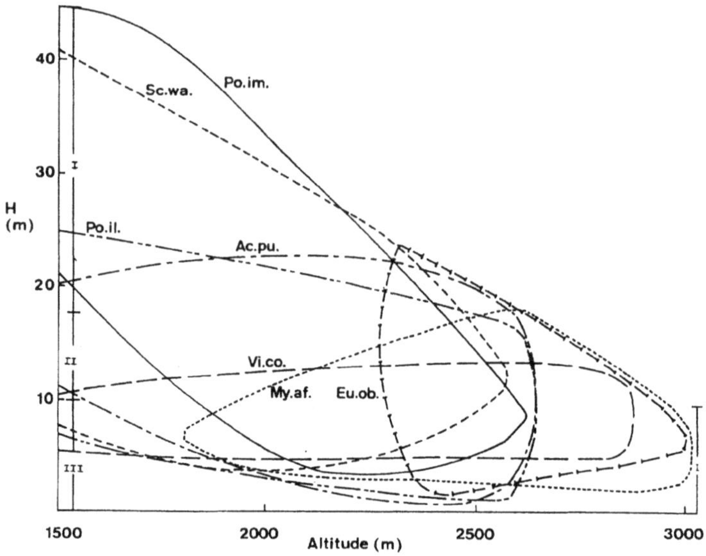

Figure 16.8 — The altitudinal distribution patterns of typical tree species viewed from the context of stratification. Note the change of species in the stratification along the altitudinal gradient. For abbreviations see Figure 16.1.

Podocarpus imbricatus and *Schima wallichii* show a wider layer range in the montane zone and always occupy the upper layer. This type can be called the MU (Montane Upper) type. The T (Transitional) type is exemplified by *Polyosma ilicifolia* and *Acronodia punctata,* which belong to the middle layer in the montane zone, and appear in the upper layer of the transitional zone. The SAU (Subalpine Upper) type refers to higher elevation species which are found only in the subalpine zone and always occupy the upper layer. And the L (Lower) type is seen in *Viburnum coriaceum* which always grows in the lowest layer, irrespective of elevation. *Saurauia pendula,* too, has a similar pattern.

Figure 16.9 is a schematic figure of the distributional species groups on this mountain. This is the basic classification of the distribution pattern according to stratification.

The distribution pattern of species shown according to vertical stratification may reflect various characteristics, such as phenology, growth pattern, and leaf size of the species along the altitudinal gradient, although data about these relationships are limited.

Richards (1983) proposed the terms 'euphotic zone' for the region of the forest above the inversion surface and 'oligophotic zone' for the shaded region below it. The former receives direct sunlight and is the most productive part of

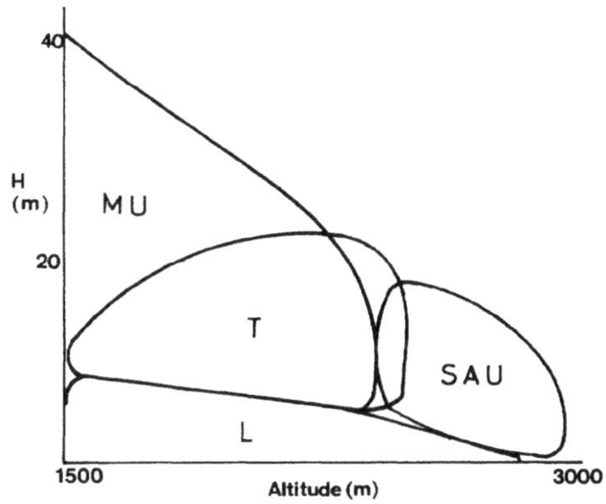

Figure 16.9 — Schematic figures of the distribution of four species groups on Mt Pangrango, West Java. Abbreviations are as follows: MU: Montane; T: Transitional; SAU: Subalpine Upper; L: Lower.

the forest, whereas the latter receives mainly light transmitted through or reflected from leaves, air movement is negligible and the general environmental conditions are constant. Richards (1952) mentioned several examples of the different patterns of phenology in the two zones.

The phenological evidence in the case of Mt Pangrango shows that species in the lower layer were not so directly related to climatic change. For instance, *Saurauia pendula,* a dominant species in the lowest layer, shows a non-seasonal leaf pattern throughout the year (Yamada 1976b). This phenomenon was also reported from Malaysia (Wong 1983) and Brunei (Yamada 1987). These are, however, studies focusing on only one site example. Hence dynamic change in the microclimate and related biological activities of species in changing patterns of stratification along altitudinal gradients have yet to be studied systematically.

Acknowledgements

The author wishes to thank the late Bapak Nurta and his sons, the late Sdr. Uden, Sdr. Idjung and Sdr. Nasurdin for helping in the survey. The Lembagu Ilmu Pengetahuan Indonesia (LIPI) and Lembaga Biologi Nasional (LBN) supported this research in various ways. Prof. Ogino gave valuable criticism and helpful discussion. Y. Imai and M. Oguchi carried out the data processing and drew the figures. The Center for Southeast Asian Studies of Kyoto University provided the financial support for this research.

References

Backer, C.A., & R.C. Bakhuizen van den Brink Jr. 1963–1968. Flora of Java. 3 vols. Groningen.

Beard, J.S. 1955. The classification of tropical American vegetation types. Ecology 36: 89–100.

Bourgeron, P.S. 1983. Spatial aspects of vegetation structure. In: F.B. Golley (ed.), Tropical rain forest ecosystems: 29–47. Elsevier, Amsterdam.

Brünig, E.F. 1983. Vegetation structure and growth. In: F.B. Golley (ed.), Tropical rain forest ecosystems: 49–75. Elsevier, Amsterdam.

Davis, T.A.W., & P.W. Richards. 1933. The vegetation of Moraballi Creek, British Guiana: an ecological study of a limited area of tropical rain forest. Part I. J. Ecol. 21: 350–384.

Hozumi, K. 1975. Studies on the frequency distribution of the weight of individual trees in a forest stand. V. The M-w diagram for various types of forest stands. Jap. J. Ecol. 25: 123–131.

Kira, T. 1978. Community architecture and organic matter dynamics in tropical lowland forests of Southeast Asia, with special reference to Pasoh Forest, West Malaysia. In: P.B. Tomlinson & M.H. Zimmermann (eds.), Tropical trees as living systems: 561–590. Cambridge Univ. Press, Camrbridge.

Monsi, M., & T. Saeki. 1953. Über den Lichtfactor in den Pflanzengesellschaften und seine Bedeutung für die Stoffproduction. Jap. J. Bot. 14: 22–52.

Ogawa, H., K. Yoda, T. Kira, K. Ogino, T. Shidei, D. Ratanawongase & C. Apasutaya. 1965. Comparative ecological studies on three main types of forest vegetation in Thailand. I. Structure and floristic composition. Nature and Life in SE Asia 4: 13–48.

Ogino, K. 1974. The forest ecological discussion on the forest vegetation and forestry in Thailand. Thesis, Kyoto Univ. (In Japanese).

Ohsawa, M., P.H.J. Nainggolan, N. Tanaka & C. Anmar. 1985. Altitudinal zonation of forest vegetation on Mount Kerinci, Sumatra: with comparisons to zonation in the temperate region of East Asia. J. Trop. Ecol. 1: 193–216.

Pompa, J., F. Bongers & J. Mearedel Castillo. 1988. Patterns in the vertical structure of the tropical lowland rain forest of Los Tuxtlas, Mexico. Vegetatio 74: 84–91.

Richards, P.W. 1952. The tropical rain forest (Reprinted with corrections, 1979). Cambridge Univ. Press, Cambridge.

Richards, P.W. 1983. The three dimensional structure of tropical rain forest. In: S.L. Sutton *et al.* (eds.), Tropical rain forest: ecology and management: 3–10. Blackwell Sci. Publ., Oxford.

Rollet, B. 1974. L'Architecture des forêts denses humides sempervirentes de plaine. Centre Techn. For. Trop. Nogent-sur-Marne.

Seifriz, W. 1923. The altitudinal distribution of plants on Mt. Gedeh, Java. Bull. Torrey Bot. Club 50: 283–305.

Smith, A.P. 1973. Stratification of temperate and tropical forests. Am. Nat. 107 (957): 671–683.

Steenis, C.G.G.J. van, A. Hamzah & M. Toha. 1972. The mountain flora of Java. Brill, Leiden.

Wong, M. 1983. Under story phenology of the virgin and regenerating habitats in Pasoh Forest Reserve, Negeri Sembilan, West Malaysia. Mal. For. 42: 197–223.

Yamada, I. 1975. Forest ecological studies of the montane forest of Mt. Pangrango, West Java. I. Stratification and floristic composition of the montane rain forest near Cibodas. Southeast Asian Studies 13: 402–426.

Yamada, I. 1976a. Forest ecological studies of the montane forest of Mt. Pangrango, West Java. II. Stratification and floristic composition of the forest vegetation of the higher part of Mt. Pangrango. Southeast Asian Studies 13: 513–534.

Yamada, I. 1976b. Forest ecological studies of the montane forest of Mt. Pangrango, West Java. III. Litter fall of the tropical montane forest near Cibodas. Southeast Asian Studies 14: 194–229.

Yamada, I. 1977. Forest ecological studies of the montane forest of Mt. Pangrango, West Java. IV. Floristic composition along the altitude. Southeast Asian Studies 15: 226–254.

Yamada, I. 1987. Report on the forest research in Negara Brunei Darussalam from 1984 through 1986. JICA.

Yamakura, T. 1987. An empirical approach to the analysis of forest stratification. I. Proposed graphical method derived by using an empirical distribution function. Bot. Mag. Tokyo 100: 109–128.

Yamakura, T. 1988. An empirical approach to the analysis of forest stratification. II. Quasi-1/2 power low of tree height in stratified forest communities. Bot. Mag. Tokyo 101: 153–162.

Yamakura, T., P. Sahunalu & Karyono. 1989. A preliminary study of changes in forest stratification along environmental gradients in Southeast Asia. Ecological Research 4: 99–116.

17 Search for phytogeographic provinces in Sumatra

Y. LAUMONIER

Summary

The flora and vegetation of Sumatra resemble those of Malaya and Borneo. However, the island is botanically undercollected. Investigations on the hierarchy of ecological criteria for vegetation classification purposes suggested that environmental subdivisions could be tested against the distribution of the flora. Quantitative vegetational studies have been carried out on floristic composition, structure and architecture of various forest types within each subzone. The altitudinal zonation of the vegetation and the correlations between plant communities and their underlying geology have been worked out, especially at lower elevations. The validity of minor floristic subdivisions on Sumatra, representing 11 'sectors' on the island, was tested against forest and habitat information from vegetation mapping surveys, botanical specimens (mainly trees and lianas), and data taken from Flora Malesiana. Results of cluster analysis on flora distribution data point out main demarcation lines delineating six floristic sectors in the lowlands and three in the mountains. The need is stressed of more intensive investigation of the Sumatran vegetation and flora, and of a more precise definition of 'lowland' areas in forest conservation strategies. The preparation of local floras or field guides should be encouraged parallel with Flora Malesiana.

Introduction

Geological and vegetational histories of Sumatra are nowadays well documented (Hamilton 1979; Audley-Charles 1987; Morley & Flenley 1987). Migration of taxa between Australia/New Guinea and Asia has occurred as far back as the Cretaceous, and during the Quaternary, changes in landscape and vegetation have been mainly attributed to fluctuations in sea level and in the altitudinal gradient of temperature. Palynological evidence of a drier climate during the Pleistocene has been pointed out for Malay Peninsula and South Borneo (Morley 1981), but such evidence is still lacking for Sumatra. Anthropogenic changes started between 6000 and 4000 years ago (Morley 1982; Maloney 1980), and Flenley (1988) places the first permanent clearings at around 2000 years BP.

The present flora of Sumatra is, however, probably only a little different from that at the end of the Tertiary. Botanists have long recognised that it largely repeats that of Malaya and Borneo in the lowlands, and that, on the other hand, it is fairly similar to that of Java in montane areas (van Steenis 1935b; Meijer 1981).

P. Baas et al. (eds.), The Plant Diversity of Malesia, 193–211.

It is noteworthy that Sumatra has fewer endemic genera (13) compared with neighbouring Peninsular Malaysia (24) and the island of Borneo (61) (van Steenis 1987), and it is often said to give an impression of harbouring a somewhat poorer flora. However, Sumatra has been surprisingly neglected by botanists, especially in the lowlands and is even less collected than Borneo and New Guinea (van Steenis-Kruseman 1973; Airy Shaw 1980). In taxonomic literature, it is often simply and arbitrarily divided into four areas (N, S, W, E).

During the process of a former ecological vegetation mapping survey (Laumonier 1983; Laumonier *et al.* 1986/87), investigations on the hierarchy of ecological criteria for the purpose of vegetation classification suggested that environmental subdivisions could be tested against flora distribution. Such maps strengthen research on eco-floristic zones, and facilitate the identification of regions of interest for further plant collection and ecological research.

To make a comparison of the proportional representation of the species within the island is probably a bit premature with the limited knowledge available at present. However, 2,500 tree species of 400 genera have been collected during our surveys, which may represent roughly 70 to 80% of the total tree flora ($\phi \geq 10$ cm) of Sumatra, and a first attempt on the tree flora could be considered in the light of recent progress in plant specimen identification.

Precise information on species distribution becomes nowadays more and more urgent for conservation policies. The present study was conducted to analyse the distribution patterns of the lowland tree flora, to compare it with the ecological classification performed for mapping, and to investigate whether any floristic sectors could be delineated in Sumatra and if so, to what extent they relate to floristic regions of Malaya and Borneo.

Method

Ecological mapping of the vegetation is partly the result of research on macro-environmental factors, mainly bioclimates, soil types and geology/geomorphology. Besides bioclimatic studies, one important objective in our vegetation mapping survey was to evaluate the relationship between natural or physiographic regions and their vegetation types. Physiography and geomorphology play an important role in sometimes subtle climatic or pedological variations: tectonic events, volcanism, local climatic changes, contributed to the formation of temporary or permanent natural barriers. Before more detailed studies on relationship of climate/vegetation or soil/vegetation can be realised, a geomorphological division may provide a first approximation of the phytogeographic zonation reflected in the structural and floristic diversity observed in the vegetation of Sumatra. To build the environmental classification, more attention has therefore been paid to the analysis and synthesis of various topographical, geomorphological and geological data in the literature.

The ecological division has been checked and validated by quantitative vegetational studies carried out on floristic composition, structure and architecture of various forest types within each subzone. The altitudinal zonation of the vegetation and the correlations between plant communities and their underlying geology have been worked out, especially at lower elevations (< 800 m). Special attention has been given to the distribution of species with poor fruit dispersal, like Dipterocarps.

The floristic affinity of the tree flora of Sumatra to each of the surrounding islands and countries in Indo-Malesia, Asia, Australia and the Pacific has then been assessed by a cluster analysis using data of Flora Malesiana, Tree Flora of Malaya and a few recent taxonomical monographs (c. 2300 species). Van Steenis' generic check list (1987) and Whitmore and Tantra's tree species check list (1986) provide necessary complementary information. Average linkage method of clustering (Sokal & Michener 1958) is known to be probably the most suitable for our purpose (Orlòci 1978; Pielou 1984). We compared it also with other amalgamation methods like the single linkage method (Hartigan 1975) and the complete linkage method.

In a second step, and according to the previous analysis of environmental factors and the resulting ecological classification, Sumatra was later divided into 11 parts, and the distribution of 1000 Malesian tree species, whose distribution was sufficiently documented, has been analysed using average linkage clustering (excluding endemics). For a matter of convenience, these sectors have been often given the names of provinces (Figure 17.1). This does not always correspond to real administrative boundaries in the present study.

Results and Discussion

Ecological classification

The detailed description of the ecological classification process and results are to be published in full elsewhere (Laumonier, in prep.), but comments can already be made here.

Bioclimatic considerations

Temperature variations and altitudinal zonation

Altitudinal zonation for the vegetation in Sumatra was dealt with by van Steenis (1935b), Jacobs (1958), Meijer (1961) and Oshawa *et al.* (1985), and this always concerns more precisely submontane and upper zones. 1,000 m altitude is generally recognised as an important, albeit approximate boundary which differentiates particular domains, after which absolute temperature minima become real limitations for certain animal and plant organisms.

In agreement with Symington's results for Malaya (1943), 800 m rather than 1,000 m seems to be a very important altitudinal boundary in the same way as another one between 300 m and 400 m. The dense, tall Dipterocarp forest always occurs below the latter boundary. As new plots were being established, an important fact began to show up in the field: another boundary between 150 m and 200 m was appearing which certain species never or rarely cross. Here, there is of course no question of a temperature limit, but in view of what we know at present, no satisfactory explanation of this fact can be given.

In a rectangle stretching from the centre of the Jambi plain to the Kerinci massif, observations concerning the Dipterocarpaceae family (Table 17.1) have demonstrated how some species are strictly confined to the very low altitude zone (< 150–200 m), such as *Anisoptera megistocarpa, Dipterocarpus lowii, D. crinitus, Hopea ferruginea, Shorea macroptera, S. pauciflora, S. singkawang*. Other species such as *Shorea* cf. *dasyphylla* and *S.* cf. *lumutensis* are

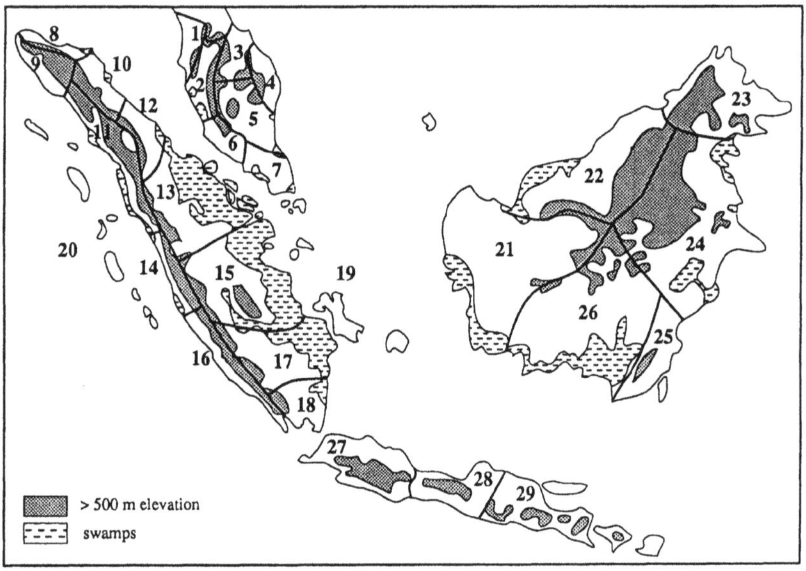

Figure 17.1 — Subdivisions used for cluster analysis: 1 = Kedak, 2 = Perak, 3 = Kelantan, 4 = Trengganu, 5 = Pahang, 6 = Selangor-Negeri Sembilan-Malaka, 7 = Johore, 8 = North Aceh, 9 = West Aceh, 10 = Langsa Aceh, 11 = Sibolga-Northwest coast, 12 = Asahan-Langkat, 13 = Riau, 14 = West Sumatra, 15 = Jambi, 16 = Bengkulu Southwest, 17 = South Sumatra, 18 = Lampung, 19 = Eastern islands, 20 = Western islands, 21 = West Kalimantan-West Sarawak, 22 = Sarawak, 23 = Sabah-North Borneo, 24 = East Kalimantan, 25 = South Kalimantan, 26 = Central Kalimantan, 27 = West Java, 28 = Central Java, 29 = East Java.

Table 17.1 — Altitudinal zonation of Dipterocarp species in Central Sumatra, below 1000 m elevation.

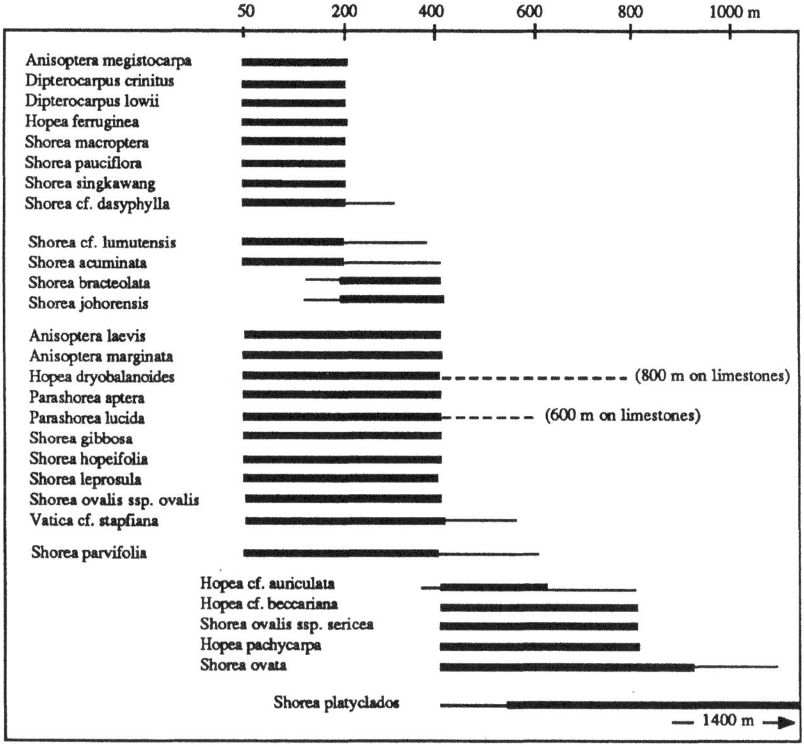

very well represented at these altitudes, but they can also occasionally be seen at an elevation of up to 400 m. There is an inverse tendency for *S. bracteolata* and *S. johorensis,* which occur occasionally in the lowlands and are more abundant in the foot hills (200–400 m).

A great majority of species show a larger ecological amplitude, which are distributed between sea level and 400 m. The most common are *Anisoptera laevis, Anisoptera marginata, Hopea dryobalanoides, Parashorea aptera, P. lucida, Shorea gibbosa, S. hopeifolia, S. leprosula, S. ovalis* subsp. *ovalis* and *Vatica* cf. *stapfiana.*

At higher altitudes, in the 400 m to 800 m hill zone, the most frequent species representatives of this level are *Hopea* cf. *auriculata, H.* cf. *beccariana, H. pachycarpa* and *Shorea ovalis* subsp. *sericea. Shorea ovata* also appears towards 400 m, but it has a greater amplitude, since it is often found in a zone up to 1,000 m. *Shorea platyclados* is an important component of submontane forest. It has a large ecological amplitude of (400–)500–1,200(–1,400) m.

The proposed zonation is given in Table 17.2. It is very close to that of Symington (1943).

Table 17.2 — Altitudinal zonation of the vegetation in Sumatra.

0–150 m	lowlands
150–400/500 m	low altitude hills
400/500–800/900 m	medium altitude hills
800/900–1,300/1,400 m	submontane
1,300/1,400–1,800/1,900 m	montane
1,800/1,900–2,400/2,500 m	upper montane
> 2,500 m	tropical subalpine

Intensity of the dry season

Ninety percent of the surface of Sumatra is subjected to very humid bioclimates, which means that the intensity of the dry season seems at first glance to be of little importance. Only the north-east coast of Banda Aceh has three to four dry or subdry months (mean monthly rainfall less than 60 mm). The east coast of the Malacca Straits, part of Lampung and some of the Barisan valleys, may have up to two dry months. There are also submontane bioclimates which have a dry season of between one and two months, with a mean temperature for the coldest month of 15–20° C.

A study of the variability of climates within Sumatra (Laumonier 1981) has shown that there is a real dry season, although it is somewhat irregular in many parts of the island. There are particularly dry years when forest fires occur even in the middle of the swamp area of Palembang and even near Padang on the equator (Braak 1925). A biologically dry season does exist at least for the south and the north-east during which the potential evapotranspiration is higher than the rainfall. Few studies have dealt with the former, due to the lack of available data concerning factors such as temperature, duration of sunshine, humidity and winds. In the same analysis, our results agree with those of Hanson and Koesoebiono (1978) who mentioned dry periods of 90 days and an exceptionally dry year approximately every four years for the region of Palembang.

Finally, it should be remembered that in areas with convection phenomena, there is a greater variability in local rainfall distribution and that the erratic character of rainfall in Sumatra is a supplementary factor which should not be ignored in any evaluation of the real impact of the dry season.

There is some doubt however, as to whether the intensity of the dry season, which is longer in northern, north-eastern and south-eastern regions of the island, is sufficient to influence the vegetation. Ashton (1972) noted a resemblance between the flora of Lampung and that of the north coast of Aceh, i.e., between the two regions where the dry season is most pronounced. Unfortunately, the vegetation of these regions has been much disturbed by man. The greater part of the east coast of Aceh and all of East Lampung are completely deforested and it is more difficult to analyse the secondary flora for environmental indicators, since many of its species have a greater ecological amplitude than those of primary forests.

On the east coast of Aceh, *Heteropogon contortus* and *Bothriochloa glabra* grasslands are very different from the humid savannas with *Imperata cylindrica,* occurring elsewhere.

Likewise, in the Padanglawas region to the south of the Toba plateaus, a region well-known for its drying winds, the grasslands and savannas are characteristic, dominated by *Arundinella setosa, Imperata cylindrica* and *Eremochloa ciliaris.* They resemble the savannas described by Corner (1940) for Peninsular Malaysia. The 'Tembesu' (*Fagraea fragrans*) is found as isolated individuals and the myrtaceous shrub (*Baeckea frutescens*), which is characteristic of nutrient-poor soil or of more constraining ecological conditions, is abundant.

It is difficult to imagine what the original vegetation in east Lampung was like. Scraps of lowland primary forest exist in the reserve of Way Kambas, as well as further west near the piedmonts. Way Kambas itself is probably just an individual case of this kind of forest, since the area consists of a mosaic of drained soil and marshes on the edge of the sea. The non-marshy part was exploited between 1968 and 1974 and was overrun by fire in several consecutive years (Santiapillai & Suprahman 1986).

Further west, observations were made on a forest which was being exploited in 1979 and which has since completely disappeared. *Dipterocarpus gracilis, Shorea guiso* and *Anisoptera costata* were abundant and seemed to grow gregariously. This characteristic undeniably differentiates them from forests further north, together with the presence of *Hopea semicuneata, Xanthophyllum erythrostachyum* and *Chrysophyllum lanceolatum.* Other species common in the canopy were *Dipterocarpus kunstleri, Shorea hopeifolia* and *Hopea mengerawan.* Although many species characteristic of these forests can be seen further north in the Palembang region (especially individuals of the lower sets), it would seem that as well as having a different structure (smaller trees, greater density of medium-sized trees per hectare, smaller forests), there were individual floristic components specific to these forests. In addition to the Dipterocarpaceae mentioned above, amongst the most important were canopy species such as *Osmelia philippina, Dillenia excelsa, Triomma malaccensis, Engelhardia ser-*

rata, Schima wallichii, or from lower strata *Knema rubens, Diplospora singularis, Aquilaria microcarpa, Ardisia lurida* and *Chrysophyllum lanceolatum*.

The presence of *Schima wallichii* and *Engelhardia serrata* as canopy trees in these lowland forests is quite remarkable. A member of the Theaceae common in the Barisan south of Lake Toba, *Schima wallichii*, is also a major component in forest regrowth after slash and burn agriculture practices in Lampung and in some hills in West Java. We have sometimes found *Engelhardia serrata* in swamp forest, but it is rare in the lowlands, and is usually considered as a mountain species.

In the Lampung lowlands forest, the distribution of these two species alone could be enough to characterise the boundaries of this floristic subregion. It is striking that these species are no longer found in primary lowland forest, nor in secondary vegetation to the north of a line connecting Palembang and Baturadja. Nor can they be considered as indicators of the length and the intensity of the dry season, since they are also found in hill and sub-montane zones as far as the Himalayas. They may, however, be considered as indicators of nutrient poor soils developed on acid tuffs where water retention is problematic. These cover large areas in Lampung. In the understorey the abundance of *Tristaniopsis merguensis* and *Ilex cymosa*, species characteristic of nutrient-poor acid substrata like peat soils, is another indication in the same direction.

It cannot really be said that these formations have a deciduous habit, since of all the species mentioned above, only *Dillenia excelsa* and *Engelhardia serrata* shed their leaves during the dry season. However, despite its more ubiquitous nature, the secondary vegetation of the area can provide some indications of the semi-deciduous tendency. Amongst the principal secondary species seen in the very degraded secondary formations dominated by *Schima wallichii* typical of Lampung and South Sumatra (the 'Seru' forests of van Steenis, 1935a), we have collected few which have affinities with the flora of northern Peninsular Malaysia and Burma, where the dry season is more marked (three to four dry months according to Gaussen *et al.* 1967). The most common are: *Cassia nodosa, Cratoxylum cochinchinensis, Dillenia obovata, Ficus lamponga, Heterophragma adenophyllum, Ilex cymosa, Lagerstroemia floribunda, Peltophorum dasyrachis*, and *Terminalia bellirica*. Very similar formations have been described for northern Peninsular Malaysia (Symington 1943), and it may be suggested that former primary formations in Lampung, although capable of retaining a certain atmospheric humidity, had, at least in certain parts, a tendency to semi-deciduousness, and that the climatic subdivision can be maintained.

Geosystems, geology and soils

There is an obvious relationship between soils and vegetation, as far as swamps are concerned. It is the same for plant formations on sandy coastal soils and to a lesser extent for those on limestones. The grasslands of Padang-lawas, mentioned above, lie over very sandy soils (the 'Zandsteppen' described by Frey-Wyssling, 1933) and may at one time have supported individual forests, similar to the 'Kerangas' forests of Borneo. These edaphic formations on tropical podzols have only been described for Borneo (Richards 1936; Brünig 1974; Kartawinata1978) and the islands of Bangka and Belitung (van Steenis 1935a; Whitten *et al.* 1984). As far as we know, they do not exist in Sumatra, although certain secondary formations in the centre of the island (Koorders 1895) may be reminiscent of the state of degradation known as 'padang' in Malay. We observed very localised secondary formations dominated by *Ploiarium alternifolium* on sandy soils in Riau foot hills.

But beside swampy or sandy areas, what is the exact role of the substrates in the distribution of vegetation types in Sumatra and especially, what is the nature of the relationship between vegetation and geology in plant distribution? Generally, botanists working in the Malay Peninsula disagree that there is any kind of relationship between soils and the distribution of forest species (Poore 1968; Kwan & Whitmore 1970), while for Borneo (Sarawak, Brunei, Sabah) there is, on the contrary, a tendency to show the existence of close links between soils and vegetation, or between geology and vegetation (Ashton 1972, 1982).

It has been impossible to test satisfactorily the nature of these links in Sumatra, simply because the level of detail of pedological analysis required would have been well beyond the objectives of the present study. However, first-hand observations on certain plots allow us to make some comments.

To investigate probable geology/vegetation correlations, we tried comparing some forests on different geological formations and especially two low altitude forests in the lowlands and the foot hills in the centre of Jambi Province, one growing on sediments and the other on older granite. Although the overall structures of these two forests are fairly similar, there are interesting floristic differences.

Briefly, if the overall floristic diversity seems very similar from one place to another (Simpson's index[1] respectively 0.96 and 0.90), there are differences for certain families or taxa. On granite substratum there is a relative floristic poverty in certain families which are better represented on sedimentary rocks. This is the case for Dipterocarpaceae, Guttiferae, Leguminosae and, to a less extent, for the Burseraceae from lower strata. Conversely, we noticed the presence of many Meliaceae on these granitic formations.

We recorded more than 20 species of Dipterocarpaceae for the plot on sediments for a sampled area of about three hectares. In comparison, the equivalent

---

1) Simpson's index, $D = 1 - \Sigma p_i^2$, where $p_i = Ni/N$ with Ni being the number of individuals in the ith species and N the total number of individuals of all species in the community.

surface area at Batang Ule (granite) has less than half this number. The dominant species on granite is *Parashorea lucida,* but one could just as well mention the strange rarity of the genera *Hopea* and *Dipterocarpus* as opposed to sedimentary soil where these genera are well represented.

In the hill zone, another sampling led us to compare the flora of metamorphic cliffs dating from the Tertiary, with that on neighbouring Quaternary volcanic rock (Merangin valley, Sungaipenuh–Kerinci district). There, too, extensive floristic inventories show significant differences at similar altitudes (main zone examined between 500–800 m), and specifically, greater floristic richness on volcanic substrate (Simpson's index 0.89 against 0.80 for those on metamorphic rock).

Thus, Dipterocarpaceae, which are very abundant on volcanic formations (*Hopea* cf. *beccariana, H. pachycarpa, Shorea ovalis, S. platyclados*) are practically non-existent on metamorphic terrain, and their ecological niche is mainly occupied by large Meliaceae, particularly *Dysoxylum acutangulum* and *Chisocheton ceramicus,* by Sapindaceae such as *Pometia pinnata,* and Anacardiaceae such as *Dracontomelon costatum. Lansium domesticum, Artocarpus anisophyllus* and *Celtis rigescens* occupy the lower strata. The latter species seems to be restricted to this environment, whereas *Casuarina nobilis* is abundant on scree and on steep slopes where rock is crumbling. The understorey is also different and we observed a great diversity of species of *Mallotus* (*M. oblongifolius* and *M.* cf. *miquelianus* being the most common), together with the presence of Marantaceae such as *Donax canniformis* which disappear in the volcanic zone. In these environments we also observed, for the first time, the very particular Araliaceae, *Trevesia burckii,* and discovered a *Rafflesia* hitherto only found further north, *Rafflesia hasseltii.*

The differences in flora could as well be attributed to a difference in age of the geological formations, and not only to lithological factors. The comparison of the flora of an old eroded Tertiary metamorphic massif (Tigapuluh Mts) with neighbouring sedimentary Quaternary plains, showed a flora which was more specific to this very interesting massif, the most obvious plants being *Shorea peltata* or the palm *Johannesteijsmannia altifrons,* which is found as pure stands in the undergrowth.

To conclude, it seems there really are indirect correlations between certain geological formations and the types of forest growing on them. Nevertheless, these results should be examined with caution, bearing in mind what is known at present about the flora of the region. In this same area there are numerous volcanic, metamorphic, granitic or sedimentary geological formations, and it would be interesting to compare their flora in detail. It would be particularly interesting to do this type of study in the framework of a larger scale mapping project in Kerinci National Park.

In the light of these results, links between physiographic regions, geosystems, vegetation types and flora have also been worked out. The geomorphological history of a landscape may influence the distribution of taxa. Using recent geological maps and the works of Verstappen (1973), van Bemmelen (1949) and Hamilton (1979), we have looked for geological accidents, which may have played a role in the isolation of plant communities. Within large physiographic regions, different geosystems whose boundaries have been geographically classified according to the dominant volcano, larger rivers or if necessary, by the name of a neighbouring large town, have been identified, the complete description of which will be published elsewhere (Laumonier, in prep.).

By doing so, we were able to differentiate, within the rather homogeneous climatic zones, other sectors of interest for analysis of the chorology of the region. They represent, below 800 m elevation and beside swamps, four sectors on the west coast, five on the eastern foot-hills and hills, and three in plains.

The Quaternary volcanic central west part acts like a very important barrier separating old geological formations of south and north Sumatra. The Talamau sector is obviously most important for floristic delineation, being for instance the southern limit of *Dryobalanops aromatica*. In other parts, mountains reaching the coast and swampy depressions act as other possible barriers. The last sector of West Aceh is set apart for its huge limestone formations.

The eastern blocks are separated also on the basis of the age of their geological formations. The south is mainly composed of volcanic products alternating with granitic cores. The area of Singkarak lake is very important for the fact that volcanos practically disappear north of it, and are replaced by large metamorphic and karstic ranges, and that this zone is in between two drier enclaves (East Singkarak and Padanglawas). In the northeastern regions, another sector is characterised by vast expanses of Toba ignimbrites and ejections from Sinabung. To the north, at the foot of the Serbolangit mountains, Tertiary hills with softer rock reappear. Tigapuluh and Duabelas mountains, sedimentary massifs which were folded and uplifted above the plain of Jambi are considered separately. Their geological history links them to the Sunda shelf which outcrop in the Riau and Lingga archipelagoes or at Bangka and Belitung.

In the plain at very low elevation (< 150 m), the three sectors are: the east Lampung block, the peneplains from Palembang to Jambi, south of the Kwantan, and the plains north of the Kwantan to the river Barumun. In the region between the confluence of the Merangin and the Tembesi and the Upper Kwantan lies an important boundary for lowland flora. It is in that area that there are often swampy depressions between the plain and the piedmonts. At the interglacial period of maximum flooding, there would have been only a narrow corridor the length of the piedmonts. In this part of the island the Batanghari itself runs through a depression parallel to the Barisan, to the west of the Tigapuluh Mountains. North of Kwantan, there are no such swampy depressions

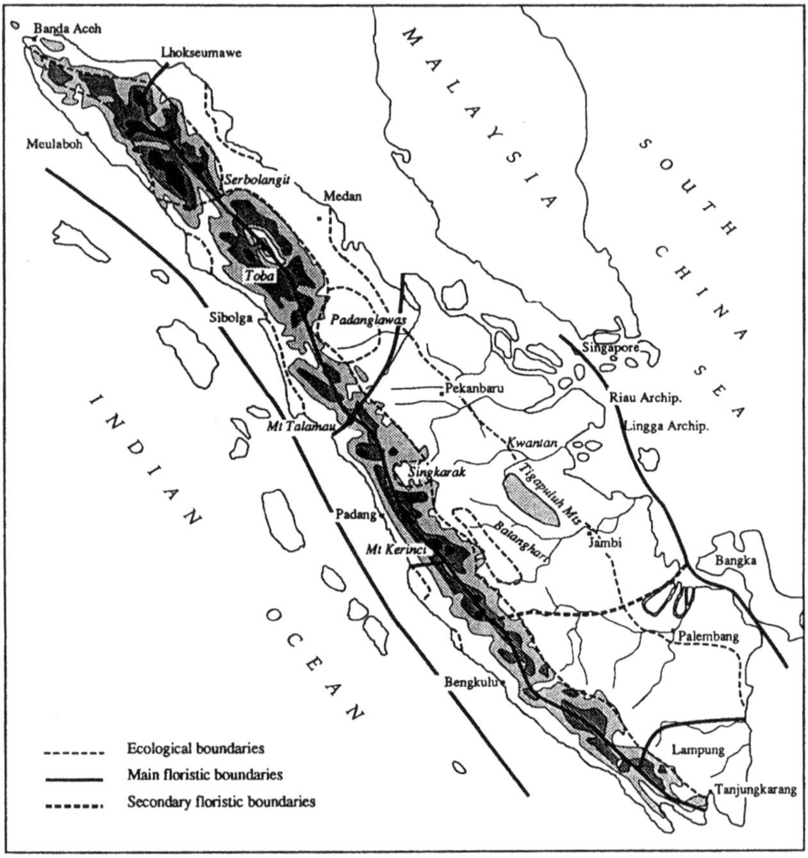

Figure 17.2 — Main ecological and floristic subdivisions in Sumatra.

between plain and mountain. The boundary of this sector is at the river Barumun, which is also the southern boundary of the climato-edaphic Padanglawas plain sector.

The combination of ecological sectors identified above allows us to consider a simple ecological classification which was used for the legend of the vegetation map. That ecological division is given on Figure 17.2.

Floristic analysis

The results of various trials for the cluster analysis show that for our data, no striking differences were observed between single, complete and average linkage methods of clustering. Single linkage has a tendency to produce longer, stringier clusters compared to complete linkage, but a few units grouped differently. Single linkage was also more affected by 'chaining', and complete linkage produced more compact clusters, often difficult to interpret. Average linkage was retained as the closest to ecological division.

Including Indo-Malesia, Australia and the Pacific, the results of the hierarchical classification at species level (Figure 17.3) show, as expected, the greatest

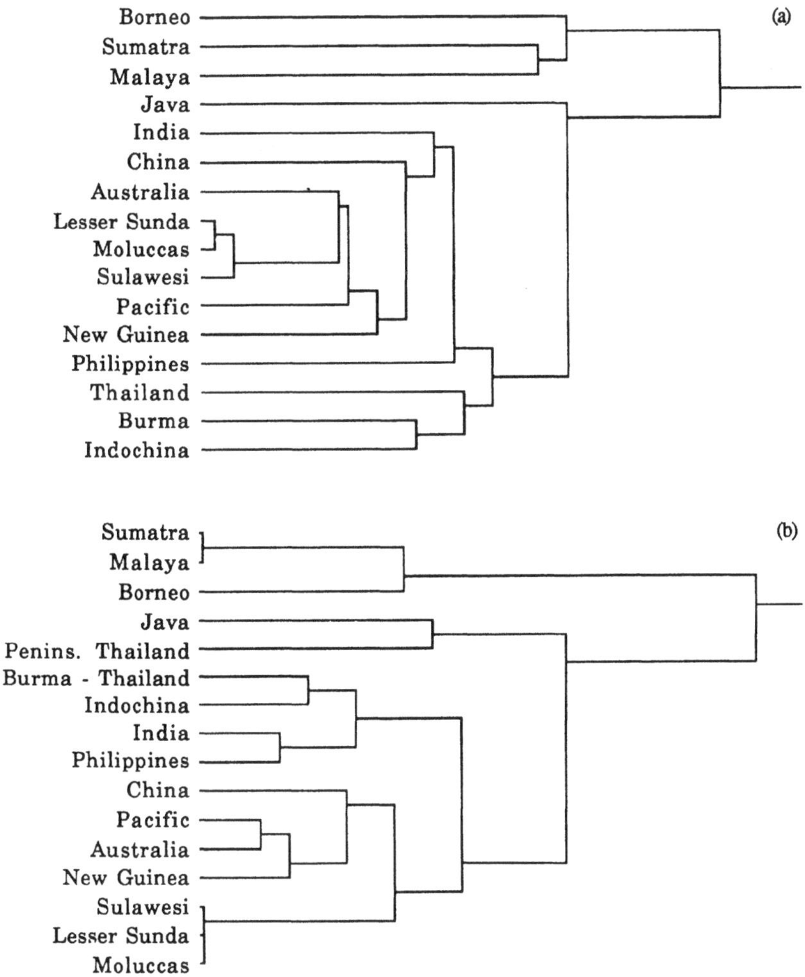

Figure 17.3 — Dendrogram showing floristic affinities of the Sumatran flora with Malesian and surrounding floristic regions, (a) including endemics, (b) excluding endemics.

floristic affinity between Sumatra and Malaya which group together and then with Borneo. Borneo, Sumatra and Malaya form a coherent subregion of Malesia. In decreasing hierarchy, they are then related to Java, and then to the group Indochina, Burma, Thailand. The Philippines comes next, which does not agree with the fact that they are usually considered as a part of West Malesia (van Steenis 1950). In the three linkage methods we tested, they appear always rather far from the group of Malaya, Borneo, Sumatra. This also agrees with the findings of van Balgooy (1987) which show that at the generic level, the Philippines group first with Sulawesi, and at the species level with Java and the Lesser Sunda Islands (excluding endemics). It is with Sulawesi that Sumatra has the least affinity, together with the Moluccas and the Lesser Sunda Islands. There are still more New Guinean, Australian or Pacific elements than from that part of Malesia.

Regarding now the subdivisions defined thanks to the search for ecological sectors, the results show that each island within the West Malesian floristic subregion remains as a coherent unit; there is no sector from any island grouping with a sector of another island (Figure 17.4).

Within Sumatra, in the lowlands, 6 main floristic sectors can be recognised which are also shown in Figure 17.2. They appear in the dendrograms of the three linkage methods, which is also a token of their validity. The classification necessitates a few remarks.

The lowlands of Riau, Jambi and South Sumatra are well apart, and so is the northeastern coastal plain. It is quite remarkable that Aceh, Western Aceh and Sibolga-Talamau, which group together as a rather coherent unit, also have floristic affinities with Lampung. The fringe of islands which runs the length of the west coast is closest to that group. From Simeuleu in the north to Enggano in the south, the level of endemism is known to be high there. This could be a relic flora (paleo-endemic), the vestige of an ancient Indo-Malesian flora. All together they will cluster then with the west central part of the island and Java.

Of the 2,300 species listed, a little less than half are strictly Malesian: 16% of these Malesian species are shared with Malaya only, whereas 11% are shared with Borneo and only 5% with Java. The level of endemic tree species can be estimated at around 20%, and Riau pocket flora elements at 6%.

A general look at the subdivision shows that it is rather close to the ecological one. Main divisions occur between old and recent geological formations with the well isolated young volcanic sector of central West Sumatra. All sectors related to the Riau pocket flora (Corner 1978) appear at the same node level, which confirms the existence of that phytogeographic unit, but not as an individualised region. The boundary of the Riau pocket flora in Sumatra remains vague. At present, one finds elements of that flora scattering in the eastern lowlands. The highest concentration of species appears however in the upper Indragiri and Tigapuluh mountain range area at the border between Jambi

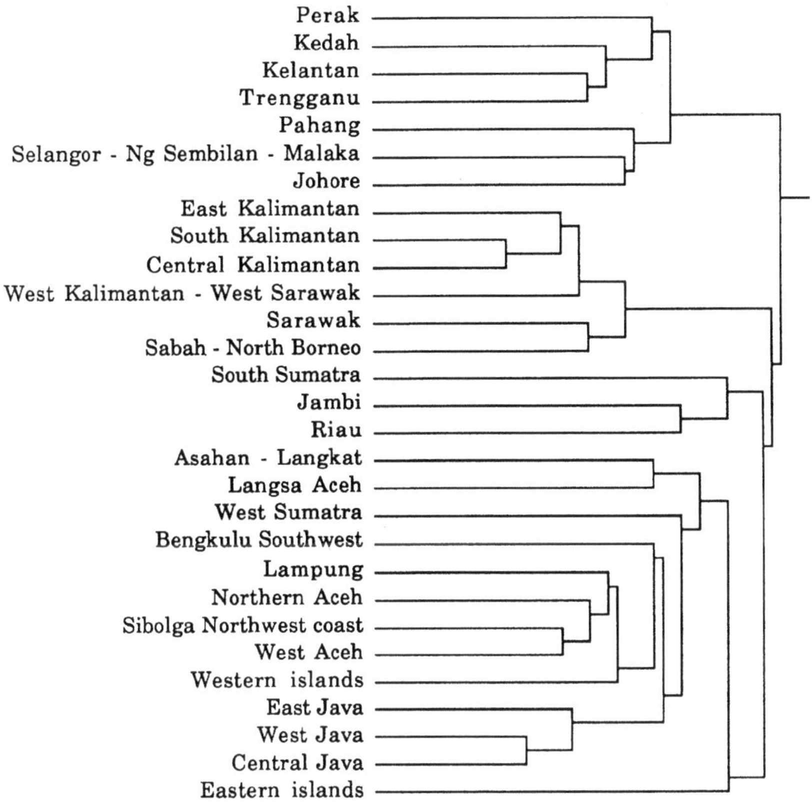

Figure 17.4 — Dendrogram showing floristic affinities of the Sumatran ecological subdivision against surrounding units in Malesia (Malesian element only).

and Riau provinces. The most common lowland species outside the swamps in that area are *Shorea lumutensis, S. gibbosa, S. peltata, Dipterocarpus lowii, Dryobalanops oblongifolia, Castanopsis schefferiana, C. johorensis, Lithocarpus bancanus, L. urceolaris* in the canopy and *Goniothalamus malayanus, Cyathocalyx ramuliflorus, Dacryodes macrocarpa, Dillenia albiflos, Lophopetalum pachyphyllum, Ptychopyxis kingii, Calophyllum flavo-ramulum, Ctenolophon parvifolius, Talauma bintuluensis, T. villosa, Artocarpus lanceifolius,* or *Xanthophyllum chartaceum* in the lower storeys.

No cluster analysis was made for the mountain flora. A brief look at the Ericaceae suggests that three additional sectors can be retained if elevation above 800 m is taken into account.

Conclusions

In summary, concerning environmental classification and vegetation mapping in an equatorial environment like the one in Sumatra, there are inadequate contrasts in climatic factors, and apart from altitudinal zonation, they cannot explain most variations observed in the vegetation and flora distribution. Similarly, apart from coastal and swamp areas or karstic formations, no striking relationship between soil and vegetation was found. Although some links may exist between the age of geological formations and their flora, no evidence has been found, in our study, for a direct relation to the substratum or lithology. Further investigation and research are needed to understand interactions between vegetation, geology and soil in such a humid tropical environment. At the present time, it makes sense to consider only relationships between physiography, geomorphology and vegetation, instead of dealing merely with intrinsic soil factors.

The present study corroborates also that, in vegetation mapping, it is fundamental to cover the entire area first to be able to observe all communities and biotopes which might have been missed by incidental observations. It prevents a subjective emphasis on favoured communities or sites. That is why the recommendations of van Steenis in 1958 of starting vegetation mapping at a 1/1,000,000 scale in the region, is perfectly valid today, when nothing has been done in that field during the past 30 years. These efforts on vegetation mapping and related studies have helped define ecological guidelines for conservation and identify floristically undercollected areas. At present, the methodology is being developed to cover more areas in South East Asia, and at the same time trials on larger scale vegetation maps are being conducted for sites selected from the previous survey.

There are strong demarcation lines within Malesia, an important one being between Sumatra and Java with about 20% of Sumatran species which do not occur in Java. Minor demarcation lines apparently also exist within the island itself, the strongest one corresponding to the existence of Lake Toba and the immense expanse of tuffs linked to its formation. This gigantic event probably influenced the distribution of flora by providing an ecological barrier to dissemination and exchanges between these two parts of the island. The south of this region is also bordered by the plain of Padanglawas, the region distinguished earlier for its sandy soils and drier climate, and to the south west by the Quaternary volcanic products of the central Sumatra Minangkabau highlands. Another important line separates Lampung from areas further north. The Barisan range itself is a considerable ecological barrier. However, at regular intervals there are corridor valleys at relatively low altitude (± 400 m) through which migrations may have occurred. This again, shows the importance of precise research into the altitudinal zonation of the flora, which should not be confined to the mountain zone only. It is not acceptable to be satisfied with the 1000 m elevation boundary, mainly faunistic, to claim that a reserve harbours 'lowland' forest. Very important floristic differences can appear as low as 150–200 m above sea level.

The floristic subdivisions identified during the course of the present study confirm the statement of Ng (1983) that differences between the various land masses in West Malesia are sufficiently important to stress that major conservation efforts have to be made separately on every land mass. They show also that within each landmass, conservation could not be made at random.

Collecting efforts should be made in Sumatra especially in the lowlands and hills, Tigapuluh mountain range, Kerinci and Leuser national park, limestone areas in Northwest Aceh and central part of the island, and East Lampung or North Aceh. Hill forest is presently the least collected. The last stand of relatively undisturbed mangroves lies in Sungai Sembilang, north of Palembang.

The problems of plant identification experienced since 1979 are far from being solved. The field experience of the present author has shown the usefulness of local treatments like the Tree Flora of Malaya, field guides and local floras which should be encouraged in the future, parallel with the more scientific Flora Malesiana, in order to stimulate interest of young local scientists towards plants.

Acknowledgements

For floristic aspects, much help in the field was provided by Pak Idjun Junaedi, and for proper plant identification I am much indebted to botanists in Leiden (M.M.J. Van Balgooy, H.P. Nooteboom, P.W. Leenhouts), Kew (J. Dransfield and the late H.K. Airy Shaw), Oxford (C. Pannell, D.J. Mabberley), Harvard (P.S. Ashton, P.F. Stevens) and Bogor (the late Pak Nedi). Annie Collier improved the English, an assistance greatly acknowledged.

References

Airy Shaw, H.K. 1980. The Euphorbiaceae of Sumatra. Kew Bull. 36: 239–374.

Ashton, P.S. 1972. The Quaternary geomorphological history of Western Malesia and lowland forest phytogeography. Trans. II Aberdeen Hull. Symp. Mal. Ecol. Misc. ser. no. 13.

Ashton, P.S. 1982. Dipterocarpaceae. Flora Malesiana I, 9^2: 237–552.

Audley-Charles, M.G. 1987. Dispersal of Gondwanaland: relevance to evolution of the angiosperms. In: T.C. Whitmore (ed.), Biogeographical evolution of the Malay Archipelago. Oxford Monographs on Biogeography 4: 5–25.

Balgooy, M.M.J. van. 1987. A plant geographical analysis of Sulawesi. In: T.C. Whitmore (ed.), Biogeographical evolution of the Malay Archipelago. Oxford Monographs on Biogeography 4: 94–102.

Bemmelen, R.W. van. 1949. The geology of Indonesia. Martinus Nijhoff, The Hague. Repr. 1970. 2 vols.

Braak, C. 1925. Het klimaat van Nederlands Indië. 2, 1: Sumatra. Kon. Magnet. en Meteor. Observ., Verh. 8. Batavia.
Brünig, E.F. 1974. Ecological studies in the kerangas forests of Sarawak and Brunei. Borneo Literature Bureau. Kuching
Corner, E.J.H. 1940. Wayside trees of Malaya. 2 vols. Repr. Malayan Nature Society, 1988.
Corner, E.J.H. 1978. The freshwater swamp-forest of South Johore and Singapore. Grdn's Bull. Suppl. No. 1. Botanic Gardens Parks, Singapore.
Flenley, J.R. 1988. Palynological evidence for land use changes in South-East Asia. J. of Biogeography 15: 185–197.
Frey-Wyssling, A. 1933. Over de zandsteppen van Kita Pinang. Trop. Natuur 22: 69–72.
Gaussen, H., P. Legris & F. Blasco. 1967. Bioclimats du Sud-Est Asiatique. Inst. Fr. Pondichéry Travx. Sec. Sci. Tech., Tome III, Fasc. 4.
Hamilton, W. 1979. Tectonics of the Indonesian region. Geological Survey Professional Paper 1078. US Gov. print. off. Washington.
Hanson, A.J., & Koesoebiono. 1978. Settling coastal swamplands in Sumatra. In: C. McAndrews & L.S. Chia (eds.), Developing economics and the environment: The Southeast Asia experience. MacGraw-Hill, Singapore.
Hartigan, J.A. 1975. Clustering algorithms. John Wiley & Sons, New York.
Jacobs, M. 1958. Contribution to the botany of Mount Kerinci and adjacent area in West Central Sumatra. Annales Bogoriense 3: 45–79.
Kartawinata, K. 1978. The 'kerangas' heath forest in Indonesia. In: J. Singh & B. Gopal (eds.), Glimpses of ecology: 145–153. International Science Publications, Jaipur.
Koorders, S.H. 1895. Losse schetsen der vegetatie van equatoriaal Sumatra. In: J.W. IJzerman, J.F. van Bemmelen, S.H. Koorders & L.A. Bakhuis, Dwars door Sumatra. Tocht van Padang naar Siak. De Erven F. Bohn, Haarlem.
Kwan, W.Y., & T.C. Whitmore. 1970. On the influence of soil properties on species distribution in Malayan lowland Dipterocarp rain forest. Malay. For. 33: 42–54.
Laumonier, Y. 1981. Ecological classification of South Sumatra forest types. BIOTROP unpubl. report. BIOTROP, Bogor.
Laumonier, Y. 1983. International Map of the Vegetation 1 : 1 000 000. Southern Sumatra. ICIV, Toulouse and BIOTROP, Bogor.
Laumonier, Y., *et al.* 1986/87. International Map of the Vegetation 1 : 1 000 000. Central Sumatra and Northern Sumatra. ICIV, Toulouse and BIOTROP, Bogor.
Maloney, B.K. 1980. Pollen analytical evidence for early forest clearance in North Sumatra. Nature 287: 324–326.
Meijer, W. 1961. On the flora of Mount Sago near Payakumbuh, Central Sumatra. Penggemar Alam 40: 3–13.
Meijer, W. 1981. Sumatra as seen by a botanist. Indonesian Circle 25.
Morley, R.J. 1981. Development and vegetation dynamics of a lowland ombrogenous peat swamp in Kalimantan Tengah, Indonesia. J. Biogeogr. 8: 383–404.
Morley, R.J. 1982. A palaeoecological interpretation of a 10 000 year pollen record for Danau Padang, Central Sumatra, Indonesia. J. Biogeogr. 9: 151–190.
Morley, R.J., & J.R. Flenley. 1987. Late Cainozoic vegetational and environmental changes in the Malay Archipelago. In: T.C. Whitmore (ed.), Biogeographical evolution of the Malay Archipelago. Oxford Monographs on Biogeography 4: 50–59.
Ng, F.S.P. 1983. Ecological principles of tropical lowland rain forest conservation. In: S.L. Sutton, T.C. Whitmore & A.C. Chadwick (eds.), Tropical rain forest, ecology and management: 359–375. Spec. Publ. n° 2 of the British Ecol. Soc., Blackwell, Oxford.

Orlòci, L. 1978. Multivariate analysis. Junk, The Hague.
Oshawa M., P.H.J. Nainggolan, N. Tanaka & C. Anwar. 1985. Altitudinal zonation of forest vegetation on Mount Kerinci, Sumatra: with comparisons to zonation in the temperate region of East Asia. J. Trop. Ecol. 1: 193–216.
Pielou, E.C. 1984. The interpretation of ecological data. John Wiley & Sons, New York.
Poore, M.E.D. 1968. Studies in Malaysian rain forest I. The forest on Triassic sediments in Jengka forest reserve. J. Ecol. 56: 143–196.
Richards, P.W. 1936. Ecological observations on the rain forest of Mt. Dulit, Sarawak. Part I & II. J. Ecol. 24: 1–37, 340–360.
Santiapillai, C., & H. Suprahman. 1986. The ecology of the elephant (Elephas maximus), in the Way Kambas Game Reserve. WWF/IUCN Report, Bogor, 95 pp.
Sokal, R.R., & C.D. Michener. 1958. A statistical method for evaluating systematic relationships. Univ Kansas Sci. Bull. 38: 1409–1438.
Steenis, C.G.G.J. van. 1935a. Maleische vegetatieschetsen I & II. Tijdschr. Kon. Ned. Aardr. Genoot. 52: 25–67, 171–203, 363–398 + map.
Steenis, C.G.G.J. van. 1935b. On the origin of the Malaysian mountain flora II. Altitudinal zones, general considerations and renewed statement of the problem. Bull. Jard. Bot. Buitenzorg III, 13: 289–417.
Steenis, C.G.G.J. van. 1950. The delimitation of Malaysia and its main plant geographical divisions. Flora Malesiana I, 1: lxx–lxxv.
Steenis, C.G.G.J. van. 1958. Vegetation map of Malaysia, scale 1 : 5 000 000. UNESCO, Paris.
Steenis, C.G.G.J. van. 1987. Checklist of generic names used for Spermatophytes in Malesian botany. Flora Malesiana Foundation, Leiden.
Steenis-Kruseman, M.J. van. 1974. Malesian plant collectors and collections. Cyclopaedia of Collectors, Suppl. 2. Flora Malesiana I, 8^1.
Symington, C.F. 1943. Foresters' manual of Dipterocarps. Mal. For. Rec. n° 16.
Verstappen, H.T. 1973. A geomorphological reconnaissance of Sumatra and adjacent islands (Indonesia). I.T.C., Enschede.
Whitmore, T.C., & I.G.M. Tantra. 1986. Tree flora of Indonesia. Check list for Sumatra. Forest Research and Development Center, Bogor.
Whitten, A.J., S.J. Damanik, J. Anwar & N. Hisyam. 1984. The ecology of Sumatra. Gadjahmada Univ. Press, Jogjakarta.

18 The secondary forest of Tanah Grogot, East Kalimantan, Indonesia

SUKRISTIJONO SUKARDJO

Summary

The structure and floristic composition of a 0.25 ha plot in a young secondary forest at Tanah Grogot, East Kalimantan, Indonesia, is described. The vegetation is a *Vitex pinnata–Geunsia furfuracea* community, with 26 species of trees, 47 species of saplings, and 56 species of ground vegetation. The (extrapolated) densities per ha are 592 trees, 4,736 saplings, and 130,400 individuals of ground vegetation. Compared with other lowland secondary forests previously studied the species diversity of trees is low, but the number of individuals is high.

Introduction

Covering the largest logging area in Indonesia, and with a potential commercial timber volume of c. 900 millions cubic metres, the forests of East Kalimantan have been intensively exploited in recent years. Moreover, traditional shifting cultivation by the inhabitants is common throughout the region. A serious degradation of the forest resource caused by both mechanical logging and indiscriminate cultivation has been the result (Kartawinata *et al.* 1977, 1978, 1979). Thus, of the 17.3 million ha covered by natural forest, already in the early seventies 2.4 million ha is now covered by secondary forest (Darjadi *et al.* 1976; Wirakusumah 1976); the proportional areas of primary and secondary forests have meanwhile shifted dramatically towards the secondary type.

Despite their growing importance throughout East Kalimantan, only few detailed studies have been made of secondary forests. This study, although limited to a single plot of 0.25 ha, is aimed at describing secondary forest in terms of floristic composition, soil type and structure from Tanah Grogot, and compare the results with literature information on other forests.

Study area

The secondary forest studied is located at Batukajang, about 180 km southwest of Balikpapan (Figure 18.1), at 350–450 m altitude and forms the lower part of the mountain forest of Mt Melihat (alt. 1,050 m). The soils have

P. Baas et al. (eds.), The Plant Diversity of Malesia, 213–224.

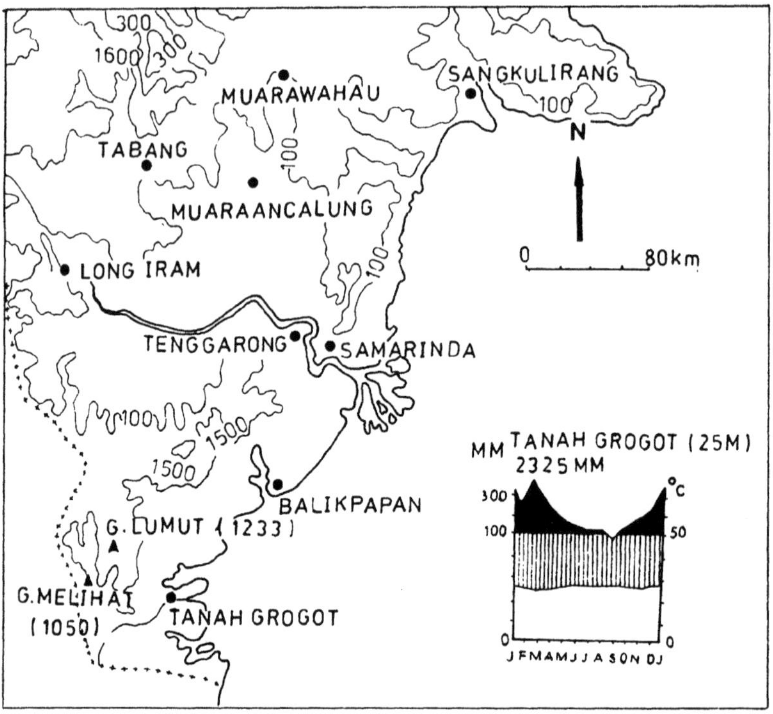

Figure 18.1 — The approximate location of the study area and a climate diagram for the meteorological station at Tanah Grogot.

been described as alluvial deposits and sedimentary rock (Anonymous 1981). The climate of the area (Figure 18.1) has been ranked as type A with a proportion of 0–9.8% dry months (Schmidt & Ferguson 1951) and an annual rainfall of c. 2,325 mm (Berlage 1949).

The forests in the region have a high species diversity, as indicated by earlier explorations (Endert 1927; Iwatsuki *et al.* 1980; Mogea *et al.* 1981); they are lowland mixed dipterocarp forests with *Vatica subcordata, Shorea ovalis, S. platyclados, Dipterocarpus cornatus, Durio* spp., *Palaquium* spp., *Castanopsis* spp., *Pentace* spp., and *Koompassia excelsa* as dominant elements of the canopy. The understorey is mostly dominated by species of *Baccaurea, Aporusa, Polyalthia,* and *Aglaia.* Parts of the forests are extensively used for shifting cultivation, while other areas have well developed secondary forest. Structurally the forest ranges from a single stratum type to a more complex, older, three-layered secondary forest.

Methods

A plot of 0.25 ha (50 × 50 m) was established and subdivided into 25 plots of 10 × 10 m. All trees over 10 cm dbh were measured and mapped. Their height was measured with a Haga altimeter. Saplings (diameter of 2.0–9.99 cm) were sampled in subplots of 5 × 5 m; seedlings and herbs were recorded from 1 × 1 m subplots. Subplots were located in each 10 × 10 m plot. All saplings were identified and their diameters measured. The percentage cover of each species of seedlings and herbs was estimated and the number of individuals of each species was counted. Vegetation data were analysed using the methods proposed by Cox (1967). Profile diagrams were prepared from selected plots of 50 × 10 m. Soil samples were collected in each 10 × 10 m plot. These were analysed by the Department of Natural Sciences, Bogor Agricultural University. Herbarium specimens were collected and deposited in Herbarium Bogoriense.

Table 18.1—The commonest species of trees and saplings in terms of their Importance Value (I. V.), Density per hectare (D) and Basal Area per hectare (BA) of the young secondary forest at Tanah Grogot.

Species	Importance Value		D		BA		H
	T	S	T	S	T	S	
Callicarpa pedicellata	29.7	15.2	68	176	3.1	1.1	25
Debregeasia capitellata	17.0	31.4	36	528	1.6	2.1	23
Diospyros borneensis	–	10.2	–	240	–	0.5	–
Ficus variegata	10.6	–	12	–	2.5	–	23
Geunsia furfuracea	46.4*	11.9	108	128	6.2	0.8	28
Glochidion sp.	10.7	–	16	–	1.3	–	23
Micromelum sp.	17.1	13.8	36	176	1.6	0.8	25
Millettia sericea	–	9.8	–	176	–	0.4	–
Peronema canescens	9.5	38.1**	16	720	0.8	2.6	20
Scolopia spinosa	–	20.4	–	304	–	1.0	–
Semecarpus heterophylla	–	9.2	–	176	–	0.3	–
Urophyllum glabrum	0.4	–	12	–	1.6	–	28
Vitex pinnata	78.8**	31.9*	180	528	12.0	2.1	21
Xanthophyllum sp.	8.3	–	8	–	1.9	–	31

** = dominant species; * = codominant species; T = tree with dbh ≥ 10 cm; S = sapling with dbh 2–10 cm; H = maximum height (trees only).

Table 18.2 — Basic statistics of the studied plot of young secondary forest at Tanah Grogot.

Vegetation attribute	Trees	Saplings	Ground vegetation (seedlings, herbs, ferns, grasses, shrubs, lianas, etc.)
Density / ha	529	4,736	130,400
Basal Area (m^2) / ha	40.6	16.0	–
Number of species	26	47	56
Number of genera	21	40	51
Number of families	14	22	27

Table 18.3 — Comparison of density and basal area in the Tanah Grogot forest and other forests.

Locality	Density (tree/ha)	Basal Area (m^2/ha)
Tanah Grogot	592	40.6
Wanariset*	541	29.7
Lempake*	445	37.5
Ketambe*	460	40.5
Jaro*	399	39.1

* Source: Kartawinata *et al.* 1981.

Results and Discussion

Characteristics of the vegetation

The young secondary forest at Kasungai, Tanah Grogot comprises 57 species of trees and saplings (Table 18.1) and 56 species of ground vegetation including tree seedlings (Table 18.2, Appendix 2). *Vatica* sp., with a density of 48 per ha (relative frequency = 0.72%) is the only member of the Dipterocarpaceae found among the tree and sapling species (Appendix 1).

The young secondary forest is dominated by three pioneer species: *Vitex pinnata* (importance value = the sum of relative frequency, density and dominance, I.V. = 78.4 and 31.9), *Geunsia furfuracea* (I.V.= 46.4 and 11.9) and *Peronema canescens* (I.V. = 9.5 and 38.1). The importance values (I.V.) between brackets refer to trees and saplings respectively. A total of 14 species have an I.V. of over 8.0 for the sapling or tree stages (Table 18.1).

For the importance values of saplings and/or trees of other species see Appendix 1. Six species appear to have an I.V. of 5.0–8.0 for the tree stage; six other species show this range of I.V. values for the sapling stage. Ten species

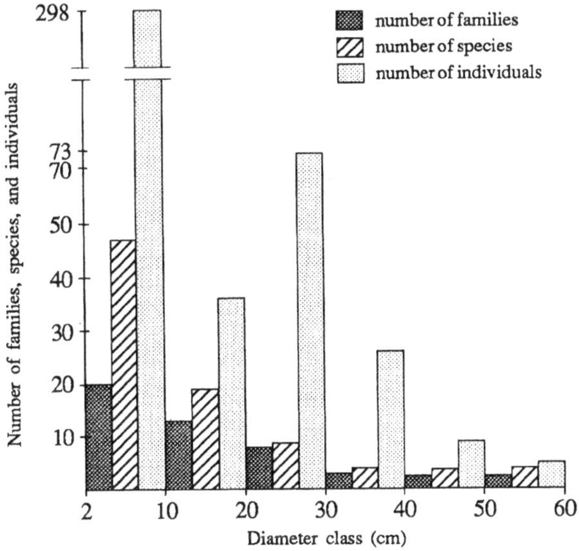

Figure 18.2 — Diameter class distribution in the Tanah Grogot plot.

of trees and 16 species of saplings have importance values between 1.0–5.0. Sixteen other species of saplings are below 1.0 in their I.V. The vegetation can be classified as a *Vitex pinnata–Geunsia furfuracea* community (cf. Table 18.1).

The statistics based on the inventory of the 0.25 ha plot are given in Table 18.2. The figure for tree density (592 per ha) is higher than in other forests in Indonesia compared in Table 18.3. However, the number of species is lowest. Both the Wanariset and Lempake forests are older than the secondary forest of Tanah Grogot. The differences in the number of species is probably at least partly due to the difference in plot size: 0.25 ha compared with 1.6 ha for the other plots listed in Table 18.3. The higher tree density probably reflects the abundance of short-lived nomad species in the plot. Soil properties might also be a factor, because the soils in Lempake and Wanariset contain more sand and are poorer in humus (personal, unpublished observation).

The floristic similarity between plots of the forest studied is relatively high as indicated by Jaccard's community coefficient of only 78.6% (cf. Greig-Smith, cited in Müller-Dombois & Ellenberg 1974). This means that the forest type is homogeneous in terms of its species composition. This seems to be typical of young lowland tropical rain forest in East Kalimantan where usually there is no single species with a high degree of dominance. Almost all individuals (90.4%) have a diameter of less that 10 cm, and only 3 species are represented by trees with a diameter over 50 cm (Figure 18.2).

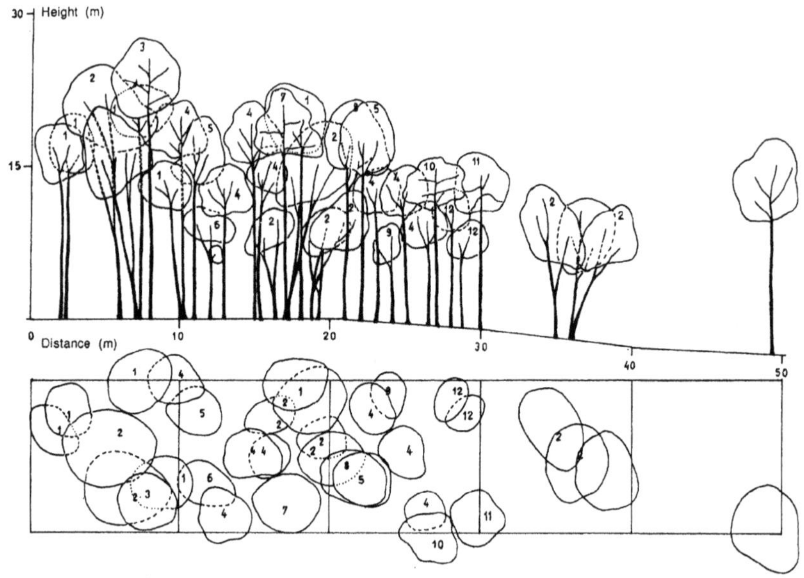

Figure 18.3 — Forest profile diagram and canopy projection showing the natural gaps of the young secondary forest. Only trees of ≥ 10 cm dbh are shown. — 1. *Callicarpa pedicellata*; 2. *Vitex pinnata*; 3. *Duabanga moluccana*; 4. *Geunsia furfuracea*; 5. *Glochidion* sp.; 6. *Debregeasia capitellata*; 7. *Terminalia microcarpa*; 8. *Premna foetida*; 9. *Ficus* sp.; 10. *Urophyllum glabrum*; 11. *Casearia* sp.; 12. *Ficus* sp. 1.

Based on Ogawa *et al.*'s method (1965) the secondary forest can be classified as unstratal, shown by the overlapping of the crowns and the crowding of boles of each species (Figures 18.3 & 18.4) and the diameter class distribution (Figure 18.2). No emergent trees occur. *Xanthophyllum* sp. has with 31 m the highest individual tree in the plot (Table 18.1).

Forest rejuvenation and habitat

The forest floor is characterised by a dense mat of grasses, sedges, herbs, and rattans. *Selaginella plana* (I.V. = 56.8) and *Scleria laevis* (I.V. = 35.4) are the dominant species (Appendix 2). Altogether 56 species, including 24 species of tree seedlings (43%) were recorded. This indicates that almost all tree species in the forest are rejuvenating themselves (Figure 18.2; Table 18.4). Only several of the commonest tree species occur at the adult stage (cf. Tables 18.1 and 18.4). Typical pioneering species represented by seedlings are *Peronema canescens*, *Debregeasia capitellata*, *Vitex pinnata*, and *Scolopia spinosa*. They all have very dispersed saplings or seedlings (Table 18.5).

Figure 18.4 — Forest profile diagram and canopy projection showing the natural gaps of the young secondary forest. Only trees of ≥ 7.5 cm dbh are shown. — 1. *Vitex pinnata*; 2. *Peronema canescens*; 3. *Debregeasia capitellata*; 4. *Xanthoxylum* sp.; 5. *Diospyros borneensis*; 6. *Callicarpa pedicellata*; 7. *Cratoxylum formosum*; 8. *Cratoxylum* sp.; 9. *Scolopia spinosa*; 10. *Omphalea grandifolia*; 11. *Artocarpus elasticus*; 12. *Micromelum* sp.; 13. *Glochidion* sp. 2; 14. *Duabanga moluccana*; 15. *Rhysotoechia acuminata*; 16. *Paranephelium xenophyllum*; 17. *Vatica* sp.; 18. *Endospermum moluccanum*; 19. *Semecarpus heterophylla*; 20. *Actinodaphne glomerulata*; 21. *Ficus variegata*; 22. *Paranephelium* sp.; 23. *Eugenia jamboloides*; 24. *Mallotus oblongifolius*; 25. *Glochidion* sp. 1.

It is interesting to note that although both the C/N ratio and percentage of organic C of the plot are lower than recorded in the primary and secondary forest of Mt Hondje, West Java (Table 18.6), seedlings and saplings of pioneers are more numerous in the Tanah Grogot forest. The abundance of the seedlings and saplings of pioneer species is probably a response to light (cf. Stoutjesdijk 1961) as gaps occupy 0.5 to 27.5% of this plot (Figures 18.3 & 18.4).

References

Anonymous. 1981. Peta jenis tanah Kalimantan Timur. Pemerintah Daerah Kalimantan Timur.

Berlage, H.P. 1949. Rainfall in Indonesia. Verhand. Kon. Magnet. & Meteorol. Obs. no 37.

Cox, G.W. 1967. Laboratory manual of general ecology. Dubuque.

Darjadi, I., E.S. Martadiwangsa & P. Wiroatmodjo. 1976. Tegakan hutan Indonesia. Buku I: Kalimantan. Direktorat Bina Program. Pengumuman no. 10.

Table 18.4 — Densities (per ha) of seedlings (S), sapling (Sp, diam. 2–10 cm) and trees (T, dbh > 10 cm) in the studied plot.

Species	Density per hectare		
	S	Sp	T
Agelaea borneensis	4,400	128	–
Aglaia sp.	3,600	32	–
Aporusa sp.	800	–	–
Aporusa elmeri	400	–	–
Casearia sp.	2,400	32	8
Cratoxylum sp.	400	–	4
Duabanga moluccana	1,600	–	8
Geunsia furfuracea	400	128	108
Leea sp.	400	–	–
Mallotus oblongifolius	800	16	4
Micromelum sp.	1,600	–	–
Millettia sericea	1,600	48	–
Omphalea grandifolia	400	16	–
Polyalthia sp.	1,200	–	–
Peronema canescens	6,000	720	16
Rhodamnia cinerea	400	16	–
Rhysotoechia acuminata	400	96	–
Scolopia spinosa	2,000	304	4
Vitex pinnata	1,200	528	180
Species of Annonaceae 1	800	–	8
Annonaceae 2	1,600	–	–
Euphorbiaceae 1	800	–	–
Euphorbiaceae 2	2,400	–	–
Meliaceae	1,600	–	–

Table 18.5 — Relative frequency (FR, %) and density (DR, %) of saplings and seedlings.

Species	Saplings		Seedlings	
	FR	DR	FR	DR
1. *Agelaea borneensis*	3.6	2.7	2.2	3.4
2. *Debregeasia capitellata*	7.2	11.1	–	–
3. *Callicarpa pedicellata*	5.0	3.7	–	–
4. *Geunsia furfuracea*	4.3	2.7	0.7	0.3
5. *Micromelum* sp.	5.0	3.7	–	–
6. *Millettia sericea*	3.6	3.7	1.5	1.2
7. *Peronema canescens*	6.5	15.5	4.4	4.6
8. *Rhysotoechia acuminata*	3.6	2.0	0.7	0.3
9. *Semecarpus heterophylla*	3.6	3.7	–	–
10. *Scolopia spinosa*	7.9	6.4	3.0	1.5
11. *Vitex pinnata*	7.9	11.1	1.5	0.9

Table 18.6 — Physical and chemical soil properties of the studied plot and of Mt Hondje, Java (data from Soerianagara 1970). All values represent averages.

	Tanah Grogot	Mt Hondje (after Soerianagara 1970) I	II	III	IV
1. Soil depth (cm)	20	20	20	20	20
2. Texture (%)					
– sand	43.18	29.27	30.30	12.57	20.40
– silt	35.92	42.80	30.80	38.40	53.37
– clay	20.90	17.93	38.90	49.03	24.23
3. Textural class	loam	loam	clay loam	clay	silt loam
4. pH		5.70	5.80	4.87	5.07
– H_2O	3.60	–	–	–	–
– KCl	3.29	–	–	–	–
5. Moisture content (% H_2O)	3.68	–	–	–	–
6. Saturated acid (%)	1.83	–	–	–	–
7. Exchangeable cation (me/100 gr)					
– K	0.86	0.04	0.08	0.05	0.02
– Na	0.97	–	–	–	–
– Ca	55.58	0.20	0.04	0.10	0.14
– Mg	7.30	–	–	–	–
8. CEC (me/100 gr)	36.36	–	–	–	–
9. Organic matter (%)	5.99	–	–	–	–
10. C organic (%)	3.48	1.47*	3.90***	3.97***	4.31***
11. N total (%)	0.35	0.15**	0.27***	0.27***	0.28***
12. C/N ratio	10.08	10.00	13.50	14.00	15.00
13. NO_3 (ppm)	12.16	–	–	–	–
14. NH_4 (ppm)	254.08	–	–	–	–
15. Available P (ppm)	30.70	0.03**	0.05***	0.04**	0.02**

I = abandoned field, II–IV = profile number. — *** = moderate; ** = low; * = very low.

(References continued)

Endert, F.H. 1927. Floristisch verslag. In: Midden-Oost Borneo expeditie: 200–285. Indisch Committee voor Wetenschappelijk Onderzoek Weltrevreden.

Iwatsuki, K., J.P. Mogea, G. Murata & K. Kartawinata. 1980. A botanical survey in Kalimantan during 1978–1979. Acta Phytotax. Geobot. 31: 1–23.

Kartawinata, K., R. Abdulhadi & T. Partomihardjo. 1981. Composition and structure of a lowland Dipterocarp forest at Wanariset, East Kalimantan. Malay. For. 44: 397–406.

Kartawinata, K., S. Adisoemarto, S. Riswan, R. Abdulhadi, J. Rahayuningsih, W. Angraitoningsih, S. Sukardjo, H. Sudjito & J. Suprijatna. 1978. Laporan kemajuan periode April 1977–Maret 1978 pelaksanaan Proyek MAB no. 1 di Kalimantan Timur: Suksesi sekunder dan perubahan biologi/ekologi lainnya di hutan tropik setelah perusakan oleh manusia. Unpubl. mimeogr. report, Herbarium Bogoriense, LIPI. 6 pp.

Kartawinata, K., S. Riswan, H. Abdulhadi, J. Rahajuningsih, H. Sudjito, S. Sukardjo & J. Suprijatna. 1979. Laporan kemajuan periode April 1978–Januari 1979 pelaksanaan Proyek MAB no. 1 di Kalimantan Timur: Suksesi sekunder dan perubahan biologi/ekologi lainnya di hutan tropik setelah perusakan oleh manusia. Unpubl. mimeogr. report, Herbarium Bogoriense, LIPI. 36 pp.

Kartawinata, K., S. Riswan, H. Sudjito & S. Sukardjo. 1977. Laporan kemajuan periode Januari 1976–Maret 1977 pelaksanaan Proyek MAB no. 1 di Kalimantan Timur: Suksesi sekunder dan perubahan biologi/ekologi lainnya di hutan hujan tropik setelah perusakan oleh manusia. Unpubl. mimeogr. report, Herbarium Bogoriense, LIPI. 24 pp.

Mogea, J.P., S. Sukardjo & A. Ma'Roef. 1981. Mengumpulkan tumbuhan di Kabupaten Pasir Kalimantan Timur. Alam Kita no 6: 1–8.

Müller-Dombois, D., & H. Ellenberg. 1974. Aims and methods of vegetation ecology. John Wiley & Sons, New York.

Ogawa, H., K. Joda, T. Kira, K. Ogino, T. Shidei, D. Ratanawongse & P. Apasutaya. 1965. Comparative ecological study on three main types of forest vegetation in Thailand. I. Structure and floristic composition. Nature and Life in South East Asia 4: 13–48.

Schmidt, F.H. & J.H.A. Ferguson. 1951. Rainfall types based on wet and dry period ratios for Indonesia with western New Guinea. Djawatan Meteorologi dan Geofisika. Verhand. no 42.

Soerianagara, I. 1970. Soil investigation in Mt Hondje Forest Reserve, West Java. Rimba Indonesia 15: 1–16.

Stoutjesdijk, P. 1961. Micrometeorological measurements in vegetation of various structure. Proc. Kon. Ned. Akad. Wet. Amsterdam, C 64: 171–207.

Wirakusumah, R.S. 1976. Toward a rational utilization of the tropical humid forest resources of East Kalimantan. In: S. Adisoemarto (ed.), East Kalimantan: 23–28. Proc. Workshop Indonesian MAB Project no 1 in East Kalimantan, LIPI, Jakarta.

Appendix 1 — A list of tree and sapling species with an Importance Value (I.V.) of less than 8.34 and their densities per ha (D) of the secondary forest in Tanah Grogot.

Species	Trees*		Saplings	
	I.V.	D	I.V.	D
1. *Actinodaphne glomerulata*	–	–	1.2	16
2. *Agelaea borneensis*	–	–	7.9	128
3. *Aglaia* sp.	–	–	2.2	32
4. *Alseodaphne* sp.	–	–	3.0	64
5. *Artocarpus elasticus*	–	–	3.1	32
6. *Casearia* sp.	6.9	8	1.9	32
7. *Cordyline fruticosa*	–	–	1.2	16
8. *Cratoxylum* sp.	2.3	4	–	–
9. *Cratoxylum formosum*	5.8	8	1.9	32
10. *Dalbergia* sp.	–	–	1.5	16
11. *Duabanga moluccana*	5.5	8	–	–
12. *Endospermum diadenum*	–	–	1.3	16
13. *Eugenia jamboloides*	2.6	4	6.6	112

(Appendix 1 continued)

14. *Euphorianthus* sp.	–	–	1.1	16
15. *Ficus* sp. 1	2.2	4	4.5	48
16. *Ficus* sp. 2	3.3	8	3.5	64
17. *Ficus michelii*	–	–	2.5	32
18. *Ficus septica*	–	–	7.9	112
19. *Ficus variegata*	–	–	6.4	96
20. *Glochidion* sp. 1	–	–	1.5	16
21. *Glochidion* sp. 2	2.8	4	7.5	128
22. *Glochidion rubrum*	3.2	4	2.8	32
23. *Macaranga trichocarpa*	6.0	8	–	–
24. *Mallotus oblongifolius*	2.4	4	1.2	16
25. *Mangifera pajang*	2.5	4	1.6	16
26. *Melanochylla elmeri*	–	–	1.6	32
27. *Millettia* sp.	–	–	2.7	48
28. *Omphalea grandiflora*	–	–	1.9	16
29. *Paranephelium xenophyllum*	–	–	3.2	48
30. *Premna foetida*	6.5	12	1.6	16
31. *Rhodamnia cinerea*	–	–	1.4	16
32. *Rhysotoechia acuminata*	–	–	6.2	96
33. *Saurauia* sp.	–	–	1.2	16
34. *Scolopia spinosa*	2.5	4	–	–
35. *Terminalia microcarpa*	2.4	4	–	–
36. *Uncaria gambir*	–	–	1.4	16
37. *Urophyllum glabrum*	–	–	1.6	32
38. *Vatica* sp.	–	–	2.0	48
39. *Vernonia* sp.	–	–	1.8	16
40. *Zanthoxylum* sp.	–	–	4.2	48
41. Species of Annonaceae	–	–	1.2	16
42. Species of Euphorbiaceae	–	–	2.7	48
43. Species of Lauraceae	5.8	8	–	–

* I.V. less than 8.3.

Appendix 2 — A list of species of ground vegetation and their importance value (I.V.) and average cover (A.C. %) of the secondary forest in Tanah Grogot.

Species	I.V.	A.C.
I. Ferns:		
1. *Diplazium* sp.	1.2	2.0
2. *Diplazium repandum*	12.7	3.7
3. *Lycopodium cernuum*	5.2	0.7
4. *Selaginella plana*	56.8**	4.2

(Appendix 2 continued)

II. Grasses:		
5. *Dendrocalamus* sp.	2.6	3.3
6. *Leptaspis urceolata*	3.9	2.0
7. *Panicum* sp.	18.5	4.0
8. *Paspalum conjugatum*	2.8	12.0
9. *Pennisetum* sp.	11.2	3.4
III. Sedge:		
10. *Scleria laevis*	35.4 *	3.8
IV. Herbs:		
11. *Amomum* sp.	1.4	4.0
12. *Borreria* sp. 1	5.1	3.0
13. *Borreria* sp. 2	1.5	5.0
14. *Donax* sp.	2.4	5.6
15. *Donax canaeformis*	4.4	8.6
16. *Commelina* sp.	1.1	1.0
17. *Lysianthus* sp.	3.9	8.3
18. *Phrynium repens*	1.9	10.0
19. *Strobilanthus* sp.	1.4	4.7
20. *Strobilanthus cernua*	3.6	1.2
21. *Zingiber* sp.	2.9	5.0
V. Shrubs:		
22. *Ixora breviloba*	3.0	5.0
23. *Leea indica*	7.2	3.4
24. *Millettia* sp.	1.4	4.0
25. Species of Leguminosae	3.5	2.3
VI. Woody climbers:		
26. *Alyxia stellata*	1.5	5.5
27. *Connarus macrophyllus*	2.7	3.0
28. *Dalbergia* sp.	1.3	3.0
29. *Uncaria gambir*	1.6	5.0
30. *Phanera fulva*	3.4	3.7
31. *Smilax* sp.	1.5	4.0
VII. Palm:		
32. *Calamus* sp.	1.8	8.0

** dominant species; * codominant species.

19 The fern flora of Seram

MASAHIRO KATO

Summary

Well over 700 fern species are estimated to occur on the island of Seram, Moluccas, judging from recent collections. This species richness may be ascribed to such non-biotic characteristics as a semi-everwet climate, mountainous topography, and limestone-rich geology, all providing fern species with a variety of habitats, and to geographical proximity to New Guinea with the world's richest fern flora, and the rest of Malesia. Seram has the strongest floristic affinity to New Guinea.

The diversity of Seram fern species includes such ecological extremes as: species of subalpine tree fern forests, which are characteristic of the vegetation of New Guinea; the obligate xeric calcicolous genus *Phanerosorus* with vegetative proliferation in both long-lived sporophytes and gametophytes; and a submerged aquatic ecotype of *Asplenium unilaterale*. *Diplazium flavoviride* is unique in its phytogeographic affinity to eastern North America.

Introduction

Seram (Ceram) is an island of c. 18,000 km^2 in the Moluccas, and located near the western tip of New Guinea (Figure 19.1). Some fern groups of Seram have been revised in studies of Flora Malesiana based on previous collections (Holttum 1959–1982). However, our knowledge of the fern flora of Seram is still quite fragmentary and far from complete, partly due to an insufficient number of specimens collected. According to van Steenis (1950), by that time 5350 plant specimens had been collected in Seram. Since only few additional specimens have been collected during these four decades, new extensive collecting explorations are urgently needed. We have carried out three botanical expeditions to Seram as well as to Ambon during 1983, 1984/1985, and 1986, and collected in total more than 11,000 specimens in field numbers of vascular plants and also 5,000 bryophyte specimens mainly from Seram. Most of our collecting trips were made to Manusela National Park in central Seram, but we also covered scattered sites from west to east Seram (Figure 19.2). This paper reports on the flora and phytogeography of Seram ferns, and on some species of particular vegetation type or life form.

P. Baas et al. (eds.), The Plant Diversity of Malesia, 225–234.

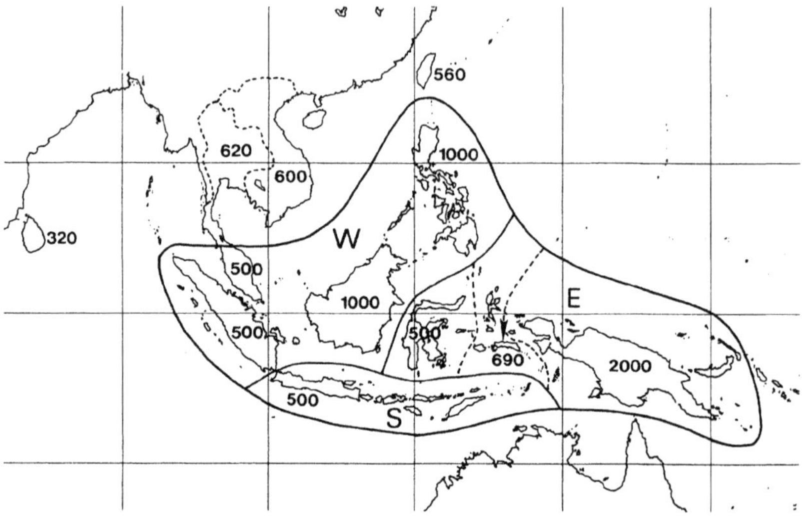

Figure 19.1 — The numbers of fern species in Malesian islands and subregions, and adjacent regions and islands. Arrowhead indicates Seram. The delimitation and subdivision of Malesia follow van Steenis (1950). E = East Malesia; S = South Malesia; W = West Malesia.

Species richness

The pteridophytes we collected ourselves from Seram reached a number of 690 species. It seems reasonable to estimate that the Seram fern flora is composed of much more than 700 species. A very rough estimation of pteridophyte species numbers in regions of Malesia is as follows: 2000 species in New Guinea (894,000 km^2), 1000 each in Borneo (739,000 km^2) and in the Philippines (290,000 km^2), and 500 each in the Malay Peninsula (132,000 km^2), Sumatra (479,000 km^2), Java (132,000 km^2) and Sulawesi (Celebes; 182,000 km^2) (Figure 19.1). Some of these numbers are surely underestimated, due to poor collections. In Southeast Asia outside Malesia, 620 species are known from Thailand (514,000 km^2), 600 from Indo-China, 560 from Taiwan (36,000 km^2), and 320 from Sri Lanka (65,000 km^2). Pacific oceanic islands isolated from continents, such as Fiji (18,000 km^2), New Caledonia (16,000 km^2), and the Hawaiian Islands (16,000 km^2), have 300 species or less. As far as I know, the most species-rich island comparable in size to Seram is Jamaica, West Indies, in which 600 species occur (Proctor 1985). Jamaica is only 2/3 as large as Seram and, like Seram, it has geological complexes with plenty of limestone, diversity of soil types, and varied topography. Taking the surface area of islands or regions into account, it can be concluded that the fern flora of Seram is one of the richest among insular floras of the world.

Non-biotic and biotic characteristics are assumed to be involved in the species richness of the Seram fern flora. The non-biotic characteristics include climate, topography, and edaphic conditions. Seram has a semi-everwet climate without a pronounced dry season, and is situated at the west end of the east big core of Malesian rain forests (van Steenis 1979). The annual precipitation ranges from 2100 mm to 4500 mm, and the smallest monthly precipitation is only slightly below 100 mm. Such a climate enables tropical rain forests to thrive on Seram. Rain forests can provide a diversity of habitats suitable for shade-requiring and epiphytic species of ferns.

Secondly, Seram is a mountainous island with many mountains of various heights including Mt Binaya of approximately 3000 m, the highest in the Moluccas. It is known that shade ferns are most diversified in various kinds of habitats in montane zones (Page 1979; Jacobs 1988). It is noteworthy that less than 300 fern species occur in the Amazon basin probably because of the absence of a great variety in environment, while very rich fern floras exist in surrounding mountain regions (Tryon & Conant 1975; Tryon 1986). Parris (1985) pointed out that the fern flora of New Guinea is twice as rich as that of Borneo due to its more mountainous topography, although both large islands are comparable in size.

Thirdly, Seram is geologically characterised by an abundance of limestone rocks, which are inhabited predominantly by ferns. Many calcicolous fern species, such as *Phanerosorus major, Cyclopeltis presliana, Heterogonium ceramense, Tectaria brooksii, Gymnocarpium oyamense, Selaginella calcicola,* occur in various microhabitats on and near limestone, while only few seed plants and bryophytes are found there.

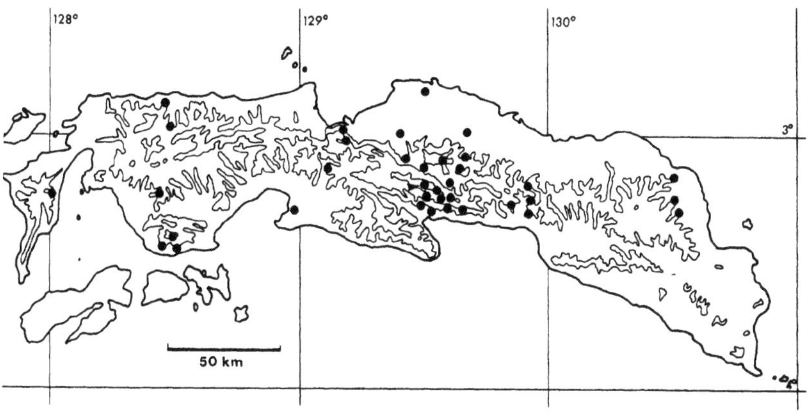

Figure 19.2 — Places explored (marked by solid circles) during the Indonesian-Japanese botanical expeditions to Seram. Arrowhead indicates Mt Binaya. Contour lines: 300 m, 1000 m, 2000 m.

Geographically Seram is close to New Guinea, a large island with the richest fern flora in the world. The species number of small islands depends on the floristic richness and geographical isolation of the nearest continent or large islands, which are the sources (Tryon 1970). The closer and the richer the flora of a continent or large island, the richer the flora of neighbouring small islands. Seram is such an island, and in this feature it is equivalent to a near continental island.

Floristic affinity

Van Steenis (1950) placed the Moluccas including Seram in the centre of the three subregions (Sulawesi, Moluccas, New Guinea) of East Malesia, on the basis of the distribution patterns of flowering plant genera (Figure 19.1). Van Balgooy (1987) argued that the flowering plant flora of Sulawesi has a closer affinity at species level to the Moluccan flora as well as the Lesser Sunda Island and Philippine floras than to the floras of Borneo and New Guinea.

Using our recent collections, and those already worked up, I classified distribution types of 256 Seram fern species of 20 groups with different systematic affinities and ecology, which are 37% of the total species collected (Psilotaceae, Lycopodiaceae, Selaginellaceae, Equisetaceae, Ophioglossaceae, Marattiaceae, Osmundaceae, Plagiogyriaceae, Gleicheniaceae, Schizaeaceae, Matoniaceae, Cyatheaceae, Dicksoniaceae, Culcitaceae, Lindsaeaceae, Davalliaceae, Oleandraceae, Lomariopsidaceae, Tectarioideae, and the *Taenitis* group). Of these, 93 (36%) species are of the East Malesia type, 66 (26%) species of the East and West Malesia type, and 89 (35%) species of the Pan-Malesia type, while only 8 (3%) species are confined to East and South Malesia (Table 19.1). Twenty-three species are endemic to Seram or the Moluccas. A preliminary study of the remaining species shows roughly similar distribution patterns.

It is evident that most of the Seram fern species have relatively wide distribution ranges and only few are endemic to Seram. This distribution pattern of the Seram fern flora differs from the high endemism of oceanic islands such as the Hawaiian Islands.

Another point is that the affinity of Seram with South Malesia is markedly weak. Such weak affinity may be ascribed to climate difference and insularity. South Malesian islands show a strongly marked wet and dry seasonal climate, which makes their flora poorer and more xeric. South Malesia, consisting of the Lesser Sunda Islands and Java, like the Moluccas, is composed of relatively small islands. These small islands, floristically poor and different from the Moluccas, have not functioned as an efficient source area to produce many species of a common East and South Malesia type. This is a marked contrast with the strong floristic affinity of angiosperms between Sulawesi and the Lesser Sunda Islands (van Steenis 1979; van Balgooy 1987).

Table 19.1 — Distribution types of 256 Seram fern species.

Distribution type	Species number	
East Malesia type	93	(36%)
E Malesia	77	
E Malesia/Oceania-Australia	16	
East-West Malesia type	66	(26%)
E & W Malesia	49	
E & W Malesia/Oceania-Australia	10	
Continental Asia/E & W Malesia	6	
Continental Asia/E & W Malesia/Oceania	1	
East-South Malesia type	8	(3%)
E & S Malesia	5	
E & S Malesia/Oceania-Australia	2	
Continental Asia/E & S Malesia	1	
Pan-Malesia type	89	(35%)
Malesia	24	
Pantropics	22	
Continental Asia/Malesia/Oceania-Australia	17	
Continental Asia/Malesia	14	
Malesia/Oceania-Australia	12	

I compared floristic affinities of Seram to New Guinea and to Sulawesi, two of the three neighbouring subregions of East Malesia (van Steenis 1950). Of the 106 out of 256 Seram fern species which range to either island (not both), 79 species are in common with New Guinea and only 27 with Sulawesi. To further clarify the affinity between Seram and New Guinea within East Malesia, I treated 65 species of the East Malesia distribution type, excluding widely distributed species. Of these, 57 species are in common with New Guinea, and only eight species with Sulawesi. Thus, as far as fern species are concerned, the floristic affinity of Seram at species level is much stronger to New Guinea than to Sulawesi. The main reason for this is the distinct difference in flora size between the two islands: 2000 species in New Guinea and 500 in Sulawesi. Species-rich New Guinea seems to have, as a source area, strong effects on the floristic composition of surrounding small islands such as Seram. Sulawesi, like New Guinea, is mountainous and has higher mountains than Seram. And, although Sulawesi is farther away from Seram than New Guinea is, there are intermediate islands linking Sulawesi and Seram. Hence, the much weaker affinity of Seram to Sulawesi than the affinity to New Guinea is not mainly due to the topography and geography of Sulawesi.

The Seram fern flora includes a few species with non-Malesian affinity. Some species of southern hemisphere genera are found among the species of the East Malesia/Oceania-Australia type: such as *Tmesipteris oblanceolata* (Braithwaite 1973, 1986; Kato 1988a), *Leptopteris alpina* (Kato 1989), and *Austrogramme boerlageana* (Hennipman 1975; Kato 1988b). *Diplazium flavo-viride* of Seram and New Guinea is very closely related to the eastern North American *D. pycnocarpon,* and they are a vicarious pair disjunct in East Asia and eastern North America (Kato & Darnaedi 1988).

Notes on particular vegetation types and life forms

The Seram fern species occur in a great variety of habitats. While most species prefer mesic environments, some others are adapted to particular or unique habitats, which are products of the limestone geology of the island. The following are notable ecological extremes which are found on the fringe of this rich flora.

Mt Binaya in central Seram is geologically characterised by predominant limestone formations. Since most places of the summit zone (2500–3000 m in altitude) are exposed with limestone rocks and stones, and a thin soil is present in only small places, *Vaccinium*-dominant scrubs are found in small patches.

Figure 19.3 — Subalpine tree fern forest near Mt Owae Puku.

Figure 19.4 — *Phanerosorus major* on limestone cliff. A = sporophytes (centre), × 0.25; B = gametophyte colony, × 3.1.

Instead, tree fern-dominant forests with mixed stands of *Vaccinium retusifolium* and *Dacrycarpus cinctus* occur on or near ridges between Mt Binaya and Mt Owae Puku (Figure 19.3). Tree fern shrublands are also typical of New Guinea subalpine zones (Holttum 1963; Hope 1980).

Two species were found in the subalpine zone of Mt Binaya. *Cyathea binayana* is dominant in *Vaccinium* or *Dacrycarpus*-mixed forests on soil-poor slopes near the ridges. *Cyathea pukuana* grows as sparse tree fern shrubs along the ridges. The former species is closely related to *C. percrassa,* which occurs at 3000–3500 m altitude in east New Guinea, while the latter is closest to *C. tomentosa* at and over 2000 m in Java and Flores, and less closely related to *C. magma* at 1750–2750 m in New Guinea (Holttum 1963; Kato 1990). Thus the two subalpine species have different phytogeographic affinities, one with East Malesian, and the other with South Malesian relatives.

Phanerosorus (Matoniaceae) is an obligate calcicolous genus. *Phanerosorus major* is distributed in Seram and Waigeo, Misool and Aru west of New Guinea, while *P. sarmentosus* is found in Sarawak in Borneo (Kato & Iwatsuki 1985a). This genus is characterised by the indeterminate growth of leaves with adventitious pinna-buds at the base of pinnae. The sporophytes and gameto-

Figure 19.5 — Aquatic ecotype of *Asplenium unilaterale* in Makariki pond. Dim white spots are air bubbles reflecting camera flashlight. F = *Ficus* root.

phytes almost always coexist on dry steep limestone cliffs (Figure 19.4). The gametophytes are slender ribbon-like and reproduce vegetatively to make colonies by proliferation from marginal cells. At maturity they produce massive antheridia and archegonia of a primitive type, and fertilisation takes place to give rise to sporophytes with a characteristic heteroblastic leaf development (Yoroi & Kato 1987). Such vegetative proliferation in both long-lived sporophyte leaves and gametophytes seems to be an adaptation to the xeric saxicolous habitat.

Seram has underground streams, which in some places flow out in ponds. One of them is Makariki pond 15 m in diameter and 2 m deep, at 350 m altitude, from which water flows down into the Makariki River. In the pond there are two stands of *Ficus* with submerged root systems. Although *Asplenium unilaterale* is almost always a lithophyte or geophyte, a submerged aquatic ecotype occurs attached to the *Ficus* root system and also on submerged tree branches or roots of other plants at the edge of the pond and among stones on the bottom of the pond (Kato & Iwatsuki 1985b). It grows submerged, 50–70 cm below the water surface in rather slow flowing water with numerous air bubbles (Figure 19.5). The aquatic ecotype cannot reproduce by air-dispersed spores, due to its constant submerged life, but instead it has vegetative reproduction by leaf-borne adventitious buds. In the water flow, gemmiferous and gemmaling-bearing pinnae get within an entangled root-rhizome system of the fern or between stones and trees, where new plantlets grow.

In Seram there are four other variants of the species which are characterised by habitat preference and reproductive mode as well as morphological features. A taxonomic study (Kato & Iwatsuki 1986) showed that the aquatic ecotype is a close derivative from an agamosporous geophyte of *Asplenium unilaterale* var. *unilaterale*. In contrast to the so-called water ferns which are distinct morphologically and taxonomically, the aquatic ecotype still retains the morphology of a land form, although stomata show partial reduction, and it is considered to be at an early stage of speciation towards the aquatic habit.

Acknowledgements

I am grateful to Drs. M.G. Price and K. Ueda for reading the manuscript and giving useful comments. This study was in part supported by Grant-in-Aid 62043018 for the Monbusho International Scientific Research Program.

References

Balgooy, M.M.J. van. 1987. A plant geographical analysis of Sulawesi. In: T.C. Whitmore (ed.), Biogeographical Evolution of the Malay Archipelago: 94–102. Oxford Univ. Press, Oxford.

Braithwaite, A.F. 1973. Tmesipteris in the Solomon Islands. Fern Gaz. 10: 293–303.

Braithwaite, A.F. 1986. Tmesipteris in Vanuatu (New Hebrides). Fern Gaz. 13: 87–96.

Hennipman, E. 1975. A re-definition of the Gymnogrammoid genus Austrogramme Fournier. Fern Gaz. 11: 61–72.

Holttum, R.E. (ed.). 1959–1982. Flora Malesiana, Ser. II Pteridophyta, Vol. 1. Martinus Nijhoff, The Hague.

Holttum, R.E. 1963. Cyatheaceae. In: R.E. Holttum (ed.), Flora Malesiana, Ser. II Pteridophyta, 1: 65–176. Martinus Nijhoff, The Hague.

Hope, G.S. 1980. New Guinea mountain vegetation communities. In: P. van Royen, The alpine flora of New Guinea. Vol. 1: General Part: 143–222. J. Cramer, Vaduz.

Jacobs, M. 1988. The tropical rain forest. A first encounter. Springer-Verlag, Berlin.

Kato, M. 1988a. Taxonomic studies of pteridophytes of Ambon and Seram (Moluccas) collected by Indonesian-Japanese botanical expeditions I. Fern-allies. Acta Phytotax. Geobot. 39: 133–146.

Kato, M. 1988b. Taenitis and allied genera of Ambon and Seram (Moluccas) and notes on taxonomic and phytogeographic relationships of Taenitis. J. Fac. Sci. Univ. Tokyo III, 14: 161–182.

Kato, M. 1989. Taxonomic studies of pteridophytes of Ambon and Seram (Moluccas) collected by Indonesian-Japanese botanical expeditions III. Eusporangiate and some lower leptosporangiate families. Acta Phytotax. Geobot. 40: 67–82.

Kato, M. 1990. Taxonomic studies of pteridophytes of Ambon and Seram (Moluccas) collected by Indonesian-Japanese botanical expeditions IV. Tree-fern families. J. Fac. Sci. Univ. Tokyo III, 14: 369–384.

Kato, M., & D. Darnaedi. 1988. Taxonomic and phytogeographic relationships of Diplazium flavoviride, D. pycnocarpon, and Diplaziopsis. Amer. Fern J. 78: 77–85.

Kato, M., & K. Iwatsuki. 1985a. Juvenile leaves and leaf ramification in Phanerosorus major (Matoniaceae). Acta Phytotax. Geobot. 36: 139–147.

Kato, M., & K. Iwatsuki. 1985b. An unusual submerged aquatic ecotype of Asplenium unilaterale. Amer. Fern J. 75: 73–76.

Kato, M., & K. Iwatsuki. 1986. Variation in ecology, morphology and reproduction of Asplenium sect. Hymenasplenium (Aspleniaceae) in Seram, Indonesia. J. Fac. Sci. Univ. Tokyo III, 14: 37–48.

Page, C.N. 1979. The diversity of ferns. An ecological perspective. In: A.F. Dyer (ed.), The Experimental Biology of Ferns: 9–56. Acad. Press, London.

Parris, B.S. 1985. Ecological aspects of distribution and speciation in Old World tropical ferns. Proc. Roy. Soc. Edinburgh 86B: 341–346.

Proctor, G.R. 1985. Ferns of Jamaica. British Museum (Natural History), London.

Steenis, C.G.G.J. van. 1950. Flora Malesiana, Ser. I, Vol. 1. Noordhoff-Kolff, Jakarta.

Steenis, C.G.G.J. van. 1979. Plant-geography of east Malesia. Bot. J. Linn. Soc. 79: 97–178.

Tryon, R.M. 1970. Development and evolution of fern floras of oceanic islands. Biotropica 2: 76–84.

Tryon, R.M. 1986. The biogeography of species, with special reference to ferns. Bot. Rev. 52: 117–156.

Tryon, R. M., & D.S. Conant. 1975. The ferns of Brazilian Amazonia. Acta Amazonia 5: 117–156.

Yoroi, R., & M. Kato. 1987. Wild gametophytes of Phanerosorus major (Matoniaceae). Amer. J. Bot. 74: 354–359.

20 Botanical progress in Papuasia

D.G. FRODIN

Summary

A sketch of botanical progress in Papuasia since 1934 is presented. The appearance in that year of Herman Lam's 'Materials towards a study of the flora of the island of New Guinea' is here taken to mark the onset of a 'second cycle' following the syntheses of the 1920s which closed the early exploratory era. Much progress has been made in filling the gaps suggested by H.J. Lam and, later, C.G.G.J. van Steenis, particularly during the third quarter of the 20th century. The overall collection density index rose from 12 in 1950 to about 28 by 1975 (some 50% higher in Papua New Guinea). Since 1975, however, diverse political and other developments have slowed progress, while at the same time more intensive and dynamic approaches to the study of biodiversity are influencing research questions, as noted by Stevens (1989). Attention is drawn to some seemingly extra-scientific dimensions, recognition and exploration of which might aid in reinvigorating botanical progress. Two are examined here: 1) systematic prosopographic analysis of published taxonomic literature, which should aid in identifying and documenting gaps in our knowledge and, thus, research directions; and 2) the need to foster effective political and other support for necessarily more detailed programmes of inventory, botanical research, database development, and conservation planning.

Introduction

The 1989 Flora Malesiana Symposium is an appropriate occasion for an inquiry into the recent progress and present state of Papuasian botany. It also marks the 50th and 100th anniversaries respectively of the conclusion of the Indies-American Expedition led by Richard Archbold and William MacGregor's successful ascent of Mt Victoria, both of exceptional importance for Papuasian floristics and biogeography, and comes 55 years after the first of the key surveys and analyses of the regional flora and vegetation by the late Herman J. Lam, former director of the Rijksherbarium (Lam 1934, 1935).

Several other narratives and analyses of progress have appeared since 1934, particularly in the last fifteen years (Jacobs 1974; van Royen 1980; Frodin & Gressitt 1982; Frodin 1985, 1988, in press; Stevens 1989) as what may be called the 'second cycle' of Papuasian botanical exploration and associated studies reached a plateau. The writer's aims here are twofold: 1) a brief overview of this 'second cycle', and 2) a look at some additional dimensions, appreciation of which might help direct a renewal of botanical inventory and field research and thus initiate a successful 'third cycle'.

P. Baas et al. (eds.), The Plant Diversity of Malesia, 235–247.

Background

In 1950 the late Prof. C.G.G.J. van Steenis wrote: "I expect that only through many planned and co-ordinated explorations can the Papuasian unit be explored within the coming 50 years to a collecting density comparable with that of the present stage of the Philippine Islands. ... Botanists of ... future generations must stimulate this research, for it deserves a high degree of priority. Phytogeographically New Guinea is a keystone in Pacific botany." (van Steenis 1950: cxiv).

When those words were written, the collecting density index (CDI, here given as collection numbers/100 km^2) in New Guinea and its associated islands (less the Solomons and Louisiades) was 12 (slightly higher in what is now Papua New Guinea). By way of comparison, the index for the Philippines was 62 and the Malay Peninsula 145. It may be noted, however, that, following a comparative lull, the eight years from 1933 witnessed a great renewal of collecting activity, increasing the index by perhaps a third. In spite of World War II, and especially if the programme of revisionary studies of the flora initiated by Elmer D. Merrill in the United States is taken into account, Papuasian botanical studies had already entered their 'second cycle' (cf. Symington 1943).

In the writer's opinion, Lam's 'Materials ...' (Lam 1934) and 'De vegetatie en de flora van Nieuw Guinée' (Lam 1935), already referred to, marks the opening of this 'second cycle'. Building upon all the work of the preceding cycle (approximately 1870–1930) and accounting for the key floristic syntheses and vegetation studies of its final decade (Gibbs 1917; Ledermann 1919; Diels 1921, 1930; Lam 1924, 1927–29, 1930; Lane-Poole 1925; Lauterbach 1928–30), Lam drew up the first recognisably modern discussion of the Papuasian flora, indicating also its relationship to a probably dynamic geological past. These were soon followed by van Steenis's papers on the Malesian mountain flora (van Steenis 1934–36).

Lam's papers also marked the definitive ascendancy of the Dutch in Papuasian botanical studies. Initially from their Indonesian centre at Buitenzorg (Bogor) and, after 1933, with the development of the Rijksherbarium at Leiden under the successive directorates of Lam and van Steenis and the stimulus provided by the Flora Malesiana project, Dutch botanists – whether on Dutch soil or not – came to dominate taxonomic research on the Papuasian flora over most of the 'second cycle'. Only in vegetation science – and then mostly after the 1950s – did British (along with Australian and New Zealand) botanists predominate, thus building on a tradition begun by Gibbs and Lane-Poole and continued by Leonard Brass in his many contributions as Archbold Expeditions botanist.

In the quarter-century after publication of the above-quoted remarks of van Steenis, many of the lacunae to which he referred (van Steenis 1950: cxiii–cxiv) have been more or less filled. In particular, there was extensive collecting in much of the tropical-pine belt, in Jazirah Doberai (Vogelkop), the central highlands of Papua New Guinea, and many of the associated islands and archipelagoes (including the Solomon Islands). The Markham Valley, the upper Watut Basin, and the Port Moresby region in Papua New Guinea, as well as Jazirah Doberai, have also benefited from the presence of scientific institutions. As for a host of other higher-level taxa, our understanding of the limits of distribution of the Dipterocarpaceae has improved. By 1972 the collecting indices for New Guinea, the Philippines and Peninsular Malaysia respectively were 26 (rather higher in Papua New Guinea), 69 and 175 – on the whole representing a creditable rate of progress towards van Steenis's goal.

The material resulting from such intense efforts – in addition to that in the decades before 1950 – and the associated Flora Malesiana treatments and other floristic and taxonomic papers (along with personal assistance) enabled Ronald Good and, later, Ruurd Hoogland and Max van Balgooy to compile revised phytogeographic analyses (Good 1960; Hoogland 1972; van Balgooy 1976). Improved views of the vegetation and its dynamics (e.g. Womersley & McAdam 1957; Walker 1973; Paijmans 1976; Hope 1980; Johns 1982, 1986; Walker & Hope 1982; Grubb & Stevens 1985) and ethnobotany (e.g. Powell 1976) also resulted. In 1970, work began at Lae towards the 'Handbooks of the Flora of Papua New Guinea'; two volumes have been published as of 1989 (Womersley 1978; Henty 1981). Some local floras have appeared (e.g. Johns & Stevens 1971; Streimann 1983).

Since 1975 (and earlier in the Solomon Islands and western New Guinea), however, the rate of collecting has dropped significantly. The regional CDI is here estimated to be presently no more than 30. Papua New Guinea, at 46 (Campbell 1989a: 14) is certainly more advanced, but the low level of activity elsewhere in Papuasia justifies Stevens's remark (Stevens 1989: 122): "we are seriously lagging behind his [van Steenis's] schedule." The ideal of 100, voiced by the compilers of 'Floristic Inventory of Tropical Countries' (Campbell & Hammond 1989) as well as earlier by van Steenis, may lie beyond our vision: 193 years for Papua New Guinea at current collecting rates (Campbell 1989a: 14). Even a CDI of 100 may not be high enough (Tan & Rojo 1989). Non-vascular plants and fungi, with some exceptions (bryophytes, lichens, and marine algae, which received renewed attention in the 1980s) lag even further behind.

Inventory has moreover been uneven. Based on information from the late Marius Jacobs, Prance (1977: 665) indicated that the Meervlakte in the west, the Kikori region in the east, and the Star Mountains were undercollected. Stevens (1989) has supplied a map, and listed a number of areas [further summarised by Prance & Campbell (1988) and Campbell (1989a: 19)]. Among them is the Star Mountains region, the 'Chocó' of Papuasia and an area from which the writer has collected or studied some 31 species of *Schefflera,* the highest such concentration in the world. Many localities have, on the other hand, enjoyed repeated visits. I shall indicate here only that more work is needed in the colline and lower montane zones (500–2500 m), including summits and upper slopes of smaller mountains (cf. Lewis 1971). These, moreover, encompass many important 'classical' areas – to which generally rather little new attention has been directed given the losses of World War II at Berlin which are a bane of present-day students (e.g. Symon 1985; Sleumer 1988c). [Recently, however, there has been a new expedition to the Hunstein River basin and Mt Sumset (Hunstein), an important 'classical' locality.]

Botanical progress in Papuasia has also been hampered by human factors; among them have been lack of coordination of efforts, isolationism, and sectoral interests on the part of many teams. In Papua New Guinea, there was relatively little effective integration of forest survey, ecological and floristic work. Botanists on Forest Department assessment surveys went for the most part as accessories, with a warrant largely separate from that of the foresters. Moreover, the numerous transects ('strips') surveyed (and the resulting data and reports) – though limited in coverage because of emphasis on trees with a minimum of 0.5 m dbh – went unused in compiling detailed vegetation maps (mostly the work of the CSIRO Division of Land Research and based on its own surveys). The Quaternary research programme of Donald Walker and his school at Australian National University, active in the 1960s and 1970s, also was largely independent (with much material remaining unpublished).

The recent drop in Papuasian collecting and research has been for political, social and economic rather than purely scientific reasons. The last fifty years have continued to show that the Papuasian flora is diverse and distinct with much of it being 'young' along with the landscape. In many taxa there has been great primary or secondary radiation, often more so than in western Malesia, and consequently a high level of 'neoendemism' (Lam 1934: 141), though admittedly from a narrower generic 'base' than in western Malesia (cf. Airy Shaw 1980). There may also be unexpected biotic discontinuities (Kalkman & Vink 1970; Stevens 1989). In addition to the differences between Mts Kerewa and Ambua, the two main summits of the Doma Peaks, reported by Kalkman and Vink, van Valkenburg (1987) has shown that *Nothofagus pullei* is present on Mt Kaindi but not on Mt Missim, and the writer has observed comparable differences between Mts Tangis and Talawe at the western end of New Britain as well as among the islands of Bootless Inlet. There are surely others. In like manner, interpretation of the basic units of taxonomic diversity can vary from one specialist to another (as, for example, in *Archidendron, Chisocheton, Drimys* and *Zornia*); moreover, approaches to interpretation have since the 1970s become more of a philosophical issue. Vegetation is also subject to differing interpretations, including ideas on classification and dynamics (cf. Johns 1986). These and other qualities of the flora and vegetation are gradually conveying a greater sense of floristic and phytogeographical distinctness from western Malesia (van Balgooy 1976; van Steenis 1979; Whitmore 1981) than was earlier believed, as well as a renewing of interest in the overall historical relationship between Asia and Australasia (Briggs 1987, for review). Future regional studies, taxonomic or otherwise, should be directed towards detailed analyses (cf. Campbell 1989b, for discussion of site study methods).

Accounts of taxonomic progress

Taxonomic progress accounts can serve not only as contributions to future *histoires totales* and *histoires problèmes* (Hexter 1979) but could also have heuristic value when used with an understanding of the biology of a given group or groups. George (1981) has provided a 'case history' of a key Australian genus which nicely illustrates larger trends in Australian botany, while van Steenis (1957a, b) has shown the value of comparing the state of knowledge of a sample of the Philippines flora in 1923 and following revision for Flora Malesiana through an analysis of nomenclatural and taxonomic decisions.

A combination and development of these two approaches to taxonomic historiography would be exceptionally valuable in itself and also aid us greatly in directing our research resources. The time for 'sweepstakes' field work is largely past, not only with accumulated knowledge but also because our techniques of information handling are becoming ever more sophisticated. It is nowadays possible, for instance, to return to the eastern Toricelli Mountains with a portable computer containing files of Schlechter data.

With the aid of recent revisions by Sleumer (1986, 1987a, b, 1988a, b, c), Cribb (1986) and Verdcourt (1979) the writer respectively examined Myrsinaceae, *Dendrobium* sect. *Spatulata* (Orchidaceae), and Leguminosae as models for historiographic analysis. The first-named family is of relatively low direct interest and probably for this reason students have been few. It also has a reputation for 'difficulty'. The importance of the third is more direct, but its many workers have meant that centrifugal forces are strong. The second taxon, while comparatively small, is in a family indirectly important, with a large press and many amateur students.

Myrsinaceae continue to yield a goodly number of novelties at this relatively late date; nearly 120 years after the beginning of continuous interior exploration in New Guinea (Frodin 1988). The many collections, often unisexual, left unidentified by Sleumer – in *Ardisia* 91, or 25–30% – and the many recognised species still poorly known suggest, as Sleumer (1988c) acknowledged, that to a large extent the revisions were "all but satisfactory." The loss of so much key material at Berlin in 1943 was here, as in other taxa, a serious handicap. [That *Discocalyx latepetiolata* could successfully be interpreted, given loss of most syntype material, was due to re-collection in 1966 at one of the type localities.]

Of the inland genera, only *Conandrium,* with two species – one widespread and variable – might be considered reasonably well known. For the others, a single example will suffice: the conspicuous small rosette-tree species of *Tapeinosperma* near Lae, twice collected, remains unnamed and undescribed.

Excluding *Labisia* and *Aegiceras,* 181 names in Myrsinaceae have been published for Papuasia. Sleumer has reduced these to 113 species, giving a retention ratio of 0.63. The greatest growth in names prior to Sleumer's revisions was between 1918 and 1950, with another 12 in the orophilic genus *Rapanea* being added in 1982. Prior to 1918 only 68 names had been published. The additions to a considerable extent represent plants centered in the montane zones or of relatively limited distribution. Many are small understorey trees, part of a synusia where in a broken biogeosphere taxonomic differentiation can fairly readily take place. Others are parts of specialised synusiae (including *Maesa calcarea* Sleum., a calciphile, and *M. rheophytica* Sleum., a rheophyte also described as *M.* sp. in van Steenis 1981: 306.) [Another rheophyte, *M. protracta* F. Muell., was first described in 1877 and is more extensively distributed.]

Widespread, relatively frequent species are few. The greater part were described before 1920, especially those at lower altitudes. Among them are (with dates, and places of first description if outside Papuasia) *Aegiceras corniculatum* (before 1775), *Conandrium polyanthum* (1900), *Rapanea leucantha* (1905), *Ardisia ternatensis* (1867, from Maluku), *A. imperialis* (1887) and *A. forbesii* (1914), *Maesa tetrandra, M. bismarckiana* (1922) and *M. haplobotrys* (1866, from Australia), *Fittingia tubiflora* (1922), *Discocalyx latepetiolata* (1922), *D. subsinuata* (1988), and *Embelia cotinoides* (1916). Three of the four described species of *Tapeinosperma* are widespread in the Solomons, with one also in the Bismarck Archipelago. A number of these may be found 'beyond the forest', so to speak; among them are *Conandrium polyanthum, Maesa bismarckiana,* and *Embelia cotinoides.* The species of mountain forests may at least sometimes be upper slope and ridgetop dwellers, with perhaps a more 'continuous' habitat – especially if they have a relatively large altitudinal range, as does *Fittingia tubiflora.* [Certainly in the writer's experience *F. conferta* grew on upper slopes and ridgetops.] Among these species one can often find an extensive synonymy.

Dendrobium sect. *Spatulata,* better known as the 'antelope' or 'Ceratobium' dendrobiums, is by contrast a predominantly lowland group of 46 currently recognised species. The 29 in Papuasia share 55 original descriptions, giving a retention ratio of 0.58. Of these, 8 were published before 1871, 13 from 1871 to 1900, 18 from 1901 to 1921, 4 from 1922 to 1950, and 12 from 1951 to the present (including 8 directly related to the 1986 revision). Species which have 'survived' respectively number 5, 4, 9, 2, and 9. In contrast to Myrsinaceae, the greater part of the names, and almost two-thirds of the currently accepted species, were described before 1920. Those most recently proposed include three occurring predominantly in the lower montane zone of the central highlands. Curiously, among the two names 'surviving' from 1922–50 is that of the 'Sepik Blue', *Dendrobium lasianthera* J.J. Smith (1932): indeed a 'late' discovery.

In Papuasian Leguminosae our overall level of knowledge is higher though, as in the Myrsinaceae, there are disparities. Many species are widely distributed, often ranging far beyond Papuasia, and moreover are beyond the forest. In the Detarieae (incl. Cynometreae), Ingeae *s. str.*, Dalbergieae, Millettieae, and Phaseoleae *s. lat.*, however, our knowledge of a number of key genera remains very imperfect, in the case of *Archidendron* even after five revisions. That genus, in Verdcourt's manual, has a nearly complete 'alphabet soup' of described but unnamed taxa – almost as many as those named – and the subsequent revision by Nielsen *et al.* (1984) has only partially improved on this state of affairs. Many species of *Archidendron* are small, weak trees with a more or less limited distribution. Only few range widely: *A. lucyi* F. Muell. is "one of the commonest species and readily identified" (Verdcourt 1979: 244). Some *Cynometra* species, such as *C. minutiflora* F. Muell., are also local (or do not often flower and fruit) and have rarely been collected. The large, woody vines of many species of *Mucuna* are hard to collect. Interpretation of several species, as in the *Zornia diphylla* complex (Verdcourt 1979, and personal communication) remains problematic, and tradition, or recourse to a broad delineation, is sometimes hard to counter.

Similar results might be drawn and trends discerned from a study of other recent revisions, e.g. *Solanum* (Symon 1985) or *Psychotria* (Rubiaceae) (Sohmer 1988).

Human factors

Prance & Campbell (1988: 538) have spoken of New Guinea as "a world priority for future work." But, as Peter Stevens (Stevens 1989) and I (Frodin, in press) have pointed out, to reduce effectively our level of ignorance will depend on human factors, including political support. Tan & Rojo (1989) have illustrated this well for the Philippines.

The political position of biotic inventory in less developed countries varies widely, for both short-term and long-term reasons including the basic structure of government, institutions and economies. More often than not it is assigned low priority, because it, and pure sciences in general, are not seen as providing immediate benefits. In some cases, the presence of seemingly authoritative publications, whatever their vintage, may further deter official interest and discourage the study of biodiversity. A higher status may be assigned, however, if it is seen as part of national goals (and prestige).

Papua New Guinea (and even more the Solomon Islands) have since independence followed the more customary pattern. The 1970s saw heavy losses of qualified non-nationals, themselves faced with tightening opportunities. The Papua New Guinea Scientific Society, the only general organisation for the sciences, lost its publication subsidy in 1975 and became inactive a few years later, having effectively failed to adapt to changing circumstances (Salter-Duke 1984). Efforts to establish a council or similar official or semi-official scientific body have so far been abortive. Without effective institutional structures, and faced with more attractive opportunities, Papua New Guineans only slowly came into the sciences, well short of 'manpower requirements' – and many of those that did soon, as in other less developed countries, moved into administrative careers by choice or necessity. Only three are presently on staff at the Division of Botany in Lae.

Little help could be expected from Australia, given its own needs and interests. Though Australian assistance continued to be generous, like much of such foreign aid, it was oriented towards more immediate economic and social development. Most aid to Papua New Guinea has been 'untied', and has been allocated in accordance with priorities laid down in the National Plan – wherein the environment has had only a small allocation and the sciences in their own right none at all.

Prospects

While recent events and the near future may give some cause for concern, in the longer term prospects, at least in Papua New Guinea, appear more favourable. Prospects for that nation's economy are thought to be relatively good [Economist 313 (7633), 16 December 1989: 70–72]. A reasonable diversity of institutions with botanical interests is presently in place, all active in collecting, research and publication. The smaller centers in particular have helped to compensate for the relative 'isolation', and post-independence decline, of the Division of Botany. The local stock of specimens, a basis for the 'natural diversity index' concept of Toledo (1985), is comparatively good relative to overseas holdings, and has been enriched by photographs (and duplicates) of many of the pre-World War II collections. A number of projects covering limited geographical areas or specific taxa have continued to be mounted, most recently floristic and vegetational studies of Puncak Trikora (Mt Wilhelmina; by Jean-Marie Mangen) and Mts Kaindi and Missim (by van Valkenburg, 1987).

Most of these projects, however, have been the work of individuals or small teams. Any renewal of large-scale programmes will have to have considerable coordinated support and, as Stevens (1989: 131) notes, be contingent on the place of botany in local life – including its relationship with the economy and environmental concerns. The 1989 Hunstein River Expedition (Association of Pacific Systematists Newsletter 6: 4–5. 1989) is a step in the right direction.

More exploration will follow, spurred by 1) new research tools such as geographical information systems (GIS), of which one (PNGRIS) is already being set up in the PNG Department of Agriculture and Stock; 2) new paradigms, such as the historically composite nature of Papuasian land masses (Duffels 1986; Duffels & de Boer 1990: this volume) and biota (Humphries & Parenti 1986: 83–84); 3) philosophical concerns, including a more dialectical approach to the study of taxa (cf. Gould 1989; Masur 1989); and 4) increasing public concern for the environment.

Appendix: Recent publications

The most important addition to Stevens' (Stevens 1989) coverage of publications are two students' manuals by R.J. Johns, firstly 'A Students' Guide to the Monocotyledons of Papua New Guinea' (Johns *et al.* 1981–), and more recently 'The Flowering Plants of Papuasia: dicotyledons' (Johns 1987–1988). Of the former, four parts have been published, while from the projected 10 parts of the latter two, covering the first two subclasses of dicotyledons in the Cronquist system, have appeared at this writing (August 1989). They encompass some original work not so far published elsewhere, and moreover greatly expand accessibility of botanical information. They provide a useful (and less expensive) alternative to the 'Handbooks of the Flora of Papua New Guinea' of which only two volumes have been published, the last in 1982 and together covering 361 species (of perhaps 15,000 vascular plants in Papuasia). Accounts of hepatics (Grolle & Piippo 1984) and lichens (Streimann 1986) have also appeared, and work is in hand at Helsinki on a revised enumeration of mosses.

References

Airy Shaw, H.K. 1980. The Euphorbiaceae of New Guinea. Her Majesty's Stationery Office (Kew Bull. Addit. Series 8), London. [iv] + 243 pp.

Balgooy, M.M.J. van. 1976. Phytogeography. In: K. Paijmans (ed.), New Guinea vegetation: 1–22. Australian National Univ. Press/CSIRO, Canberra.

Briggs, J.C. 1987. Biogeography and plate tectonics. Elsevier (Developments in palaeontology and stratigraphy 10), Amsterdam.

Campbell, D.G. 1989a. The importance of floristic inventory in the tropics. In: D.G. Campbell & H.D. Hammond, Floristic inventory of tropical countries: 5–30. New York.

Campbell, D.G. 1989b. Quantitative inventory of tropical forests. In: D.G. Campbell & H.D. Hammond, Floristic inventory of tropical countries: 523–533. New York.

Campbell, D.G., & H.D. Hammond (eds.). 1989. Floristic inventory of tropical countries. New York Botanical Garden, New York. [x] + 545 pp.

Cribb, P.J. 1986. A revision of Dendrobium sect. Spatulata (Orchidaceae). Kew Bull. 41: 615–692, pl. 6–13.

Diels, L. 1921. Die pflanzengeographische Stellung der Gebirgsflora von Neu-Guinea. Ber. Freien Ver. Pflanzengeogr. u. Syst. Bot. f. das Jahr 1919: 45–59. Lande, Berlin.

Diels, L. 1930. Ein Beitrag zur Analyse der Hochgebirgs-Flora von Neu-Guinea. Bot. Jahrb. Syst. 63: 324–329.

Duffels, J.P. 1986. Biogeography of Indo-Pacific Cicadoidea: a tentative recognition of areas of endemism. Cladistics 2: 318–336.

Duffels, J.P., & A.J. de Boer. 1990. Areas of endemism and composite areas in East Malesia. In: P. Baas, C. Kalkman & R. Geesink, The plant diversity of Malesia: 249–272. Kluwer Academic Publishers, Dordrecht, Boston, London.

Frodin, D.G. 1985. Herbaria in Papua New Guinea and nearby areas. In: S.H. Sohmer (ed.), Forum on systematics resources in the Pacific: 54–62. Bishop Mus. Press (Bern. P. Bishop Mus. Spec. Publ. 74), Honolulu.

Frodin, D.G. 1988. The natural world of New Guinea: hopes, realities and legacies. In: R. MacLeod & P.F. Rehbock, Nature in its greatest extent: Western science in the Pacific: 89–138. Univ. Hawaii Press, Honolulu.

Frodin, D.G. In press. Explorers, institutions and outside influences: botany north of Thursday. In: P. Short (ed.), Development of systematic botany in Australasia. Australian Syst. Bot. Soc., Melbourne.

Frodin, D.G., & J.L. Gressitt. 1982. Biological exploration in New Guinea. In: J.L. Gressitt (ed.), Biogeography and ecology of New Guinea: 87–130. Junk (Monogr. Biol. 42), The Hague.

George, A.S. 1981. The genus Banksia L. f. – a case history in Australian botany. In: A. Wheeler & J.H. Price (eds.), History in the service of systematics: 53–59. Soc Bibliogr. Nat. Hist. (Spec. Publ. 1), London.

Gibbs, L.S. 1917. Dutch North-West New Guinea. A contribution to the flora and phytogeography of the Arfak Mountains. Taylor & Francis, London.

Good, R. 1960. On the geographical relationships of the angiosperm flora of New Guinea. Bull. Brit. Mus. (Nat. Hist.), Bot. 2 (8): 203–226.

Gould, S.J. 1989. Wonderful life. Norton, New York.

Grolle, R., & S. Piippo. 1984. Annotated catalogue of Western Melanesian bryophytes, I. Hepaticae and Anthocerotae. Finnish Bot. Publ. Board, Akateeminen Kirjakauppa (Acta Bot. Fennica 125), Helsinki. 86 pp.

Grubb, P.J., & P.F. Stevens. 1985. The forests of the Fatima Basin and Mt. Kerigomna, Papua New Guinea: with a review of montane and subalpine rainforests in Papuasia. Australian Nat. Univ./ANUTech (Dept Biogeogr. & Geomorph., Research School of Pacific Studies, Publ. BG/5), Canberra. xiv + 221 pp.

Henty, E.E. (ed.). 1981. Handbooks of the Flora of Papua New Guinea 2. Melbourne Univ. Press, Melbourne. ix + 276 pp., map.

Hexter, J.H. 1979. Fernand Braudel and the Monde Braudelien. In: J.H. Hexter, On historians: reappraisals of some of the makers of modern history: 61–145. Harvard Univ. Press, Cambridge, Mass. [x] + 310 pp. [1972; originally in the Journal of Modern History 44: 480–539.]

Hoogland, R.D. 1972. Plant distribution patterns across the Torres Strait. In: D. Walker (ed.), Bridge and barrier: the natural and cultural history of Torres Strait: 131–152. Australian Nat. Univ. (Dept Biogeogr. & Geomorph., Research School of Pacific Studies, Publ. BG/3), Canberra.

Hope, G.S. 1980. New Guinea mountain vegetation communities. In: P. van Royen (comp.), The alpine flora of New Guinea 1, chapter 8: 153–222. Cramer/Gantner, Vaduz.

Humphries, C.J., & L.R. Parenti. 1986. Cladistic biogeography. Oxford Univ. Press (Oxford Monogr. Biogeogr. 2), Oxford.

Jacobs, M. 1974. Botanical panorama of the Malesian archipelago (vascular plants). In: Unesco, Natural resources of humid tropical Asia: 263–294. Unesco (Natural Resources Research 12), Paris.

Johns, R.J. 1982. Plant zonation. In: J.L. Gressitt (ed.), Biogeography and ecology of New Guinea: 309–330. Junk (Monogr. Biol. 42), The Hague.

Johns, R.J. 1986. The instability of the tropical ecosystem in New Guinea. Blumea 31: 341–371.

Johns, R.J. 1987–88. The flowering plants of Papuasia: dicotyledons. Parts 1 & 2 (of 10). Dept For., Papua New Guinea Univ. of Tech., Lae. (Part 2 also published as Christensen Research Institute Publication 1.)

Johns, R.J., & P.F. Stevens. 1971. Mount Wilhelm Flora: a checklist of the species. Div. of Bot., Dept For., Papua New Guinea (Bot. Bull. 6), Lae. [vi] + 57 pp., map. [Reissued 1974.]

Johns, R.J., *et al.* 1981–. A students' guide to the monocotyledons of Papua New Guinea. Parts 1–. Dept For., Univ. of Tech., Lae [Four parts published as of 1989.]

Kalkman, C., & W. Vink. 1970. Botanical exploration in the Doma Peaks region, New Guinea. Blumea 18: 87–135.

Lam, H.J. 1924. Vegetationsbilder aus dem Innern von Neu-Guinea. In: G. Karsten & H. Schenck (eds.), Vegetationsbilder 15 (5–7), 18 pl. Fischer, Jena.

Lam, H.J. 1927–29. Fragmenta papuana I-VII. Natuurk. Tijd. Ned.-Indië 87: 110–138, 139–180; 88: 187–227, 252–324; 89: 67–140, 291–388 (repr. separately; 360 pp.). [English translation as H.J. Lam. 1945. Fragmenta papuana. Arnold Arboretum, Harvard Univ. (Sargentia 2), Jamaica Plain, Mass. 196 pp.]

Lam, H.J. 1930. Het genetisch-plantengeografisch onderzoek van den Indischen Archipel en Wegener's verschuivings-theorie. Tijd. Kon. Ned. Aardr. Gen. 47: 553–581.

Lam, H.J. 1934. Materials towards a study of the flora of the island of New Guinea. Blumea 1: 115–159.

Lam, H.J. 1935. De vegetatie en de flora van Nieuw Guinee. In: W.C. Klein (comp.), Nieuw Guinee 1: 187–210. Molukken-Instituut, Amsterdam.

Lane-Poole, C.E. 1925. The forest resources of the territories of Papua and New Guinea. Govt Printer (Australian Parliamentary Papers), Melbourne.

Lauterbach, C. 1928–30. Die Pflanzenformationen einiger Gebiete Nordost-Neu-Guineas und des Bismarck-Archipels, I–IV. Bot. Jahrb. Syst. 62: 284–304, 452–501, 550–569; 63: 1–28, 419–476.

Ledermann, C. 1919. Einiges von der Kaiserin-Augusta-Fluss Expedition. Bot. Jahrb. Syst. 55, Beibl. 122: 33–44.

Lewis, W.H. 1971. High floristic endemism in low cloud forests of Panama. Biotropica 3: 78–80.

Masur, L.P. 1989. Stephen Jay Gould's vision of history. Massachusetts Review 30 (3): 467–484.

Nielsen, I., T. Baretta-Kuipers & P. Guinet. 1984. The genus Archidendron (Leguminosae–Mimosoideae). Nordic J. Bot. (Opera Botanica 76), Copenhagen. 120 pp.

Paijmans, K. 1976. Vegetation. In: K. Paijmans (ed.), New Guinea vegetation: 23–105. Australian National Univ. Press/CSIRO, Canberra.

Powell, J.M. 1976. Ethnobotany. In: K. Paijmans (ed.), New Guinea vegetation: 106–199. Australian National Univ. Press/CSIRO, Canberra.

Prance, G.T. 1977 (1978). Floristic inventory of the tropics: where do we stand? Ann. Missouri Bot. Gard. 64: 659–684; i-ii (Errata).

Prance, G.T., & D.G. Campbell. 1988. The present state of tropical floristics. Taxon 37: 519–548.

Royen, P. van. 1980. History of the exploration of the high altitude regions of New Guinea. In: P. van Royen (comp.), The alpine flora of New Guinea 1, chapter 10: 249–295. Cramer/Gantner, Vaduz.

Salter-Duke, B.J. 1984. A role for scientific societies in Papua New Guinea. In: J.R. Morton (ed.), The role of science and technology in the development of Papua New Guinea: the policy dimensions 2: 110–122. Fac. Science, Univ. Papua New Guinea, Port Moresby.

Sleumer, H.O. 1986. A revision of the genus Rapanea Aubl. (Myrsinaceae) in New Guinea. Blumea 31: 245–269.

Sleumer, H.O. 1987a. A revision of the genus Maesa Forsk. (Myrsinaceae) in New Guinea, the Moluccas and the Solomon Islands. Blumea 32: 39–65.

Sleumer, H.O. 1987b. The genera Embelia Burm. f. and Grenacheria Mez (Myrsinaceae) in New Guinea. Blumea 32: 385–396.

Sleumer, H.O. 1988a. The genera Discocalyx Mez, Fittingia Mez, Loheria Merr. and Tapeinosperma Hook. f. (Myrsinaceae) in New Guinea. Blumea 33: 81–107.

Sleumer, H.O. 1988b. The genus Conandrium Mez (Myrsinaceae). Blumea 33: 109–113.

Sleumer, H.O. 1988c. A revision of the genus Ardisia (Myrsinaceae) in New Guinea. Blumea 33: 115–140.

Sohmer, S.H. 1988. The nonclimbing species of Psychotria (Rubiaceae) in New Guinea and the Bismarck Archipelago. Bishop Mus. Press (Bishop Mus. Bull. in Botany 1), Honolulu. [viii] + 339 pp.

Steenis, C.G.G.J. van. 1934–1936. On the origin of the Malaysian mountain flora. Parts 1–3. Bull. Jard. Bot. Buitenzorg sér. III, 13: 135–262; 289–417; 14: 56–72.

Steenis, C.G.G.J. van. 1950. Desiderata for future exploration. In: M.J. van Steenis-Kruseman, Malaysian plant collectors and collections. Flora Malesiana I, 1: cvii–cxvi, maps 2 & 3. Djakarta: Noordhoff/Kolff. [Repr. 1985, Koeltz, Koenigstein.]

Steenis, C.G.G.J. van. 1957a. The significance of the Flora Malesiana for the critical knowledge of the Philippine flora. In: Proc. Eighth Pacif. Sci. Congr. 4 (Bot.): 493–499. Manila.

Steenis, C.G.G.J. van. 1957b. Specific and infraspecific delimitation. In: C.G.G.J. van Steenis (ed.), Flora Malesiana, I, 5: clxvii–ccxxix. Noordhoff, Groningen.

Steenis, C.G.G.J. van. 1979. Plant-geography of East Malesia. Bot. J. Linn. Soc. 79: 97–178.

Steenis, C.G.G.J. van. 1981. Rheophytes of the world. Sijthoff & Noordhoff, Alphen a/d Rijn. [xvi] + 407 pp.

Stevens, P.F. 1989. New Guinea. In: D.G. Campbell & H.D. Hammond (eds.), Floristic inventory of tropical countries: 120–312. New York.

Streimann, H. 1983. The plants of the Upper Watut Watershed of Papua New Guinea. National Bot. Gard., Canberra. iv + 209 pp., map.

Streimann, H. 1986. Catalogue of the lichens of Papua New Guinea and Irian Jaya. Cramer/Gantner, Vaduz.

Symington, C.F. 1943. The future of colonial forest botany. Empire For. J. 22: 11–23.

Symon, D.E. 1985. The Solanaceae of New Guinea. J. Adelaide Bot. Gard. 8, Adelaide. i + 171 pp.

Tan, B.C., & J.P. Rojo. 1989. The Philippines. In: D.G. Campbell & H.D. Hammond (eds.), Floristic inventory of tropical countries: 44–62. New York.

Toledo M., V. 1985. A critical evaluation of the floristic knowledge in Latin America and the Caribbean. Washington, D.C. (Report presented to the Nature Conservancy International Program.) 78, 17 pp.

Valkenburg, J.L.C.H. van. 1987. Floristic changes following human disturbance of mid-montane forest. Wageningen. [iii] + 25 pp., unpaged appendices.

Verdcourt, B. 1979. A manual of New Guinea legumes. Div. Botany, Office of For., Papua New Guinea (Bot. Bull. 11), Lae. 645 pp.

Walker, D. 1973. Highlands vegetation. Australian Natural History 17: 410–414, 419.

Walker, D., & G.S. Hope. 1982. Late Quaternary vegetation history. In: J.L. Gressitt (ed.), Biogeography and ecology of New Guinea: 263–285. Junk (Monogr. Biol. 42), The Hague.

Whitmore, T.C. (ed.). 1981. Wallace's line and plate tectonics. Oxford Univ. Press (Oxford Monogr. Biogeogr. 1), Oxford.

Womersley, J.S. 1978. Handbooks of the Flora of Papua New Guinea 1. Melbourne Univ. Press, Melbourne. xvii + 278 pp., map.

Womersley, J.S., & J.B. McAdam. 1957. The forests and forest conditions in the Territories of Papua and New Guinea. [Dept For., Territory of Papua New Guinea, Port Moresby.] iv + 62 pp. [Reprinted, 1975.]

21 Areas of endemism and composite areas in East Malesia

J.P. DUFFELS and A.J. DE BOER

Summary

East Malesia is a tectonic zone of interaction between the Asian and Australian plates. Sulawesi, Maluku and New Guinea are composite areas comprising fragments of different geological origin. Areas of endemism in East Malesia are recognised following the methods of phylogenetic taxon-analysis and area-cladistics. These methods are preferred above methods dealing with biotic similarity in the analysis of area-relationships, since they provide an instrument to recognise general and unique biogeographic patterns in the evolution of biota. The presently known area-cladistic patterns demonstrate that the biota of Sulawesi are related to Asian biota only, whilst New Guinea has biotic relationships to Asia as well as to Australia. The biogeographic relationships of Maluku need further study. An explanation for the absence of taxa of Australian affinities in Sulawesi can be found in the supposition that the Australian eastern arc of Sulawesi was under water until its collision with the subaerial western part of Sulawesi. Sulawesi and Maluku are areas of endemism, both comprising several areas of endemism of lower rank. New Guinea comprises three areas of endemism: northern New Guinea, the Vogelkop region and the central part of the island. The island cannot be regarded an area of endemism in itself. Northern New Guinea has a primary biogeographic coherence to other fragments of the Tertiary Outer Melanesian Arc (OMA) and a sister area relationship to central New Guinea, which forms a part of the Australian Inner Melanesian Arc (IMA). The Vogelkop region has a sister area relationship to OMA and Maluku. An explanation for these New Guinea patterns can be found in the coherence of the OMA and in the dispersal of Oriental OMA-groups into the IMA after the collision of both island arcs forming New Guinea. A sister area relationship spanning East Malesia and the Southwest Pacific is found between Sulawesi and Maluku, New Guinea and the other islands of the OMA. Relationships of New Guinea to fragments of the IMA and Australia are not found in cicadas, but are well-known in several groups of organisms of Australian origin.

Introduction

The occurrence of different species of animals and plants in various parts of the world has intrigued biologists now for already more than 150 years. In 1858 Sclater presented a subdivision of the earth's surface in six avifaunal regions, which he grouped in two creations. Sclater's classification was widely used, since its validity for many more groups of animals was recognised by Wallace (1860, 1876) and others. The zoological division of the world was further refined in 1876, when Wallace subdivided each region in four sub-

P. Baas et al. (eds.), The Plant Diversity of Malesia, 249–272.

regions so that a total of 24 subregions was recognised. It is very remarkable from a biogeographic point of view that the zoological subregions of Wallace correspond to a large extent to the botanical regions defined by de Candolle (1838) (cf. Nelson & Platnick 1981).

Wallace (1863, 1876) recognised two subregions in the Malay Archipelago: the Indo-Malayan subregion, west of Makassar Strait, as a part of the Oriental region, and the Austro-Malayan subregion, east of Makassar Strait, as a part of the Australian region. The nature of Wallace's zoological division has been stated clearly in his study on the Indian and Australian butterflies of the family Pieridae (1867): "Zoological divisions of the earth's surface can only be true and useful ones, in so far as they agree with the most ancient and permanent barriers to the diffusion of species."

The concept of plate tectonics provided an explanation for the borderlines between the various regions and subregions. Recent reconstructions of continental shifts (Audley-Charles 1981, 1987; Smith *et al.* 1981; Hamilton 1979) show that the interaction zone between the continental Greater Sunda Islands and the New Guinea–Australia continent corresponds with the most ancient barrier hampering exchange of flora and fauna between these continents. The complexity of biological evolution in this transition zone is already illustrated by Alfred Wallace's doubts concerning the biogeographic position of Sulawesi: "... whose central position and relations both to Asia and to Australia render it very difficult to decide in which of the primary zoological regions it ought to be placed ..." (Wallace 1880: 422). As illustration of his doubts Wallace included Sulawesi either in the Oriental region (Wallace 1863–1876) or in the Australian region (Wallace 1910), with lines of Wallace west or east of Sulawesi as a consequence. The area between the Greater Sunda Islands and New Guinea has intrigued also later biogeographers as indicated by several other lines proposed to separate the faunas of primarily Asian and Australian origin (Simpson 1977; George 1981). Recent historical biogeography studies demonstrate that Makassar Strait represents an important discontinuity in the distribution of Malesian taxa (Schuh & Stonedahl 1986; Duffels 1986). The extensions of this borderline to the north and south are less well defined because the Philippines in the north and the Lesser Sunda Islands in the south show relationships to both the western and the eastern part of Malesia.

The present study deals with the biogeography of East Malesia, which comprises Sulawesi, Maluku and New Guinea in the subdivision proposed by van Steenis (1950). Some attention is also given to the island groups east of New Guinea since these form a biogeographic entity with this area.

The aim of our biogeographic study is the search for general patterns of historical biogeographic relationships between areas of endemism. The first step in this study is the recognition of areas of endemism. By way of definition Nelson & Platnick (1981: 468) state: "An area of endemism is delimited by the more or less coincident distribution of taxa that occur nowhere else." Several of the islands and island groups in East Malesia are such areas of endemism. Some of these are composed of fragments of different geological origin. Geological composite islands may have biotas of different biogeographic affinities, representing the affinities of the original biotas of the components.

Faunistic and floristic similarities

Biotic similarities (in terms of the number of shared taxa in relation to the number of unshared taxa) between biogeographical homogeneous primary areas of Malesia have been studied by van Balgooy (1987) and Holloway & Jardine (1968). In his plant biogeographic analysis of Sulawesi, van Balgooy demarcated the primary areas of Malesia on the basis of the distribution of taxa that do not cross certain boundaries in the area; Sulawesi was divided into seven parts. The abundance of data in the volumes 4–10[1] of Flora Malesiana series I (1950–1984) formed the basis for his study. The floristic affinities of Sulawesi as a whole, as well as each part of Sulawesi, were calculated first by using Kroeber's formula and next by cluster analysis using the 'group average' method. These procedures were followed for both species and genera. At both species and genus level similarities of the Sulawesi flora were demonstrated with the north (Philippines), east (Maluku) and south (Lesser Sunda Islands). There appears to be considerably less similarity with Borneo. The analysis showed further that Sulawesi is coherent in a biogeographic sense. The floristic similarities among the different parts of Sulawesi are greater than those to neighbouring islands.

Another, classic example of using numerical methods for the analysis of biogeographic data in the Indo-Australian area is the study by Holloway & Jardine (1968) on the distribution of 870 species of butterflies, 876 species of birds and 124 species of bats. Their approach was based on the study of faunal affinities of primary areas, which were delimited by geographical criteria. Regarded as primary areas were some parts of the SE Asian mainland, Australia and the islands and island groups between these two continents. A hierarchical zoogeographic classification in well defined regions, subregions, provinces etc. was obtained after cluster analysis from a matrix of dissimilarity coefficients. The resulting maps of faunal boundaries derived for the butterflies (Holloway & Jardine 1968: fig. 5) match well with those derived for the birds (ibid.: fig. 6) but the zoogeographic classification for the bats differed markedly from those of the other two groups.

The floristic and faunistic homogeneous primary areas, as recognised by van Balgooy (1987) and Holloway & Jardine (1968) on account of the accumulation of distribution limits of genera and species groups, are for the greater part areas of endemism (cf. Schuh & Stonedahl 1986; Duffels 1986). However, endemism in primary areas is not informative to establish biotic similarity between areas of endemism, because the evolutionary relationships between endemic taxa are not considered. Biotic similarity is determined by the number of shared taxa in relation to the number of unshared taxa. Van Balgooy (1987: 99) stated frankly "endemic taxa obscure the relationships with other areas."

Holloway & Jardin regarded their studies "as a first stage in the zoogeographic study of a region. The second stage should consist of detailed studies of the patterns of variation and distribution of taxonomic groups of lower rank and attempts to relate these in detail to past and present geographical and ecological factors." After the introduction of new methods of analytical biogeography (see the next paragraph) Holloway (1984, 1987) and others concentrated on the investigation of inter-island relationships in the Indo-Australian archipelago in terms of phylogenetic relationships of taxa inhabiting these islands.

Analytical biogeography

Ball (1976) distinguished three main phases in the explanation of biogeographical patterns: an empirical or descriptive phase, a narrative phase and an analytical phase. In the descriptive phase basic data are collected on the present distributions of groups of organisms; not an easy task when organisms in undercollected areas, as most parts of the tropics, are concerned. The narrative and analytical phases are the subject matter of explanatory historical biogeography (Ball 1976). Cracraft (1988) characterised these phases as follows: "Essentially, narrative and analytical explanations were said to differ from one another in that the former are not based on corroborated phylogenetic hypotheses for the taxa involved." Phylogenetic hypotheses are nowadays regarded a prerequisite in analysing historical biogeographic patterns.

The introduction of phylogenetic systematics by Hennig (1950, 1966) provided a method to formulate hypotheses on the evolution of groups of organisms by character analysis. Only unique derived characters (synapomorphies) were accepted as convincing phylogenetic evidence to establish the monophyletic origin of taxa (cf. Hennig & Schlee 1978). The relationships within taxa of monophyletic origin are visualised in taxon-cladograms.

The application of Hennig's phylogenetic systematics (alternatively called cladistics) in historical biogeography has resulted in the method of cladistic biogeography (Platnick & Nelson 1978). The successive methodological developments in this field have been summarised by Humphries & Parenti (1986) and Wiley (1988a). Following this method biogeographical patterns are studied by substituting the names of the areas in which the species are found for the species names in the taxon-cladogram. The resulting cladogram is a taxon-area-cladogram, which shows the relationships of the areas as suggested by the phylogeny and distribution of a particular taxon. Such a reconstruction of biogeographic relationships within a certain geographic area can only be obtained from groups of organisms with restricted, endemic distributions in this area, and it is obvious that widespread taxa are uninformative in regard to such analysis. Three-area statements are the most basic units for expressing these area-relationships. Taxon-areacladograms obtained from different groups of animals and plants may show one or more congruent and unique patterns depending on the common and special factors which affected the evolution and/or distribution of these groups. Congruence of taxon-areacladograms suggests a pattern that is not random but orderly.

The theory of continental drift that was developed in the 1960's, and especially later plate-tectonic reconstructions (e.g. Smith *et al.* 1981), provided explanations for the existing biogeographic order. Partition of continents or continental fragments, the collision of areas, and the formation of island arcs played a role in the evolution of the whole biota of the area where these events occurred. Vicariant plate-tectonic events, as well as the development of common dispersal pathways (McDowall 1980; Duffels 1983a, b) may both result in congruent area-relationships for different groups of organisms. Congruent area-relationships within a certain area can develop only in groups of organisms, which respond similarly to successive manifestation and disappearance of barriers. The geographic and climatological conditions of the period in which a group evolved determined the biogeographic pattern shown by the group. Groups of organisms that evolved in different periods will therefore show different patterns of area-relationships (cf. Cracraft 1988). Unique area-relationships are found in those groups of organisms, wherein special factors determined the relationships of the areas to such an extent that no congruence to other patterns exist.

Historical biogeography aims to discriminate general patterns of area-relationships, shown by two or more groups occupying the same area, from unique distribution patterns of individual taxa. In case of congruence of area-cladistic patterns biogeographers try to relate the sequence of splitting in the general taxon-areacladogram to geological and climatological changes in the area concerned. Recently developed methods of biogeographic parsimony analysis aim to find one general area-cladogram (e.g. Wiley 1988b), which represents the most parsimonious tree of area-relationships. The search for one general area-cladogram, however, may obscure the presence of two or more general patterns of area-relationships existing among the original taxon-areacladograms (cf. Cracraft 1988; Oosterbroek, in litt.; Vane-Wright, in press).

Barriers and connections in East Malesia

The individual biogeographic patterns resulting from fragmentation of areas of distribution and occupation of new areas by dispersal of organisms are determined to a large extent by the nature of the barriers and the ecology and vagility of these organisms. Plate-tectonic events, climatological changes, sea level changes and shifts of vegetation zones may lead both to the formation of new potential barriers and the disappearance of existing barriers. The different groups of organisms forming part of the biota of an area may respond differently to sequential geographic, climatological and ecological changes. This means that different area-relationships may be expected from the phylogenetic reconstructions of different groups living side-to-side in the same area.

A first impression of the geographic barriers in Malesia can be obtained from sea-depth maps of the area. These show that the Greater Sunda Islands (Java, Sumatra, Borneo) are separated by shallow waters of less than 200 m deep, while the islands east of Makassar Strait (Sulawesi, Moluccas, New Guinea) are separated by seas of 200–7000 m deep. The geological history of the area confirms the coherence of Sundaland (the southern part of the Indochina peninsula, Malaysia, Sumatra, Borneo and Java) and the isolation of the islands of East Malesia. The eastern part of Malesia and the West Pacific evolved from a complex series of interactions of continents and island-arcs. These island-arcs were chains of islands with changing subaerial continuity and isolation. Historical biogeography of the Indo-Pacific has already demonstrated that island arcs have functioned as historic routes of dispersal for terrestrial animals and plants (Duffels 1983a & b, 1986), and for marine insects of coastal waters (Møller Andersen 1989a, b).

Sulawesi is a geologically composite area with a western arc of Southeast Asian origin and an eastern arc of Australian origin (Audley-Charles 1981, 1987). The western arc of the island probably had a long and isolated subaerial history. Land connections between West Sulawesi and Borneo either across Makassar Strait or by closure of the strait or by close juxtaposition of these islands is not supported by the biogeography of the Sulawesian mammal and insect fauna (Musser 1987; Holloway 1987; Duffels 1989, in press). The eastern Sulawesi arc formed a part of the Outer Banda Arc that ran from Buton, through Seram and Tanimbar, to Timor and began to emerge as an island during its mid-Miocene collision with western Sulawesi.

The sea level changes during the Quaternary did not change the contours of the island very much as has been the case on the Sunda and Sahul shelfs. The supposed maximum fall of 200 m (Morley & Flenley 1987) did not expose much more of Sulawesi, as the coastlines slope quite sharply. The sea level rises in the Pleistocene, which had a maximum of 25 m above the present level, separated the south-west peninsula from the rest of Sulawesi by marine transgression of the Tempe depression (Whitten *et al.* 1988). During the drier periods of the Pleistocene the area of seasonal forest expanded, while the rainforest area reduced in size. Lowering of vegetation zones during cool Pleistocene periods enabled mountain plants and animals to spread more widely (Whitten *et al.* 1988).

Maluku (= Moluccas) is an archipelago of composite geological nature. The present position of the North Maluku (Obi, Bacan, Halmahera, Morotai, Talaud etc.) and the shallow waters separating these islands from the North Vogelkop of New Guinea suggest that the North Moluccas are tectonically detached parts of New Guinea, which have moved westward, together with the Banggai and the Sula Islands (Audley-Charles 1987). The South Maluku (Buru, Ambon and Seram) formed part of the Outer Banda Arc, which continued to Kai, Tanimbar and Timor.

New Guinea and the Melanesian Arcs. The northward drifting of Australia from a late Cretaceous position as part of Gondwanaland is a major factor in the formation of East Malesia. The leading edge of the Australian plate, which later transformed into the central mountain ranges of New Guinea, was a part of the Inner Melanesian Arc (IMA), which stretched from New Guinea over New Caledonia to New Zealand. In the late Miocene (10 my BP) or early Pliocene the continental IMA collided with the Outer Melanesian Arc (OMA). The latter island arc is of oceanic derivation and developed independently of the IMA. Remnants of the OMA are North Maluku, the northern part of the Vogelkop peninsula, Biak and Japen Islands in the Geelvink Bay, the northcoast mountain ranges of New Guinea (and probably the southeastern Papuan peninsula), the Bismarck Archipelago, the Solomon Islands, Vanuatu, Fiji and Tonga. Though not a part of the OMA, the WNW trend of the Samoan Islands, running from the now exposed islands to the Combe Bank suggests an old island-chain connection with parts of the OMA (Ewart 1988). Reviews of the geological history of New Guinea and the West Pacific (Coleman 1980; Crook 1981; Holloway 1984; Duffels 1986; Ewart 1988) provide more details on composition and fragmentation of these island arcs. It is most important in a biogeographic respect that northern oceanic New Guinea and continental central and South New Guinea, which are separated nowadays by the Sepik and Ramu-Markham valleys, were situated widely apart in the Paleogene. The Early Miocene witnessed the reduction of the distance between the IMA and OMA resulting in the collision of both arcs in New Guinea in the late Miocene. The formation of the central mountains occurred in the Pliocene or later.

Changes in sea level in the Tertiary and Quaternary altered the amount of dry land at the northern margin of the Australian plate and in the chains of islands in both Melanesian arcs. A major low sea level stand of 200 m below present sea level occurred during the late Miocene. The Pleistocene saw four sea level falls of 150–200 m. Sea level rises of 100 m occurred in middle Miocene and early Pliocene, and relative small rises occurred three times in the Pleistocene (Ewart 1988).

The Vogelkop of New Guinea (= Tjendrawasih) takes an isolated position in relation to the other parts of New Guinea. During expansion of the Banda Arc in late Neogene, this peninsula rotated clockwise northward from a southern position near the edge of the New Guinea-Australia plate.

No detailed information is available on the development of the vegetation of East Malesia during the Tertiary and early Quaternary. Palynological studies in the mountains of New Guinea form the main source of our present knowledge of changes in altitudinal zonation in the late Quaternary (Hope & Peterson 1975; Flenley 1979; Morley & Flenley 1987). Evidence exists for a great depression of vegetation boundaries in New Guinea prior to 37,000 yr BP. The forest limit was as much as 1700 m lower than today (about 3800 m), but this depression was followed by a gradual rise to more that 3000 m until about 30,000 yr BP. Then followed a period of cold conditions, with a forest limit at 2200–2400 m, until 10,000 yr BP. Such shifts of vegetation zones in New Guinea caused repeated expansions and contractions of montane forest zones, resulting in formation and disappearance of barriers for forest-dwelling animals.

Areas of endemism in East Malesia

Sulawesi

The island of Sulawesi must be regarded as an area of endemism on account of the occurrence of several monophyletic groups of organisms. Especially in cicadas endemism is very striking. The Sulawesi cicadas are currently under study by de Jong (1985, 1986, 1987) and Duffels (1977, 1982, 1986, 1989), so that the results presented here are preliminary. The distribution of 77 recognised Sulawesi species are taken into consideration, but 24 of these are still awaiting description. A recent review of the biogeography of the Sulawesi cicadas (Duffels, in press) lists six monophyletic genera and species groups, which are endemic or centred in Sulawesi: *Dilobopyga* Duffels from Sulawesi (30 species), Selayar (1), Banggai (2), Sula (1), Sangihe (1) and South Moluccas (1); the genera *Brachylobopyga* Duffels (2 species) and *Prasia* Stål (7), and the *Lembeja parvula* group (2), which are restricted to Sulawesi; the *Lembeja foliata* group from Sulawesi (12 species) and Sangihe (1 species); the *Lembeja fatiloqua* group with a wider distribution in Sulawesi (6), Sumba (1), Sumbawa (1), Mindanao and Borneo (1) and New Guinea and Australia (1). These distributions suggest that the Sulawesian area of endemism probably includes Banggai and Sangihe, and possibly some other nearby islands that form part of these Sulawesi centred distributions. Other examples of major endemic radiations in Sulawesi are found in the satyrine butterflies of the genus *Lohora* (Vane-Wright, pers. comm.), in a monophyletic subgroup of the danaid butterflies of the genus *Parantica* Moore (Ackery & Vane-Wright 1984) and in the *velutina* group of the limacodid genus *Narosa* Walker (Holloway 1987: fig. 9.6).

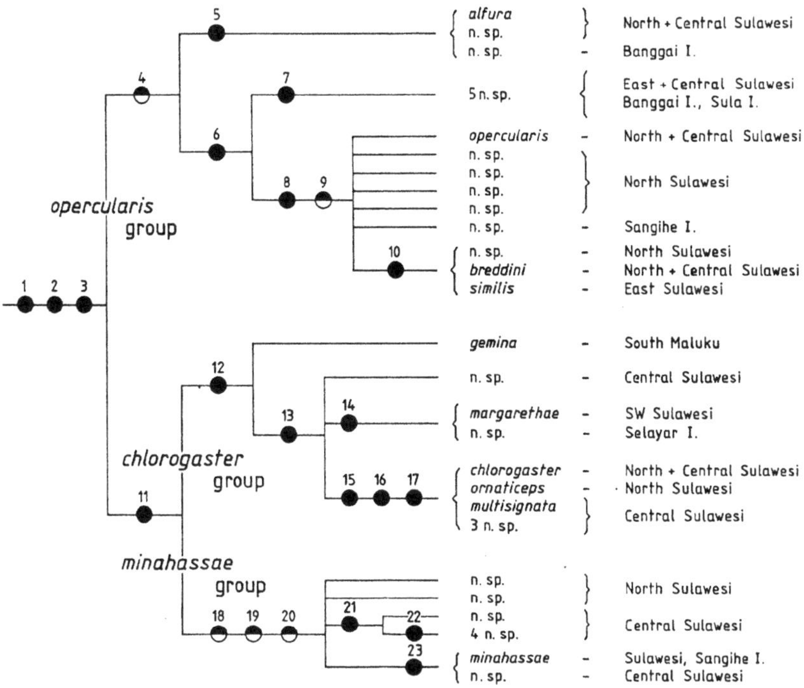

Figure 21.1 — Taxon-areacladogram of the genus *Dilobopyga*. The numbers refer to synapomorphies listed in Duffels (in press: table 3); weak synapomorphies are half black and half white.

The taxon-areacladogram of *Dilobopyga* (Figure 21.1) shows the striking endemism of the species and species groups of this largest cicada genus of Sulawesi. On the species level, endemism is found in North, Southwest, central and East Sulawesi. On a higher taxonomic level a subgroup (9 species) of the *D. opercularis* group is centred in North Sulawesi and a subgroup (5 species) of the *minahassae* group is restricted to central Sulawesi. Another monophyletic subgroup of the *opercularis* group shows a biogeographic coherence of central Sulawesi to the eastern part of the island and to the Banggai and Sula archipelagos. Southwest Sulawesi has only one endemic species of *Dilobopyga*.

The genus *Lembeja* Distant, studied by de Jong (1985, 1986, 1987), shows another example of a distribution in North and central Sulawesi in the *foliata* group. The *L. fatiloqua* group, which has a relatively wide distribution (see above) has five endemics in Southwest Sulawesi.

Allopatry of (sub)species restricted to one of the arms or the central part

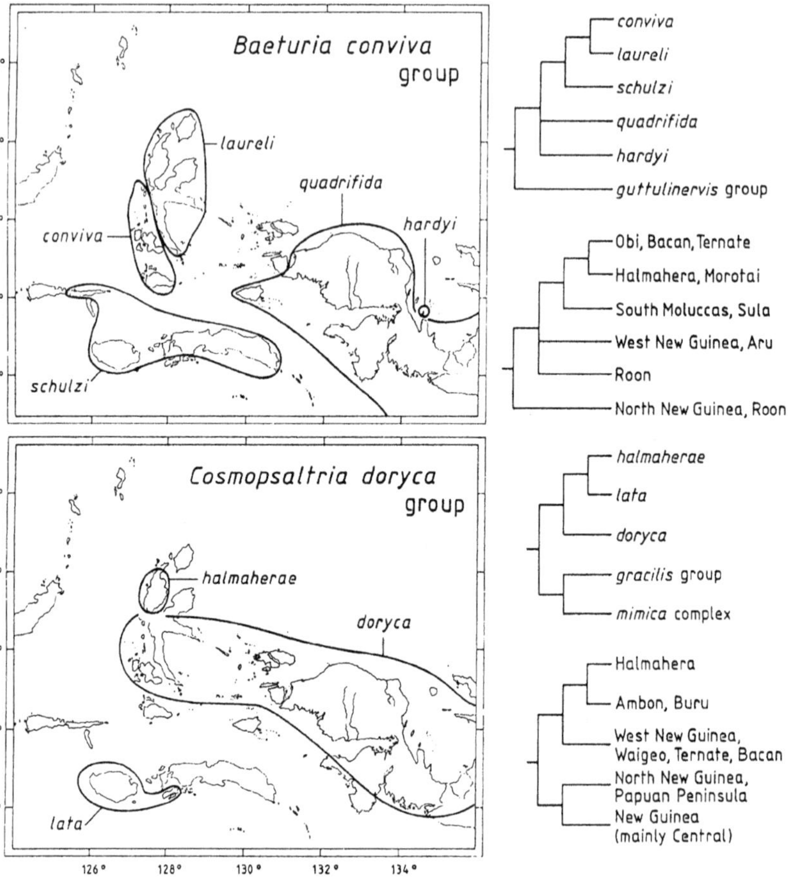

Figure 21.2 — Distribution, taxon-cladograms and taxon-areacladograms of two groups of cicadas in Maluku.

of Sulawesi are found in many other groups of organisms, e.g. macaques (Fooden 1969), waterstriders (Whitten *et al.* 1988), beetles (Kibby 1985) and carpenter bees (van der Vecht 1953).

Maluku

Several groups of cicadas have species or species groups endemic to Maluku or part of this area. The most significant biogeographic discontinuity in Maluku is found between the North Moluccas (Bacan, Halmahera, Morotai and Talaud) and the South Moluccas (Seram, Ambon and Buru); Obi appears to hold

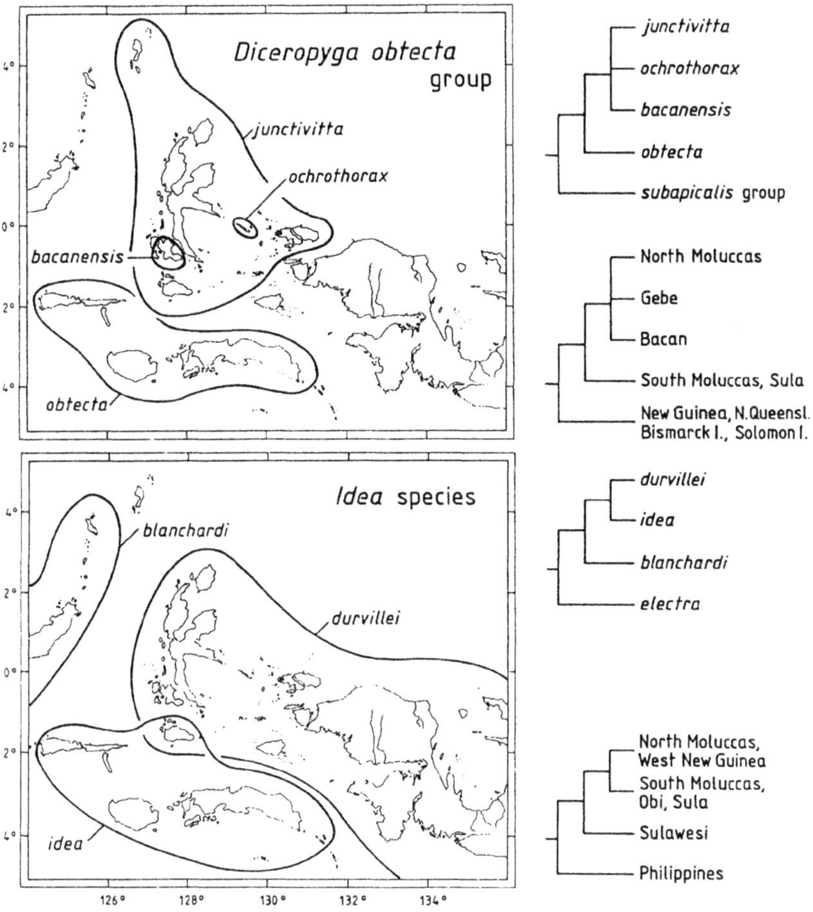

Figure 21.3 — Distribution, taxon-cladograms and taxon-areacladograms of a group of cicadas (above) and a group of danaid butterflies (below) in Maluku (data on butterflies from Kitching *et al.* 1987).

an ambiguous position. The cicadas of the *Diceropyga obtecta* group (Duffels 1988b) (Figure 21.3), the *Baeturia conviva* group (de Boer 1986) (Figure 21.2), and the *Cosmopsaltria doryca* group (Duffels 1988c) (Figure 21.2) show sister group relationships between the North Moluccas and either the South Moluccas or the South Moluccas and Sula. A discontinuity between the North Moluccas plus West New Guinea and the South Moluccas is found in the butterfly genus *Idea* Fabricius (Kitching *et al.* 1987) (Figure 21.3). Remarkable is the differing amount of endemic cicadas between North and South Moluccas. Single-island endemics are found in Halmahera (Figure 21.2), Bacan (Figure

21.3), and Gebe Island (Figure 21.3), while other species have a wider distribution within the North Moluccas. But each of the groups mentioned have only one endemic in the South Moluccas. Most striking is the similar distribution of three of the four endemics in Seram, Buru and Sula (Figures 21.2 & 21.3).

Maluku as a whole has a close but fairly complicated relationship to New Guinea, wherein two patterns can be distinguished: firstly a close relationship of Maluku to western New Guinea, including the Vogelkop area, secondly a sister group relationship of Maluku to North New Guinea and eastern parts of the Outer Melanesian Arc (OMA). The first pattern is found in the *Baeturia conviva* group and in the *Cosmopsaltria doryca* group. The Maluku species of the *Baeturia conviva* group have a monophyletic origin; the two other species of this group are recorded from Roon Island and from West New Guinea, Vogelkop and Aru Islands respectively (de Boer 1986). A group consisting of *Baeturia guttulinervis* Blöte and three related species, from North New Guinea and Roon I. form the sister group of the whole *B. conviva* group (de Boer, unpublished). The sister species of the Maluku species *Cosmopsaltria lata* (Walker) and *C. halmaherae* Duffels is *C. doryca* Boisduval, which is distributed in West New Guinea and doubtfully recorded from Ternate; the sister group of the *doryca* group is widely distributed in New Guinea and recorded from the Bismarck Archipelago (Duffels 1983a). The second relationship of Maluku to the OMA, by-passing the Vogelkop area, is found in both the genus *Diceropyga* (Duffels 1977, 1986), which includes the endemic *obtecta* group from Maluku, and the *Baeturia bloetei* group (de Boer, 1989). Other examples of OMA distribution are presented by the genera *Gymnotympana* Stål (de Boer, unpublished) and *Toxopeusella* Kirkaldy (Boulard 1981). These four groups have a wide distribution from Maluku through the northern mountains of New Guinea, to the Bismarck Archipelago and the Solomon Islands; the *B. bloetei* group extends further east into the West Pacific.

The genus *Idea* mentioned above shows a sister group relationship between two species from North Moluccas-western New Guinea and from the South Moluccas respectively and a Sulawesi species (Kitching *et al.* 1987).

A relationship between South Moluccas and Sulawesi is demonstrated by *Dilobopyga gemina* (Distant) (Figure 21.1), which is a sister species of a group of 9 *Dilobopyga* species from Sulawesi. *Dilobopyga gemina* has two endemic subspecies, one in Seram, Saparoea, Goram and Misoöl (*D. gemina gemina*) and one in Buru (*D. gemina toxopei* (Schmidt)).

New Guinea

This island has fascinated biogeographers by its very peculiar endemic biota demonstrating Asian (e.g. many groups of plants and insects) and Australian (e.g. many groups of birds and mammals) relationships. New Guinea is a composite area in a geological sense, in which at least three areas of endemism can be recognised: 1) the Vogelkop peninsula (= Tjendrawasih) including the islands Waigeo, Salawati, and Misoöl, 2) central New Guinea, 3) the northern mountain ranges.

These areas of endemism were found by phylogenetic analysis of cicadas of the subtribe Cosmopsaltriaria (Duffels 1983a), while systematic and biogeographic study of the cicadas of *Baeturia* Stål and related genera (de Boer 1982, 1986, 1989) and studies of other groups corroborate these results.

The Vogelkop area

The Vogelkop peninsula of New Guinea and adjacent islands form an area of endemism as demonstrated by the cicada genus *Rhadinopyga* Duffels (Duffels 1985, 1986: fig. 7). Two of its species are endemic to Waigeo, one is endemic to Salawati and the fourth species is recorded from Misoöl and Batjan, and Sorong on the Vogelkop. Other undescribed species are recorded from several localities in the Vogelkop area. De Boer (unpublished) recognised a new genus related to *Baeturia* with two new species. One of those is only recorded from Sorong and almost certainly an endemic of the Vogelkop, while the other species is recorded from one locality in the Vogelkop, from Roon Island in the Geelvink Bay and from Torricelli Mts, which is a part of the northcoast mountains of New Guinea. A third genus of cicadas, *Arfaka* Distant, comprises two species, endemic to the Vogelkop peninsula and the island of Misoöl respectively (de Jong, pers. comm.). It is interesting that four groups of cicadas, the genus *Diceropyga* Duffels (sister genus of *Rhadinopyga*) (Duffels 1986), the *Baeturia bloetei* group (de Boer 1989) and the genera *Gymnotympana* and *Toxopeusella*, which are both widely distributed in Maluku and the northern mountains of New Guinea, by-pass the Vogelkop area.

Central New Guinea

Central New Guinea was recognised as an area of endemism by the occurrence of the monophyletic *Cosmopsaltria mimica* complex (Duffels 1986: fig. 4). This complex comprises 17 (sub)species arranged in five species groups, which are principally distributed in central New Guinea (Duffels 1983a, 1986). Though one species group (*capitata* group) is primarily distributed in the lowland of New Guinea, the other four show central mountain range patterns with largely sympatric (*mimica* group) to largely allopatric (*meeki* group) species distribution. Another endemic group of central New Guinea and Huon peninsula is the *Baeturia nasuta* group (de Boer 1982). This group comprises 8 species; one of which, *B. parva* Blöte, does not occur in central New Guinea, but is distributed in the northern mountain ranges and lowlands of New Guinea, in New Britain, New Ireland, Admiralty Islands, Biak and Misoöl.

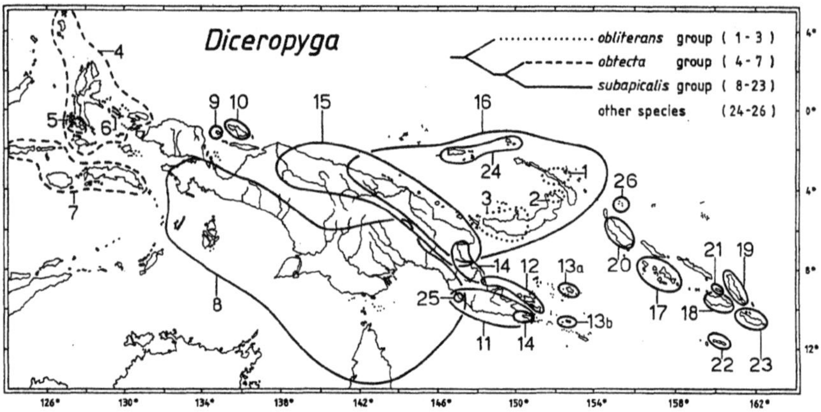

Figure 21.4 — Distribution of the species groups and species of *Diceropyga*: *obliterans* group (1–3): *noonadani* (1), *novaebritannicae* (2), *obliterans* (3); *obtecta* group (4–7): *junctivitta* (4), *bacanensis* (5), *ochrothorax* (6), *obtecta* (7); *subapicalis* group (8–23): *subapicalis* (8), new species (9), new species (10), *bicornis* (11), *auriculata* (12), *woodlarkensis* (13) subsp. *woodlarkensis* (a), subsp. *inexpectata* (b), *subjuga* (14), *bihamata* (15), *gravesteini* (16), *major* (17), *guadalcanalensis* (18), *malaitensis* (19), *bougainvillensis* (20), *tortifer* (21), *rennellensis* (22), *aurita* (23), other species: *triangulata* (24), *novaeguinae* (25), *didyma* (26).

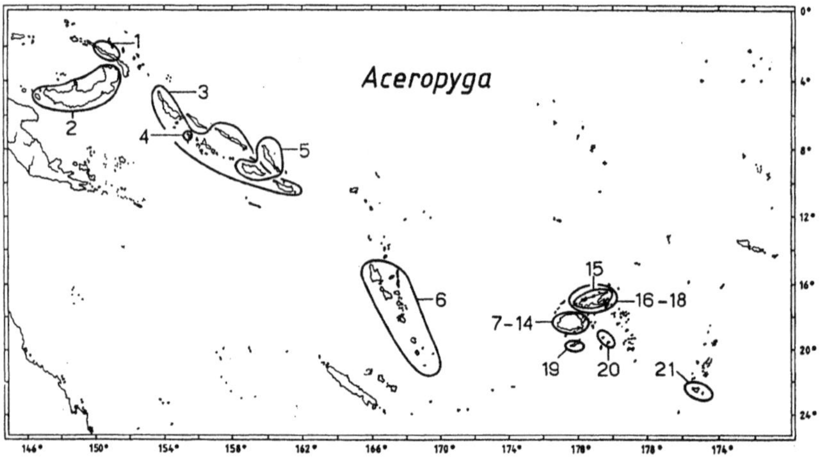

Figure 21.5 — Distribution of the species of *Aceropyga*: *novaeirelandicae* (1), *obliqua* (2), *aluana aluana* (3), *aluana minuta* (4), *aluana torquata* (5), *poecilochlora* (6), *distans distans* (7), *stuarti stuarti* (8), *corynetus corynetus* (9), *philorites* (10), *acuta* (11), *macracantha* (12), *egmondae* (13), *pterophon* (14), *huireka* (15), *distans taveuniensis* (16), *stuarti pallens* (17), *corynetus ungulata* (18), *corynetus monacantha* (19), *distans lineifera* (20), *albostriata* (21).

The considerable Pleistocene shifts of montane vegetation zones in New Guinea, together with the geological instability of this island, account for the high species endemicity in the New Guinean mountain ranges. This is demonstrated in the *Cosmopsaltria mimica* complex (Duffels 1983a) and the *Baeturia nasuta* group (de Boer 1982), which are mainly distributed in the lower montane forest zone (1000–2650 m). Different stages of evolution are suggested by largely sympatric and allopatric distribution patterns of the species groups of the *mimica* complex. One of the species of this complex (*Cosmopsaltria signata* Duffels), which is found at relatively high altitude (1800–3300 m), has a disjunct distribution in the central mountain ranges (Duffels 1983a: fig. 89).

The Outer Melanesian Arc remnants

All areas originating from the Outer Melanesian Arc (OMA) taken together form one single area of endemism. Though the Vogelkop and Maluku are here discussed as separate areas of endemism, they should be regarded as part of this area of endemism as well. These OMA-remnants further include northern New Guinea and the Papuan peninsula, the Bismarck Archipelago and Admiralty Islands, Solomon Islands, Vanuatu, Fiji and Tonga and probably Samoa. The coherence of this area of endemism is best illustrated by the distribution of the recently revised *Baeturia bloetei* group (Figures 21.6 & 21.7), throughout this area from Maluku to American and western Samoa, but excluding the Vogelkop and Fiji Islands and with a western extension to Timor (de Boer 1989). The distribution of the monophyletic group formed by the genera *Rhadinopyga, Diceropyga* (Figure 21.4) and *Aceropyga* Duffels (Figure 21.5) is very similar to that of the *Baeturia bloetei* group. This group does not reach Samoa, but its distribution area does include Fiji and the Vogelkop. Only one species, *Diceropyga subapicalis* (Walker) (Figure 21.4), occurs outside the OMA fragments, in South New Guinea, Aru Islands and North Queensland (Duffels 1977, 1986).

Several other species groups occupy only part of the OMA; their distributions coincide in some cases with the distribution of subgroups of the taxa discussed above.

Northern New Guinea, Bismarck Archipelago and Solomon Islands — These areas together form one area of endemism for marine waterstriders; *Halovelia annemariae* and some undescribed species of *Xenobates* are restricted to that area (Møller Andersen 1989a, b). The *Diceropyga subapicalis* group, consisting of 16 (sub)species, has a very similar distribution (Figure 21.4). A less clear cut OMA distribution can be found in the butterflies of the *Graphium agamemnon* group (Saigusa *et al.* 1982). The species of this group are mainly distributed throughout Maluku, New Guinea, Queensland, Bismarck Archipelago and

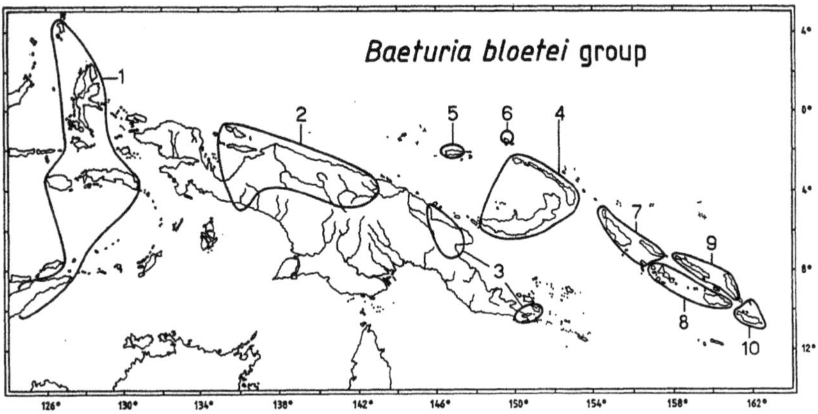

Figure 21.6 — Western part of the area of distribution of the *Baeturia bloetei* group: *bloetei* (1), *macgillavryi* (2), *papuensis* (3), *bismarckensis* (4), *manusensis* (5), *mussauensis* (6), *brandti* (7), *sedlacekorum* (8), *reijnhoudti* (9), *cristovalensis* (10).

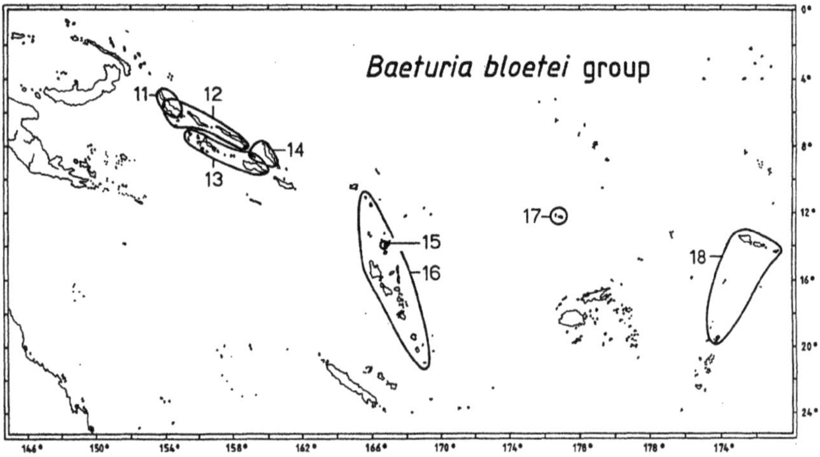

Figure 21.7 — Eastern part of the area of distribution of the *Baeturia bloetei* group: *gressitti* (11), *bilebanarai*(12), *mendanai* (13), *marginata* (14), *boulardi* (15), *edauberti* (16), *rotumae* (17), *maddisoni* (18).

Solomon Islands, with one very widely distributed species extending westward into the Oriental region. The four species with more restricted distribution areas are endemics of the Bismarck Archipelago (1 species) and Solomon Islands (3) respectively.

Bismarck Archipelago, Solomon Islands, Vanuatu, Fiji and Tonga Islands — This eastern part of the OMA is an area of endemism as suggested by the genus

Aceropyga (Figure 21.5). Two subgroups of the *Baeturia bloetei* group, which are found from the Solomon Islands eastwards, have a very similar distribution pattern. One of these subgroups includes the Solomon Islands, Vanua-Lava of the Bank Islands and Rotuma Island. The other subgroup is found in the Solomon Islands, Vanuatu, Tonga and Samoa Islands (de Boer 1989). However, the monophyly of these subgroups is not absolutely certain.

Within the OMA smaller areas of endemism can be recognised. Due to incomplete knowledge of the phylogeny of the respective species groups, the relations between these areas is not always clear. Maluku and the Vogelkop area have already been discussed as separate areas of endemism.

Northern New Guinea including the Papuan peninsula — This is a recognised area of endemism for the *Cosmopsaltria gracilis* group (Duffels 1983a). The current studies of *Baeturia* and some related genera reveal an abundance of species with very restricted distributions in northern New Guinea as well as the Papuan peninsula.

Bismarck and Admiralty Islands — The *Diceropyga obliterans* group (3 species) (Figure 21.4) and *Baeturia bismarckensis* of the *bloetei* group (Figure 21.6) are endemic to the Bismarck Archipelago, while *Aceropyga* has two island endemics in this area. The *bloetei* group has two endemics in the Admiralty Islands and *Diceropyga* one. The three species of the *bloetei* group from the Bismarck and Admiralty Islands possibly form a monophyletic subgroup, but probably do so together with *Baeturia papuensis* from North Papua New Guinea.

Solomon Islands — These islands are very rich in endemic species, all restricted to part of this island group, as demonstrated by the *Diceropyga subapicalis* group (7 endemic species) (Figure 21.4), the *Baeturia bloetei* group (8 endemic species) (Figures 21.6 & 21.7) and the genus *Aceropyga* (3 endemic species) (Figure 21.5). Further endemics of this area are the cicada genus *Heteropsaltria* Jacobi (1 species) and two species of the related genus *Nggeliana* Boulard (Boulard 1979; Duffels 1986). As yet no monophyletic taxa have been recognised that characterise the Solomon Islands as a whole as an area of endemism.

South-West Pacific — The area east of the Solomon Islands is usually characterised by a decrease in number of species. For example the *Baeturia bloetei* group has 8 endemics on the Solomon Islands, but only two on Vanuatu (one on Vanua-Lava and one widely distributed), one on Rotuma Island and one on Samoa-Tonga. *Aceropyga* with three endemics on the Solomon Islands has one endemic on Vanuatu-East Carolines and one endemic in the Tonga chain, but no less than 14 endemic (sub)species on Fiji. Further details on the biogeography of the Southwest Pacific island groups Fiji, Tonga and Samoa are not discussed here but can be read from Duffels (1988a), Ewart (1988) and de Boer (1989).

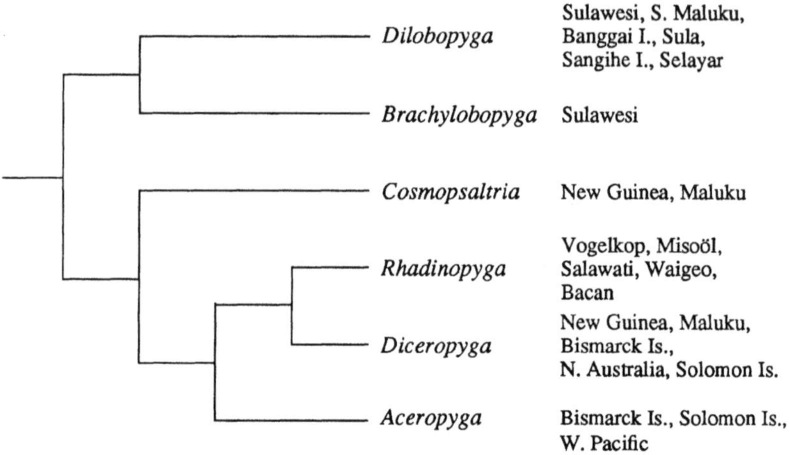

Figure 21.8 — Taxon-areacladogram of the Cosmopsaltriaria.

General biogeographic patterns

The first step in the search for general biogeographic patterns is the recognition of areas of endemism on the basis of the endemic occurrence of species and/or monophyletic groups of species. Within the geographic entities of Sulawesi, Maluku and New Guinea we have recognised several areas of endemism. The available taxon-areacladograms show that the areas of endemism within each of these geographic entities do not necessarily form one larger area of endemism. An explanation for these complex biogeographic patterns can be found in the composite or hybrid geological origin of the geographic entities of East Malesia (see paragraph on barriers and connections). We have avoided the use of the term composite or hybrid area in biogeographic sense, defined by Wiley (1988b) as an area that contains two biotas of different biogeographic affinities. In fact most areas of endemism have biotic relationships to more than one other area of endemism, and therefore it is not of much interest to indicate whether an area is composite in a biogeographic sense (Oosterbroek, in litt.). An area of endemism may have different biotic relationships as a result of large-scale vicariant events in different geological times and unique dispersal events. However, since the aim of biogeographic analysis remains to be the search for general biogeographic patterns, it seems to us that a discussion on the different biotic relationships of an area of endemism becomes meaningful only in relation to the recognition of general patterns of area-relationships and the search for their causal explanations.

As far as the cicadas of New Guinea are concerned northern New Guinea is closest related to other fragments of the Outer Melanesian Arc (OMA) (see the

Diceropyga subapicalis group and the *Baeturia bloetei* group), whilst at least one group shows relationships of northern New Guinea to central New Guinea (the sister group relationship of the *Cosmopsaltria gracilis* group and the *Cosmopsaltria mimica* complex). The Vogelkop area shows a sister group relationship to the fragments of the OMA together, including Maluku, in the subtribe Cosmopsaltriaria (Figure 21.8). These relationships of the New Guinean areas of endemism indicate that New Guinea as a whole can probably not be regarded as one single area of endemism of higher rank, though each of these areas of endemism shows relationships to the Outer Melanesian Arc. On the other hand, several other groups of insects and plants show an area-cladistic coherence of New Guinea with Australia, and either New Zealand or New Caledonia (Ross 1956; Thornton 1980; Patterson 1981; Humphries & Parenti 1986). These patterns represent area-cladistic relationships between fragments of the Inner Melanesian Arc (New Guinea, New Caledonia and New Zealand) and Australia. It would be interesting to analyse the area-cladistic relationships in these Australian IMA groups to allow further comparison of the New Guinean patterns in these groups with those available for the Oriental primary OMA groups. An explanation for the existing relationships of the animals and plants of New Guinea to both, the Oriental biota along the OMA as well as the Australian biota of Australia itself and the IMA, can be found in the geological history of New Guinea. In the Tertiary central New Guinea formed a part of the IMA, whilst northern New Guinea and the Vogelkop peninsula were parts of the OMA. Both island arcs were – initially widely – separated, until their late Miocene collision, forming New Guinea, allowed an exchange of organisms (see the paragraph on barriers and connections). Well-documented examples of Oriental groups, evolved from ancestors that migrated from the northern New Guinean parts of OMA into central New Guinea, are the cicadas of the *Cosmopsaltria mimica* complex and the *Baeturia nasuta* group. Oriental groups that dispersed from northern to central New Guinea, did not reach the other fragments of the IMA (New Caledonia and New Zealand), probably because these became widely separated from the New Guinean part of the arc at the time of collision. Some Oriental groups reached northern Australia probably during Pleistocene periods of low sea level stands. In the opposite direction Australian IMA groups probably dispersed from central New Guinea to northern New Guinea and other parts of the OMA. The taxon-areacladograms available suggest that the biotic exchange between both arcs remained limited to New Guinea.

The present juxtaposition of Sulawesi to Borneo suggests a close biotic relationship between these islands. Biogeographic data, however, give evidence of relatively recent relationships of Sulawesi to Sundaland, which are even more recent than the relationships of Sulawesi to the Philippines (Holloway 1987; Duffels, in press). The high degree of endemism on the generic level in Sulawesi indicates a long subaerial isolation of this island.

Sulawesi and some adjacent, small islands and archipelagos form one large area of endemism. Areas of endemism of lower rank are found in the northern peninsula (plus the Sangihe Islands), Southwest Sulawesi (plus Selayar), the central part of the island, and East Sulawesi (plus Banggai Archipelago and perhaps Sula). Sulawesi shows an interesting sister group relationship to Maluku, New Guinea and the islands of the OMA in the vicariance of two groups of genera of the cicada subtribe Cosmopsaltriaria (Figure 21.8), but such an old pattern has not been found between Sulawesi and Australia plus the IMA. Since Sulawesi is geological composite with a western arc of Asian origin and an eastern arc of Australian origin, we would expect to find area-relationships with Asia ànd Australia. A possible explanation for the absence of Australian connections comes from a certain consensus among geologists who assume that the western Sulawesi arc has a long and isolated subaerial history, while the eastern arc began to emerge as an island arc during its Mid Miocene collision with western Sulawesi. If the eastern component was really submerged before the collision, no ancient faunal component of Australian origin can be expected in East Sulawesi because it never existed (cf. Holloway 1987: 104).

Acknowledgements

We wish to thank our colleagues Jeremy Holloway, CAB International Institute of Entomology, Pjotr Oosterbroek, Institute for Taxonomic Zoology, Amsterdam and Dick Vane-Wright, British Museum (Nat. Hist.), London for their comments on the manuscript; Dick Langerak for preparation of the figures, and Annelies Stoel for typing.

References

Ackery, P.R., & R.I. Vane-Wright. 1984. Milkweed butterflies. British Museum (Natural History), London.

Audley-Charles, M.G. 1981. Geological history of the region of Wallace's line. In: T.C. Whitmore (ed.), Wallace's line and plate tectonics: 24–35. Clarendon Press, Oxford.

Audley-Charles, M.G. 1987. Dispersal of Gondwanaland: relevance to evolution of the angiosperms. In: T.C. Whitmore (ed.), Biogeographical evolution of the Malay Archipelago: 1–25. Clarendon Press, Oxford.

Balgooy, M.M.J. van. 1987. A plant geographical analysis of Sulawesi. In: T.C. Whitmore (ed.), Biogeographical evolution of the Malay Archipelago: 94–102. Clarendon Press, Oxford.

Ball, I.R. 1976. Nature and formulation of biogeographical hypotheses. Syst. Zool. 24: 407–430.

Boer, A.J. de. 1982. The taxonomy and biogeography of the nasuta group of the genus Baeturia Stål, 1866 (Homoptera, Tibicinidae). Beaufortia 32: 57–78.

Boer, A.J. de. 1986. The taxonomy and biogeography of the conviva group of the genus Baeturia Stål, 1866 (Homoptera, Tibicinidae). Beaufortia 36: 167-182.

Boer, A.J. de. 1989. The taxonomy and biogeography of the bloetei group of the genus Baeturia Stål, 1866 (Homoptera, Tibicinidae). Beaufortia 39: 1–43.
Boulard, M. 1979. Cigales nouvelles des Iles Salomon et des Iles Sous-le-Vent. Notes biogéographiques (Hom. Cicadoidea). Revue Fr. Ent. (n.s.): 49–60.
Boulard, M. 1981. Sur trois Toxopeusella nouvelles des collections du British Museum (Hom. Cicadoidea). Entomologist's Month. Mag. 117: 129–138.
Candolle, A.P. de. 1838. Statistique de la famille Composées. Treuttel & Würtz, Paris.
Coleman, P.J. 1980. Plate tectonics background to biogeographic development in the Southwest Pacific over the last 100 million years. Palaeogeogr. Palaeoclimat. Palaeoecol. 31: 105–121.
Cracraft, J. 1988. Deep-history biogeography: retrieving the historical pattern of evolving continental biotas. Syst. Zool. 37: 221–236.
Crook, K.A.W. 1981. The break-up of the Australian-Antarctic segment of Gondwanaland. In: A. Keast (ed.), Ecological biogeography of Australia: 1–14. Monographiae Biologicae 41.
Duffels, J.P. 1977. A revision of the genus Diceropyga Stål, 1870 (Homoptera, Cicadidae). Monografieën Ned. Ent. Vereen. 8: 1–227.
Duffels, J.P. 1982. Brachylobopyga decorata n. gen., n. sp. from Sulawesi, a new taxon of the subtribe Cosmopsaltriaria (Homoptera, Cicadidae). Ent. Ber., Amst. 42: 156–160.
Duffels, J.P. 1983a. Taxonomy, phylogeny and biogeography of the genus Cosmopsaltria, with remarks on the historic biogeography of the subtribe Cosmopsaltriaria (Homoptera: Cicadidae). Pacific Insects Monogr. 39: 1–127.
Duffels, J.P. 1983b. Distribution patterns of Oriental Cicadoidea (Homoptera) East of Wallace's Line and plate tectonics. In: I.W.B. Thornton (ed.), Symposium on biogeography and plate tectonics in the Pacific. Geojournal 7: 491–498.
Duffels, J.P. 1985. Rhadinopyga n. gen. from the 'Vogelkop' of New Guinea and adjacent islands, a new genus of the subtribe Cosmopsaltriaria (Homoptera, Cicadoidea: Cicadidae). Bijdr. Dierk. 55: 275–279.
Duffels, J.P. 1986. Biogeography of Indopacific Cicadoidea, a tentative recognition of areas of endemism. Cladistics 2: 318–336.
Duffels, J.P. 1988a. The cicadas of Fiji, Samoa and Tonga Islands, their taxonomy and biogeography (Homoptera, Cicadoidea). Entomon. 10: 1–108.
Duffels, J.P. 1988b. Biogeography of the cicadas of the island of Bacan, Maluku, Indonesia, with description of Diceropyga bacanensis n. sp. (Homoptera, Cicadidae). Tijdschr. Ent. 131: 7–12.
Duffels, J.P. 1988c. Cosmopsaltria halmaherae n. sp. endemic to Halmahera, Maluku, Indonesia (Homoptera, Cicadidae). Bijdr. Dierk. 58: 20–24.
Duffels, J.P. 1989. The Sulawesi genus Brachylobopyga (Homoptera: Cicadidae). Tijdschr. Ent. 132: 123–127.
Duffels, J.P. In press. Biogeography of Sulawesi cicadas (Homoptera: Cicadoidea). Proc. Roy. Ent. Soc.
Ewart, A. 1988. Geological history of the Fiji–Tonga–Samoan region of the S.W. Pacific and some palaeogeographic and biogeographic implications. In: J.P. Duffels, The cicadas of Fiji, Samoa and Tonga Islands, their taxonomy and biogeography (Homoptera, Cicadoidea). Entomon. 10: 15–23.
Flenley, J.R. 1979. The equatorial rain forest. Butterworths, London.
Fooden, J. 1969. Taxonomy and evolution of the monkeys of Celebes (Primates: Cercopithecidae). S. Karger, Basel.

George, W., 1981. Wallace and his line. In: T.C. Whitmore (ed.), Wallace's line and plate tectonics: 3–8. Clarendon Press, Oxford.

Hamilton, W. 1979. Tectonics of the Indonesian region. Geol. Surv. Prof. Pap. 1078: i–ix, 1–345.

Hennig, W. 1950. Grundzüge einer Theorie der phylogenetischen Systematik. Deutscher Zentralverlag, Berlin.

Hennig, W. 1966. Phylogenetic systematics. Univ. of Illinois Press, Urbana.

Hennig, W., & D. Schlee. 1978. Abriss der phylogenetischen Systematik. Stuttg. Beitr. Naturk. 319: 1–11.

Holloway, J.D. 1984. Lepidoptera and the Melanesian Arcs. In: F.J. Radovski, P.H. Raven & S.H. Sohmer (eds.), Biogeography of the tropical Pacific: 129–169. Proc. Symp. Association of Systematics Collection, Lawrence, U.S.A.

Holloway, J.D. 1987. Lepidoptera patterns involving Sulawesi: What do they indicate of past geography. In: T.C. Whitmore (ed.), Biogeographical evolution of the Malay archipelago: 103–118. Clarendon Press, Oxford.

Holloway, J.D., & N. Jardine. 1968. Two approaches to zoogeography: a study based on the distribution of butterflies, birds and bats in the Indo-Australian area. Proc. Linn. Soc. Lond. 179: 153–188.

Hope, G.S., & J.A. Peterson. 1975. Glaciation and vegetation in the high New Guinea mountains. In: R.P. Suggate & M.M. Cresswell (eds.), Quaternary Studies: 155–162. Royal Society New Zealand, Wellington.

Humphries, C.J., & L.R. Parenti. 1986. Cladistic biogeography. Clarendon Press, Oxford.

Jong, M.R. de. 1985. Taxonomy and biogeography of Oriental Prasiini 1: The genus Prasia Stål, 1863 (Homoptera, Tibicinidae). Tijdschr. Ent. 128: 165–191.

Jong, M.R. de. 1986. Taxonomy and biogeography of Oriental Prasiini 2: The foliata group of the genus Lembeja Distant, 1892 (Homoptera, Tibicinidae). Tijdschr. Ent. 129: 141–180.

Jong, M.R. de. 1987. Taxonomy and biogeography of Oriental Prasiini 3: The fatiloqua and parvula groups of the genus Lembeja Distant, 1892 (Homoptera, Tibicinidae). Tijdschr. Ent. 130: 177–209.

Kibby, G. 1985. A review of Thopeutica, a subgenus of Cicindela (Coleoptera, Cicindelidae), with a key to species and descriptions of two new taxa. J. Nat. Hist. 19: 21–36.

Kitching, I.J., R.I. Vane-Wright & P.R. Ackery. 1987. The cladistics of Ideas (Lepidoptera, Danainae). Cladistics 3: 14–34.

McDowall, R.M. 1980. Freshwater fishes and plate tectonics in the Southwest Pacific. Palaeogeogr. Palaeoclimat. Palaeoecol. 31: 337-352.

Møller Andersen, N. 1989a. The coral bugs, genus Halovelia Bergroth (Hemiptera, Veliidae). I. History, classification, and taxonomy of species except the H. malaya-group. Ent. Scand. 20: 75–120.

Møller Andersen, N. 1989b. The coral bugs, genus Halovelia Bergroth (Hemiptera, Veliidae). II. Taxonomy of the H. malaya-group, cladistics, ecology, biology, and biogeography. Ent. Scand. 20: 179–227.

Morley, R.J., & J.R. Flenley. 1987. Late Caenozoic vegetational and environmental changes in the Malay Archipelago. In: T.C. Whitmore (ed.), Biogeographical evolution of the Malay Archipelago: 50–59.

Musser, G.G. 1987. The mammals of Sulawesi. In: T.C. Whitmore (ed.) Biogeographical evolution of the Malay archipelago: 73–93. Clarendon Press, Oxford.

Nelson, G., & N. Platnick. 1981. Systematics and biogeography. Cladistics and vicariance. Columbia Univ. Press, New York.

Patterson, C. 1981. Methods of paleobiogeography. In: G. Nelson & D.E. Rosen (eds.), Vicariance biogeography: A critique: 446–497. Columbia Univ. Press, New York.
Platnick, N.I., & G. Nelson. 1978. A method of analysis for historical biogeography. Syst. Zool. 27: 1–15.
Ross, H.H. 1956. Evolution and classification of the mountain caddisflies. Univ. of Illinois Press, Urbana.
Saigusa, T., A. Nakanishi, H. Shima & O. Yata. 1982. Phylogeny and geographical distribution of the swallow-tail subgenus Graphium (Lepidoptera: Papilionidae). Ent. Gen. 8: 59–69.
Schuh, R.T., & G.M. Stonedahl. 1986. Historical biogeography in the Indo-Pacific: a cladistic approach. Cladistics 2: 337–355.
Sclater, P.L. 1858. On the general geographical distribution of the members of the class Aves. J. Proc. Linn. Soc., Zool. 2: 130–145.
Simpson, G.G. 1977. Too many lines; the limits of the Oriental and Australian zoogeographic regions. Proc. Amer. Phil. Soc. 121: 107–120.
Smith, A.G., A.M. Hurley & J.C. Briden. 1981. Phanerozoic paleocontinental world maps. Cambridge Univ. Press, Cambridge.
Steenis, C.G.G.J. van. 1950. Delimination of Malaysia and its main plant geographical divisions. Flora Malesiana ser. I, 1: lii-lxxv.
Thornton, I.W. B. 1980. Plate tectonics and the distribution of the insectfamily Philotarsidae (Order Psocoptera) in the Southwest Pacific. Palaeogeogr. Palaeoclimat. Palaeoecol. 31: 251–266.
Vane-Wright, R.I. In press. The Philippines – Key to the biogeography of Wallacea? Proc. Roy. Ent. Soc.
Vecht, J. van der. 1953. The carpenter bees (Xylocopa Latr.) of Celebes, with notes on some other Indonesian Xylocopa species. Idea 9: 57–69.
Wallace, A.R. 1860. On the zoological geography of the Malay Archipelago. J. Proc. Linn. Soc., Zool. 4: 172–184.
Wallace, A.R. 1863. On the physical geography of the Malay Archipelago. J. Roy. Geogr. Soc. 33: 217–234.
Wallace, A.R. 1867. On the Pieridae of the Indian and Australian regions. Trans. Ent. Soc., Lond. (3) 4: 301–415.
Wallace, A.R. 1876. The geographical distribution of animals. MacMillan, London.
Wallace, A.R. 1880. Island life. Macmillan, London.
Wallace, A.R. 1910. The world of life. Chapman & Hall, London.
Whitten, A.J., M. Mustafa & G.S. Henderson. 1988. The ecology of Sulawesi. Gadjah Mada Univ. Press, Yogyakarta.
Wiley, E.O. 1988a. Vicariance biogeography. Ann. Rev. Ecol. Syst. 19: 513–542.
Wiley, E.O. 1988b. Parsimony analysis and vicariance biogeography. Syst. Zool. 37: 271–290.

Index to animal genera and species

22 Biogeographical relationships of Australia and Malesia: Loranthaceae as a model

BRYAN A. BARLOW

Summary

Major interdigitation of an autochthonous Australian and a Malesian flora occurred in New Guinea and northern Australia in late Tertiary and Quaternary time, when the isolation of the two floras was weakened by geophysical and climatic events. In Australia this interaction had been preceded by early differentiation of warm and cool and of moist and dry rain forest elements from a mesic pan-Gondwanan flora, and of open forest and heathland vegetation in response to edaphic evolution. However, there is a growing perception that even in the Malesian flora there are ancient links to an early Gondwanan flora, perhaps involving India and parts of southeast Asia as Gondwanan fragments. Throughout these changes many primitive Gondwanan elements have survived in durable refugia in northeastern and eastern Australia. Thus in Australia there has developed an intricate pattern of humid forest communities which differ in composition, age, endemism, relictuality and richness. In the mistletoe family Loranthaceae, which is a classic austral group, there are various biogeographic patterns which are consistent with the suggested episodes in the development of the Australasian-Malesian flora. These are described as examples to illustrate the possible vegetation history.

Introduction

This contribution to the historical biogeography of the Malesian region is offered with two objectives. The first is to apply current views on Australian biogeography to understanding the floristic history of Malesia, because it is clear that the two are intimately linked. The Australian plate, and its associated Gondwanan fragments, should be no less a focus in such analyses than the Laurasian land mass. The second objective is to examine one family, Loranthaceae, now reasonably well known biogeographically, in order to show how its patterns of distribution and relationship fit the current models. A brief comparison is made with another mistletoe family, Viscaceae, which superficially appears to have had a different history.

Historical relationship between the Malesian and Australian floras

A summary of those aspects of the origin of the Australian flora which are relevant to Malesia will of necessity involve some of what is now becoming

P. Baas et al. (eds.), The Plant Diversity of Malesia, 273–292.

classical Gondwanan biogeography. Because Australian biogeographers work with the biota of an island continent, this subject is perhaps of more fundamental importance to them than to most others.

Australia as a Gondwanan fragment: floristic implications

Although the northward separation of Australia from Antarctica commenced about 125 million years ago (Ma), movement remained slow (4.5 mm/yr) until 55 Ma, when the rate increased to 60–70 mm/yr (Crook & Taylor 1985; Wellman & McDougall 1974). Narrow land connections between Australia, Antarctica and South America may have persisted until c. 40 Ma (Kemp 1978), and when circumpolar oceanic circulation was initiated, heat transport from equator to pole was reduced and temperature gradients increased. Thus at the same time as Australia was moving away from Antarctica the air circulation was becoming cooler and drier, and arid conditions started to overtake Australia from the south about 30 Ma (Bowler 1982). The unique scleromorphic and arid-adapted flora of Australia's open forests and heathlands developed in response to these cycles of cooling and aridity, and to associated edaphic evolution (Barlow 1981). The rain forests of Australia, now very restricted geographically, include within them the relicts of the original Gondwanan forest flora and the ancestral stocks from which this unique Australian flora evolved. Thus within the Australian flora it is usual to recognise a *Gondwanan element* comprising two subelements, a relict one and a derived *autochthonous* one (Nelson 1981; Barlow 1981).

The Coral Sea started to open about 80 Ma, initiating the separation of New Guinea from eastern Australia. The southern part of New Guinea is a continuation of the Australian plate, and most of the New Guinean uplands are very young. In addition to the emplacement of exotic terranes there has been recent vertical uplift along the axis of the island at a rate of c. 2 mm/yr (Chappell 1974; Pain & Ollier 1984). Thus whilst New Guinea is clearly part of the Australian plate with respect to floristic history, the extensive everwet uplands of the present day did not exist in this form through most of the period of angiosperm history of the region.

The intrusive Malesian component in the Australian/Papuasian flora

The Miocene collision between the northward-moving Australian plate and the westward-moving volcanic arcs of the Sunda region is clearly of major significance in understanding the origins and relationships of the flora of tropical Australia (Barlow & Hyland 1988). The likely scenario has been described and developed by many authors, notably by Audley-Charles (1981, 1987) and Audley-Charles *et al.* (1972, 1981, 1988). The northwestern margin of the

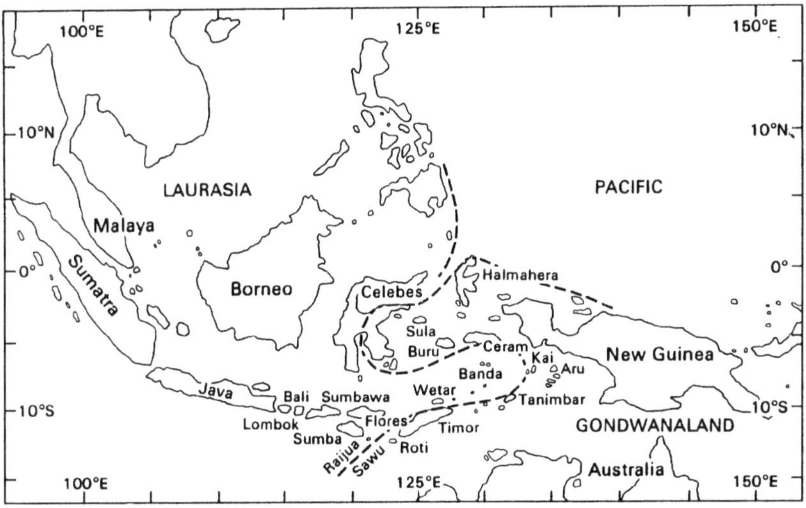

Figure 22.1 — Relationships of Australian and Sundaland plate areas at the present day (from Barlow & Hyland 1988).

Australian plate as it existed at the time of contact is now represented by the Outer Banda Arc, extending as a great deformation from Timor to Buru, the southeastern and eastern arms of Sulawesi, the Moluccas and the Vogelkop of New Guinea (Figure 22.1). Although there is increasing evidence that the blocks on the Asian side of the collision boundary are also Gondwanan fragments (Audley-Charles 1987, Audley-Charles *et al.* 1988), the Miocene contact ended a long period of isolation of differentiated Australian and Malesian floras.

Land surfaces in the contact zone increased as the convergence continued. Although the collision began in the mid-Miocene, it was probably not before latest Miocene or early Pliocene that a land migration route from Sulawesi to Australia via New Guinea was established (Audley-Charles 1981), and a strong floristic contact between the Australian and Malesian regions ensued. Prior to this, of course, a filtered exchange of taxa of higher dispersibility would have occurred (McKenna 1972). By the late Pliocene, lowered sea levels would have exposed extensive land on both sides of the contact boundary. Whilst the Makassar Strait has long existed as a seaway, and is the most robust part of Wallace's Line, Borneo and Sulawesi would have been almost continuous at its southern end (Audley-Charles 1981). By the middle Pleistocene the Lesser Sunda Islands would have been connected to Asia by continuous land (at least as far as Timor), and would have been separated from mainland Australia by only a short sea distance.

The second element usually recognised in the Australian flora is the *intrusive* one (Nelson 1981; Barlow 1981), comprising plants which have entered Australia subsequent to its separation from Gondwanaland. It is a composite of several different subelements, one of which is the *tropical* one, made up of Malesian plants of post-Miocene inception. An *Indogondwanan* subelement was defined by Barlow (1981) to accommodate plants of ancient Gondwanan derivation which have reached Australia as part of the post-Miocene intrusion. Whilst this subelement has usually been linked with the history of the Indian subcontinent as a Gondwanan fragment (Specht 1981), we now know that other parts of southeast Asia may have similar geological histories (Audley-Charles 1987, Audley-Charles *et al.* 1988). The tropical and Indogondwanan subelements may therefore be identical (Barlow & Hyland 1988), and are discussed further below.

In the Malesian region the two great blocks of tropical rain forest are separated by a great corridor of seasonally dry climates covering much of the Philippines, Sulawesi, Moluccas, Lesser Sunda and Banda Arc islands, part of Java and part of southern New Guinea (Whitmore 1981). In this corridor monsoon forest forms a mosaic with rain forest, the latter being confined to gallery and other wetter sites. The Australian occurrence of monsoon forests is an extension of this pattern, and northern Australia and some seasonally drier parts of northeastern Australia are obviously part of the same corridor.

By the late Pliocene progressive increase in seasonality of the climate would have resulted in expansion of monsoon forests at the expense of closed mesic forests. Thus at a time when the potential for migration over continuous land in the Australian/Malesian region was greatest, the predominant vegetation in the corridor may have been the monsoon forest type (Barlow & Hyland 1988). As well as contributing directly to the monsoon forest flora of northern Australia, this intrusive stock has also been significant in the evolution of an Australian arid zone flora (Burbidge 1960; Barlow 1981).

The monsoon corridor may have provided some geographic isolation between the western and eastern Malesian rain forest blocks. The eastern Malesian rain forest flora which occupied New Guinea may have retained some integrity because of this isolation. However its impact on the stable mesic forests of northeastern Australia was weaker than in New Guinea, and these have retained many relictual elements, including the highest concentration of primitive angiosperm families in the world (see below).

Australian rain forest stability in the last 100,000 years

The uplands of eastern Australia have apparently existed at least since the beginning of the Tertiary period (Ollier 1982, 1986), and have continuously provided habitats favourable for survival of the ancient Gondwanan rain forests

(Barlow & Hyland 1988). The other old uplands in Australia have all been overcome by major climatic cold and arid cycles, and the present day uplands of New Guinea and especially New Zealand are too young to have become major refugia. Thus the old eastern uplands of the Australian plate may be the only large part of the entire Australasian/Pacific region where Gondwanan warm mesic forests have persisted continuously since Cretaceous times. However, these rain forests have obviously been influenced by floristic exchange, and a young intrusive Malesian subelement is easily recognisable in the flora of the Australasian wet tropics.

The integrated Gondwanan-intrusive rain forests in Australia have been moulded in a dynamic pattern of shifting vegetation mosaics in the Quaternary (Webb & Tracey 1981; Kershaw 1981; Barlow & Hyland 1988). Boundaries between rain forests and open forests have advanced and retreated, and rain forest areas have been fragmented and coalesced. Of special interest in this context is the extent to which recent floristic exchange may have occurred between

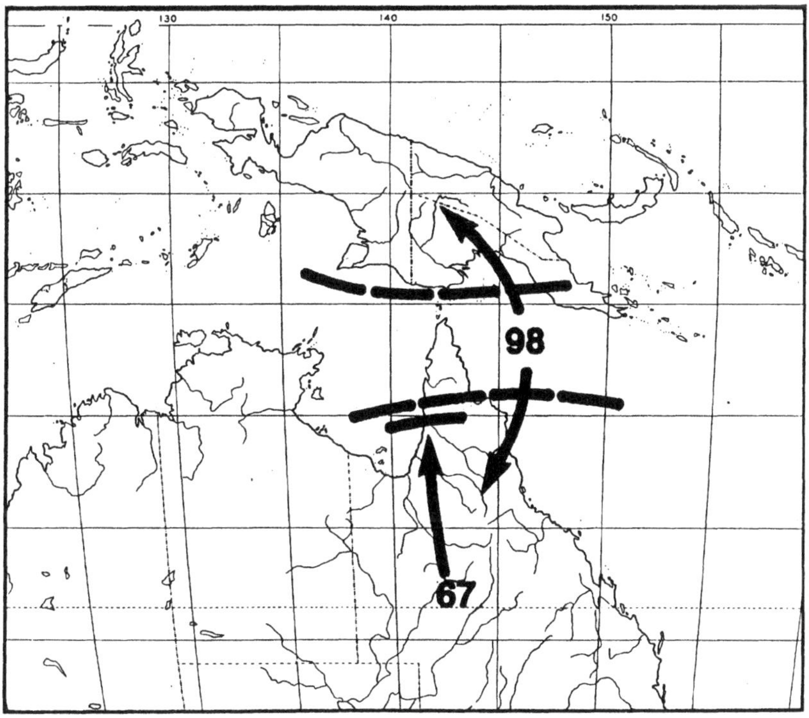

Figure 22.2 — Generic limits and disjunctions in Australian rain forest genera in Cape York Peninsula, Queensland. See text for explanation.

New Guinean and Australian rain forests during phases of rain forest expansion. Whilst the two lands have been broadly connected above sea level through much of Quaternary time, the low-lying Torresian land surface may have been mostly unsuitable for humid forest colonisation. Because conditions favouring rain forests in the Quaternary were seldom better than they are at present, the Torres Strait region may have been an effective ecological barrier to floristic exchange (Walker 1972).

Closed forests may have extended almost continuously along a narrow Torresian isthmus from New Guinea to Queensland 8,000 years ago (Nix & Kalma 1972). More protracted contact may have occurred around 100,000 years ago when tropical forests had a more durable tenure. Conversely during periods of maximum cooling and aridity the mesic forests may have disappeared almost completely from the lands of low relief in the Torres Strait region. Such an event may have occurred most recently 20,000–17,000 years ago (Nix & Kalma 1972).

The present rain forests of northern Cape York Peninsula thus represent the limited results of recolonisation, from both New Guinean and Queensland refugia, which probably reached maximum momentum about 8,000 years ago. These forests are relatively depauperate, and have low endemism (Barlow &

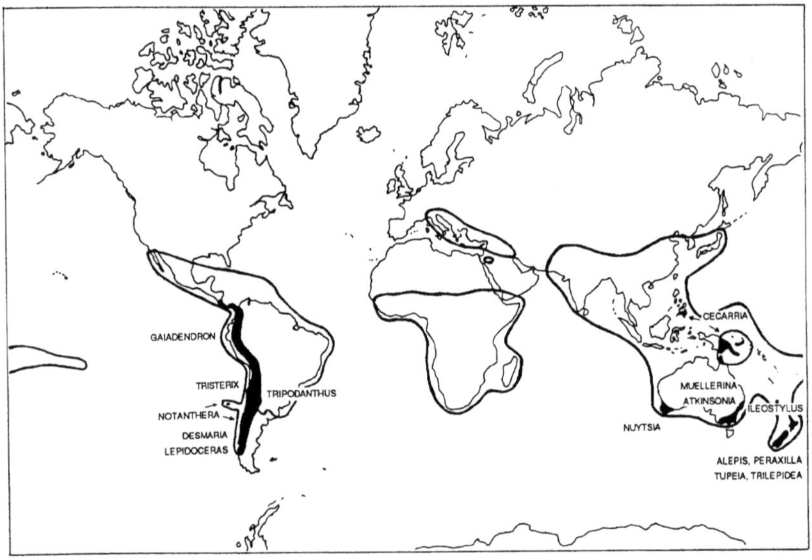

Figure 22.3 — World distribution of Loranthaceae. Occurrence of relictual genera is shown in black (from Barlow 1983).

Hyland 1988). Of 165 rain forest genera which occur in north Queensland but not in northern Cape York Peninsula, about 40% are Gondwanan, reaching their northern range limits in north Queensland. Almost all of the remainder are further distributed in New Guinea and/or the Malesian region, and their absences from the northern area simply represent disjunctions in their distributions (Figure 22.2). This is consistent with the idea that the northern region has undergone more pronounced floristic perturbations in the Recent than the areas immediately to the north and south, which are more stable and more refugial in character. Critical systematic studies support the view that Recent colonisation of the northern Cape York Peninsula has involved New Guinea as a source area (Whiffin & Hyland 1986).

Case study - Loranthaceae

Global biogeography

Although fossil evidence is lacking, the biogeographic history of the Loranthaceae is clearly indicated by morphological, systematic and karyological data (Barlow & Wiens 1971; Barlow 1983; 1987). Dispersibility of loranths is low, and their strongly continental distribution allows high confidence in the correlation of phylogeny and migration. The most primitive extant genera are all small in size, with relatively restricted areas, confined almost exclusively to Gondwanan land surfaces, and mostly in temperate habitats (Figure 22.3). They appear to be relictual endemics, and their scattered occurrence in the southern lands strongly suggests that the Loranthaceae are an old southern family.

The cytogeography of the family is a good indicator of geographic history (Barlow 1983). There is considerable genomic stability, and cytogeographic data are useful indicators of change at higher taxonomic levels. The primary basic number, $x = 12$, occurs in several of the small, primitive relictual genera as well as in a group of genera distributed from southeast Asia to Australia and the South Pacific. The basic numbers $x = 11$ and 10 occur only in a few of the small southern relictual genera. The basic number $x = 9$ occurs in the relictual *Cecarria* and in all of the remaining derived and advanced genera of the Old World. The basic number $x = 8$ occurs in the relictual *Tripodanthus* and in all of the derived and advanced genera of the New World. Probably an ancestral stock with $x = 12$ was widespread in the southern lands, and dysploid reduction to $x = 9$ and 8 has preceded massive independent evolutionary radiations in the tropics of the Old and New Worlds respectively. The occurrence of all basic numbers among the relictual southern genera strongly suggests that this dysploid reduction occurred before the fragmentation of Gondwanaland (Figure 22.4). Not only is the family very old (despite its specialised growth habit), but its karyological differentiation occurred very early in its history (Barlow 1987).

The loranths presumably originated in the mesic, warm to mild, closed forests of Gondwanaland, and the parasitic habit presumably did not arise in response to water stress but to competition for nutrients in complex ecosystems. Four main lines of evolution occurred independently in the Afro-Indian, Indian-Indosinian, Australasian-Papuasian and South American regions of Gondwanaland, from genomically different basal stocks, and were isolated by the fragmentation of the supercontinent.

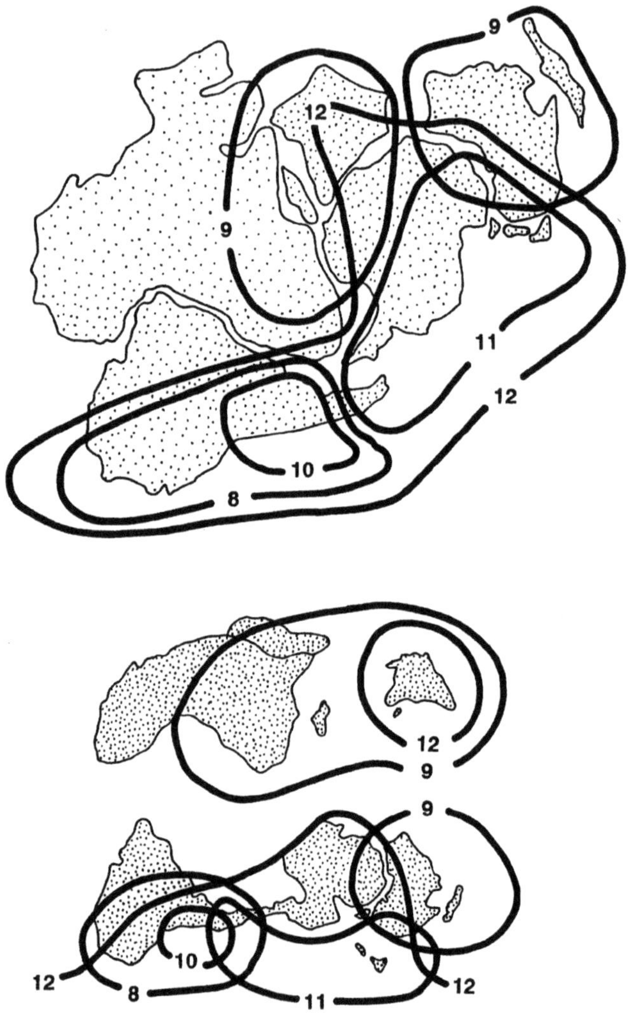

Figure 22.4 — A model for karyological differentiation of Loranthaceae in Gondwanaland prior to fragmentation, and the consequent geographic separation of different phyletic stocks.

Loranths in Australia and Malesia

Analysis of the distribution, centres of species richness and variation in genera of Loranthaceae occurring in the Malesian/Australian region reveals different patterns highly relevant to interpretation of the complex biogeography. These can be considered in the context of the elements of the Australian flora described above, and are summarised in Table 22.1.

Table 22.1 — Status of genera of Loranthaceae with respect to the elements of the Australian flora.

Element Subelement (component)	Genus	Chrom. No. x	No. of spp.	Distribution
Gondwanan				
relict	*Cecarria*	9	1	Papuasia/Philippines
autochthonous (old austral)	*Nuytsia*	12	1	W Australia
	Atkinsonia	12	1	E Australia
	Muellerina	11	4	E Australia
(Tertiary tropical)	*Dactyliophora*	9	4	Australia/Papuasia
	Amyema	9	c. 80	Australia/Papuasia/Malesia
	Diplatia	9	3	Australia
(young Papuasian endemics)	*Distrianthes*		1	New Guinea
	Papuanthes	9	1	New Guinea
	Tetradyas		1	New Guinea
	Sogerianthe	9	4	New Guinea/Solomons
Intrusive				
Indogondwanan (old Laurasian/ Malesian)	*Macrosolen*	12	c. 40	India-New Guinea
	Lepeostegeres	12	c. 15	Borneo-New Guinea
	Decaisnina	12	c. 30	Malesia-Pacific
	Dendrophthoe	9	c. 40	Asia-Australia
(young Papuasian/ Australian endemics)	*Amylotheca*	12	4	New Guinea/Australia/Pacific
	Lysiana	12	8	Australia
	Benthamina	9	1	Australia
(young Malesian endemics)	*Cyne*		1	Philippines
	Lampas		1	Borneo
	Trithecanthera		1	Borneo
	Kingella		1	Malay Peninsula
	Loxanthera		1	Sumatra-Borneo
	Lepidaria		c. 12	Sumatra-Philippines

Gondwanan element

The primitive genus *Cecarria* (n = 9) occurs in the rain forests of the Papuasian region, from the Philippines to north Queensland (Figure 22.3), thus sharing its area with several primitive angiosperm families. Although it has apparently crossed Wallace's Line into the Philippines, it is probably of ancient origin on the Australian plate and thus belongs with the relict subelement of the Gondwanan element of the Australian flora.

There are also primitive, relictual small genera with small areas in southern Australia. These include the monotypic root parasitic genera *Nuytsia* and *Atkinsonia* (x = 12), and also *Muellerina,* with x = 11 and many plesiomorphic character states, in the open forests of southeastern Australia. In view of their adaptation to open forest habitats, these genera are clearly part of the autochthonous subelement.

There are several other genera of Loranthaceae which are either endemic in Australia or are strongly centred in the Australian/Papuasian region (Figure 22.5), and which have almost certainly originated there. They can therefore be identified as part of the Gondwanan element. Like *Cecarria,* they all have x = 9.

The largest of these genera is *Amyema,* with c. 80 species, of which c. 60 occur in Australia and New Guinea. The genus has extended its range westwards across Wallace's Line after the Miocene collision event, but has not

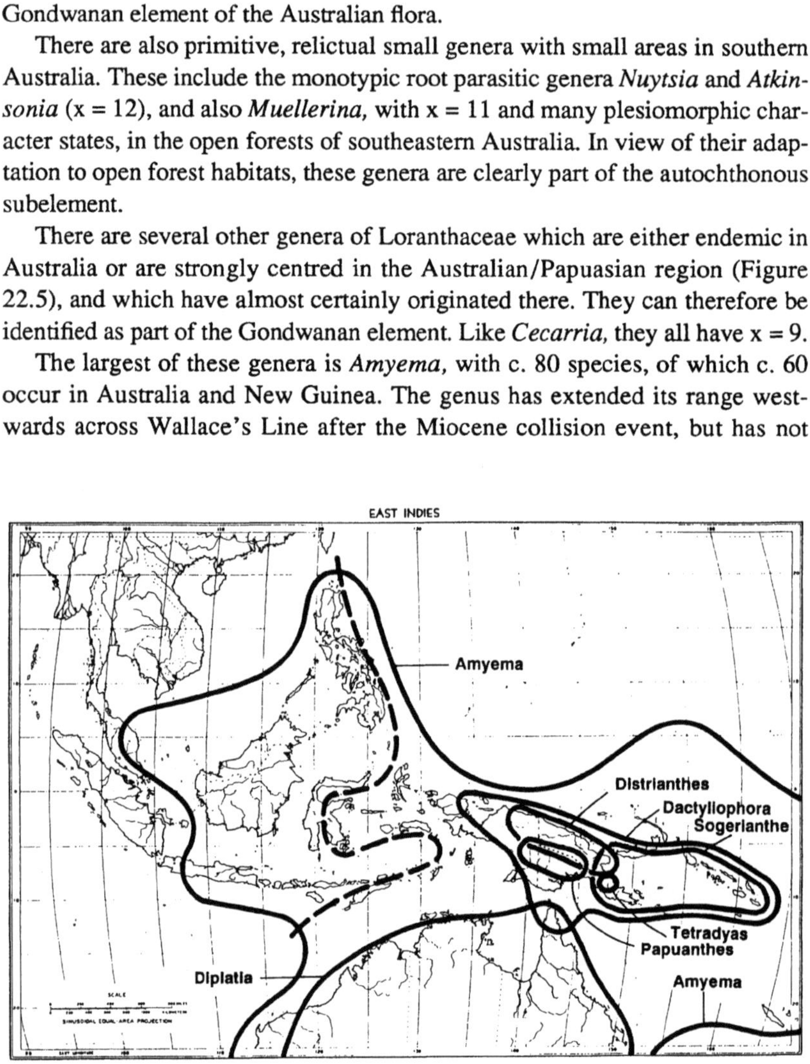

Figure 22.5 — Distribution of autochthonous Gondwanan Australian genera of Loranthaceae.

occupied extensive territory on the Asian mainland. In Australia it is well represented in all communities from closed mesic forests to arid shrub steppe. However, there is greater morphological diversity in New Guinea, including a greater retention of plesiomorphic characters. The origin of the genus was probably in mesic forests at the northern margin of the Australian plate, followed by radiation into areas of seasonal, arid and subalpine climates as they appeared. This likelihood is supported by the occurrence in New Guinea and north Queensland of *Dactyliophora* (4 spp.), a very closely related small genus in mesic forests which is possibly relictual and near the stem of the *Amyema* stock. In contrast, the small genus *Diplatia* (3 spp.) is clearly derived from *Amyema,* and occurs widely in open seasonal communities in inland Australia. The relatively few species of *Amyema* which occur to the west of Wallace's Line are presumably derived from stocks which reached the area from a New Guinean source.

Young Papuasian endemics. — Some other genera of loranths highlight the importance of not overlooking New Guinea as part of the Australian phytogeographic region. There are a few small, morphologically specialised genera in New Guinea which are probably no older than the vertical uplift which has followed the emplacement of exotic terranes on the New Guinean margin of the Australian plate. Whilst perhaps younger than the Miocene collision, and not in fact represented in present mainland Australia, they are Gondwanan in derivation and belong with the autochthonous subelement of the Australian flora. They include *Distrianthes* (1 sp.), *Papuanthes* (1 sp.), *Sogerianthe* (4 spp.) and the poorly known *Tetradyas* (1 sp.). Those genera known karyologically have x = 9 and are phyletically close to *Amyema.*

Loranthaceae as Indogondwanan intrusives

There are a few loranth genera which are centred in southeast Asia or Malesia and which have only a limited representation in Papuasia and/or the Australian tropics (Figure 22.6). *Lepeostegeres* (c. 15 spp.; x = 12) is strongly centred in Malesia, having most of its species in Borneo and the Philippines. The genus is represented in New Guinea by only a single species, and thus presents a simple example of the tropical Malesian intrusive element in the Australian/New Guinean flora.

Two other loranth genera are centred in southeast Asia and are represented by only one or a few species in Papuasia and/or the Australian tropics. *Macrosolen* (x = 12) comprises c. 40 spp., mostly in southeast Asia and western Malesia, and only one of these reaches New Guinea (Figure 22.6). Similarly the richness and diversity in *Dendrophthoe* (x = 9) attenuates to the east, and there are only a few species in New Guinea and northern and eastern Australia (Figure 22.7). Like *Lepeostegeres,* these genera are clearly intrusives into the Australian

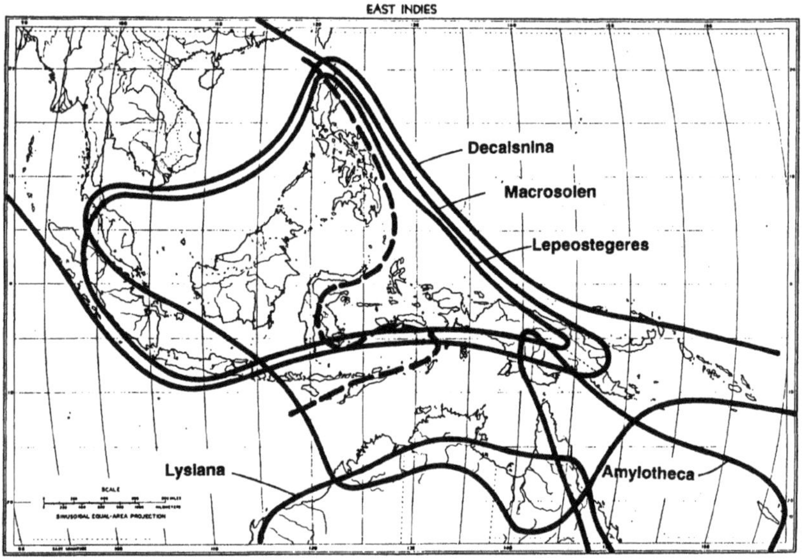

Figure 22.6 — Distribution of genera of Loranthaceae with x = 12 which are intrusive into the Malesian/Australian region.

region, having attained limited extensions eastwards across Wallace's Line following the Miocene collision.

The patterns in *Macrosolen* and *Dendrophthoe,* which are surely genera of pre-Miocene establishment in Asia, raise a specific question about their history. In view of the apparent Gondwanan origin of the family, Loranthaceae must have dispersed from Gondwanaland to Laurasia prior to the Miocene collision. There are phyletic links between African, Madagascan and Asian loranths (Barlow 1983), and the likely explanation involves the geophysical history of India and parts of southeast Asia as Gondwanan fragments. During the time India was rafting northwards towards collision with Asia in the middle Eocene, its climates may have been continuously equable. India may therefore have carried a differentiated flora of Gondwanan origin to Laurasia at the same time as another subset of the Gondwanan flora was differentiating in the newly isolated Australian fragment. Similarly, the earlier separation of what is now Burma and Malaya from the northern margin of Gondwanaland (Audley-Charles *et al.* 1988) may have been of major significance in delivering angiosperms from their Gondwanan source to Laurasia. Indeed, this event can be reconciled with the conclusions of Takhtajan (1969) about the site of origin of the angiosperms (Martin & Dowd 1989). *Macrosolen, Lepeostegeres* and *Dendrophthoe* thus appear to be excellent examples of an indogondwanan subelement in the intrusive floristic element which reached the Australian plate after the Miocene collision. Other loranthaceous genera such as *Elytranthe, Helixanthera, Barathranthus, Taxillus* and *Scurrula* have similar histories in the Asian region, but have not penetrated Malesia as far as Australia.

Recognising this biogeographic pattern may be the key to understanding the history of three Australian genera which have x = 12. *Decaisnina* (c. 30 spp.) is well represented on both sides of Wallace's Line, and is therefore very interesting biogeographically (Figure 22.6). Most of its diversity is found to the east of Wallace's Line, from Sulawesi to northern Australia and Pacific islands. To the west of Wallace's Line it is well represented in the Philippines, where significant speciation has occurred. The genus is extremely difficult taxonomically at the species level. Like *Amyema*, it may have originated in mesic forests at the northern margin of the Australian plate, and later radiated into areas of monsoonal climate as they appeared, but unlike *Amyema* it may be younger than the Miocene contact of the Australian and Sunda floras. Because of the strong

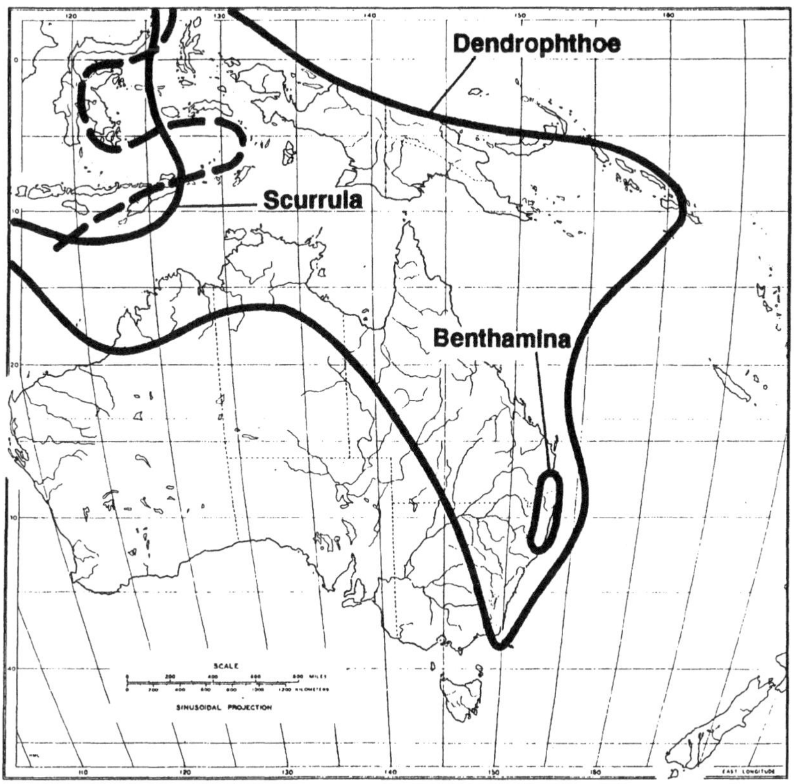

Figure 22.7 — Distribution of genera of Loranthaceae with x = 9 (or 8) which are intrusive into the Malesian/Australian region.

development of genera with x = 12 in the Indomalesian region, it is probable that *Decaisnina* is derived from an intrusive stock even if its origin was on the Australian plate.

This conclusion can be extended to the other two genera in this group, which are more specialised than *Decaisnina* but probably derived from the same ancestral stock. *Amylotheca* (4 spp.) occurs in New Guinea, Australia and New Caledonia, mostly in rain forests (Figure 22.6). *Lysiana* (8 spp.) is widespread in open forest communities in Australia, but the least specialised species are tropical and subtropical, again indicating an origin near the northern margin of the Australian plate (Figure 22.6).

In previous biogeographical analyses of Loranthaceae (Barlow 1983, 1987), I concluded that the last three genera were evidence of a radiation of a stock with x = 12 from the Australian fragment of Gondwanaland into the Indomalesian region. If they are now correctly interpreted as local derivatives of an intrusive stock, it follows that there is no extant evidence of an autochthonous group of mistletoes with this genome on the Australian plate other than the relictual root parasites *Nuytsia* and *Atkinsonia*. The entire phyletic line of stem-parasitic loranths with x = 12 is therefore Asian in derivation, and may have been endemic to the Indian and/or Indosinian fragments of Gondwanaland (Figure 22.4). In this respect the loranth flora of the Indian-Indosinian part of Gondwanaland represents a fourth phyletic line of Gondwanan Loranthaceae, different from the Afro-Indian, Australasian and South American lines previously recognised.

Perhaps the most striking occurrence of a likely Indogondwanan intrusive in Australia is *Benthamina* (x = 9). This monotypic genus occurs in closed forests in a small area in eastern Australia (Figure 22.7), but appears to belong to the Asian-centred *Taxillus* alliance. The extent of the disjunction is shown by the geographically nearest related genus, *Scurrula,* which reaches Java, so that *Benthamina,* if correctly placed phyletically, is the only genus in its group to occur to the east of Wallace's Line.

Malesian endemics

An outline of the floristic history of loranths in the region is completed by mention of the few remaining endemic Malesian loranth genera not represented to the east of Wallace's Line. There are only four genera, all apparently monotypic, which can be considered to be young Malesian endemics by virtue of their morphological specialisation and small areas. They include *Cyne* (Philippines), *Lampas* and *Trithecanthera* (Borneo) and *Kingella* (Malay Peninsula). Genera with wider ranges, and therefore perhaps older, include *Loxanthera* (1 sp.) and *Lepidaria* (c. 12 spp.). These two genera have similar ranges including Sumatra, Malay Peninsula and Borneo, with *Lepidaria* also in the Philippines.

If the occurrence of apparently young endemic genera of Loranthaceae in western Malesia is an indicator of the extent of local evolutionary diversification, then it has been relatively limited. It is similar in magnitude to that which has occurred in the northern part of the Australian plate on the other side of Wallace's Line. There has clearly been much speciation since the present geography and climate have been established, but much of it has occurred in older genera of Asian and Australian origins which have penetrated the region as the land configurations have allowed.

Loranths and Wallace's Line

Different lineages of Loranthaceae were clearly well established on both sides of Wallace's Line at the time of the Miocene contact. As shown above, distributional patterns in the Loranthaceae with respect to Wallace's Line range from localised species transgressions to differentiation and diversification of new genera from intrusive stocks which have crossed the Line. This suggests that there has been prolonged exchange, perhaps continuously since migration first became possible after the Miocene contact. In view of their biology as parasites, this is most probable. Loranths are not directly dependent on soil or rainfall, and in the region under discussion are mostly of low host specificity. Because their habitat requirement is essentially a susceptible host tree, new habitats may have been readily available and dispersal of loranths may have quickly followed the Miocene contact.

When assessed on the basis of numbers of genera or derived species involved, extensions of Loranthaceae in the two directions across Wallace's Line have been approximately equal. *Cecarria, Amyema* and *Decaisnina* have crossed from the east, and *Macrosolen, Lepeostegeres, Dendrophthoe* and *Helixanthera* from the west, along with the progenitors of *Amylotheca, Lysiana* and *Benthamina*. About 30–40 derived species are involved in each case.

When assessed on the basis of territorial expansion, however, the intrusive loranth element which entered the Australasian region has clearly been more successful. It is best represented in the monsoon woodland belt. Five of the six Australian species of *Dendrophthoe* are found in this zone, as well as four of the six Australian species of *Decaisnina* and half of the species of *Lysiana*. Intrusive loranths have undergone significant speciation in Australia, although their total impact on the older autochthonous loranths has been limited.

Loranths as Recent recolonisers in the Torres Strait area

Loranths provide several examples of a distribution pattern which is becoming more common as botanical exploration in the northern part of Cape York Peninsula continues. Some common and widespread mistletoes of the lowland rain forests of New Guinea have been discovered with apparently very local occurrences in the northern Peninsula area. Notable are *Amyema friesianum, A. seemenianum, Dactyliophora novaeguineae* and *Decaisnina hollrungii* (Barlow 1984), which present a different mistletoe facies in the area from their Australian counterparts. They indicate recolonisation of the area by New Guinean closed forest elements over the last few thousand years (Barlow & Hyland 1988). Even more notable is the occurrence of the relictual, monotypic *Cecarria obtusifolia* in only one known locality in the northern Cape area in addition to its several disjunct occurrences in Papuasia and the Philippines. However, the status of this species as a recent waif or an old relict is less certain, because although it occurs in a botanical region with a young, depauperate rain forest flora virtually devoid of endemics or relicts, it is localised in the most substantial and diverse rain forest block in the area.

Amyema mackayense and *Dendrophthoe glabrescens* are widespread in Australia, and also have very local occurrences on the northern shores of Torres Strait. They probably indicate very recent exchange in the opposite direction, even though they are parasites of mangroves and thus limited in their distribution by the shorelines of the recent past.

Other loranths may provide examples of earlier Quaternary exchanges between Australia and New Guinea. The extremely polymorphic *Dendrophthoe falcata* is widespread from Asia to New Guinea, but only has a local occurrence in the rain forests of north Queensland, and may not be genetically close to its Australian congeners of the monsoon belt (see above). *Amyema queenslandicum* also has a small area in north Queensland, and appears to be most closely related to the widespread *A. strongylophyllum* of New Guinea. *Amyema conspicuum* is widespread in Australia but has a subspecies with a very local occurrence in northern Papua New Guinea.

The distributions of these loranths indicate colonisation in both directions between New Guinea and north Queensland in Quaternary times. They strongly confirm the conclusion of Walker (1972) that the filter which regulates such exchanges is not the existence of Torres Strait *per se* as a water barrier, but the ecological opportunities in northern Cape York Peninsula provided by the climatic cycles of the Quaternary.

Case study - Viscaceae

Global biogeography

Fossil data for Viscaceae are also lacking, and the interpretation of morphological and karyological polarities is more difficult than in Loranthaceae, so that the biogeographic history is less clear (Wiens & Barlow 1971; Barlow 1983, 1987). Within the family the seven genera have very different distribution patterns. *Arceuthobium* occurs in north temperate areas in both the Old and New Worlds. *Phoradendron* and *Dendrophthora* are confined to the New

World, with centres of richness in the South American tropics. *Ginalloa* and *Notothixos* are more or less confined to the Malesian/Papuasian tropics. *Viscum* is widespread in the Old World, with major centres of species richness in southern Africa and India/southeast Asia. The last genus, *Korthalsella,* is exceptional among the mistletoes in its high dispersibility and its wide range over remote islands of the Indian and Pacific Oceans, in addition to its occurrence in a small area in eastern Africa and more widely in India/Malesia/Papuasia.

Whilst there is a general absence of relictual occurrences in the southern lands and a paucity of south temperate taxa, the distribution of Viscaceae in Africa, India, South America and Malesia/Papuasia, involving 5 of the 7 genera, is predominantly on Gondwanan land surfaces. In the Viscaceae, like the Loranthaceae, there may have been separate lines of evolution from a Gondwanan stock which have radiated more or less independently in the southern land masses (Barlow 1987).

Viscaceae in Malesia and Australia

Four genera of Viscaceae, *Viscum, Notothixos, Ginalloa* and *Korthalsella,* occur in the Malesian/Australasian region. All have centres of richness in southeast Asia or Malesia, and in all except *Korthalsella* the distributions become strongly attenuated in the Australasian area (Figure 22.8). The Australian/

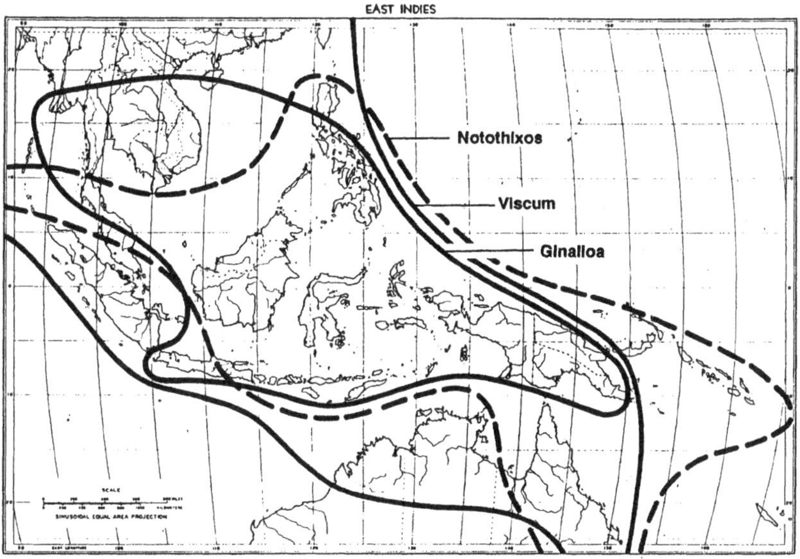

Figure 22.8 — Distribution of genera of Viscaceae (excluding *Korthalsella*) which are intrusive into the Malesian/Australian region.

Papuasian block therefore does not represent a primary source area of Viscaceae, at least for extant taxa. These genera are clearly part of an intrusive element with respect to the Australian flora, and if the family is ultimately of Gondwanan derivation then they belong in the Indogondwanan subelement. The Indian and Indosinian blocks may again be of significance in explaining the present world pattern of the family, and of the Malesian/Papuasian region in particular. The present occurrences of *Viscum, Notothixos, Ginalloa* and *Korthalsella* in Australasia may represent extension from a southeast Asian secondary centre in late Tertiary time, and be linked with the Miocene collision event.

Viscaceae as Recent recolonisers in the Torres Strait area

Like the loranths, a few species of Viscaceae show local occurrences across Torres Strait which are indicative of recent exchanges. *Notothixos cornifolius* has an extensive north-south range in eastern Australia, and a very local occurrence in New Guinea. Conversely, *Notothixos leiophyllus* is widespread in eastern Malesia, and occurs in Cape York Peninsula. *Viscum ovalifolium* ranges from Burma and southern China throughout Malesia, and has localised occurrences in Australia in the Kimberley region and in Cape York Peninsula. Because of their much greater dispersibility, *Korthalsella* species may be less dependent on habitat continuity for migration, but 3 of the 6 Australian species occur on both sides of Torres Strait.

Conclusion

Developments in Gondwanan geophysics continue to resolve residual problems in Australasian and Malesian biogeography, and indeed in angiosperm biogeography generally. Whilst the Malesian region conserves many primitive angiosperms, it seemed unlikely to be the centre of angiosperm origin because of its turbulent geological history. If parts of southern Laurasia are fragments of northern Gondwana (Audley-Charles *et al.* 1988) then the area from Assam to Fiji nominated by Takhtajan (1969) as the cradle of the angiosperms emerges as a contiguous part of northern Gondwana, and is more realistically a putative centre of origin (Martin & Dowd 1989). The early dispersal of angiosperms from Gondwana is more easily explained, and even today most of the world's rain forests occur on Gondwanan land surfaces. It is more obvious now that the two floristic elements interdigitating from both sides of Wallace's Line are both of common Gondwanan derivation, as visualised by Webb and Tracey (1981). In explaining the origins of the Australian flora, it is not necessary to distinguish between intrusive Malesian and Indogondwanan subelements.

The Loranthaceae provide a good case study of a family with clear Gondwanan derivation, an early diversification in isolated stocks on Gondwanan fragments, and the eventual Miocene reunion at Wallace's Line. In Australia some loranth genera are derived from the intrusive stocks, thus completing a migration and evolution cycle which began in the Mesozoic.

References

Audley-Charles, M.G. 1981. Geological history of the region of Wallace's Line. In: T.C. Whitmore (ed.), Wallace's Line and Plate Tectonics: 24–35. Clarendon Press, Oxford.

Audley-Charles, M.G. 1987. Dispersal of Gondwanaland: relevance to evolution of the angiosperms. In: T.C. Whitmore (ed.), Biogeographical evolution of the Malay Archipelago: 5–25. Clarendon Press, Oxford.

Audley-Charles, M.G., P.D. Ballantyne & R. Hall. 1988. Mesozoic-Cenozoic rift-drift sequence of Asian fragments from Gondwanaland. Tectonophysics 155: 317–330.

Audley-Charles, M.G., D.J. Carter & J.S. Milsom. 1972. Tectonic development of eastern Indonesia in relation to Gondwanaland dispersal. Nature Phys. Sci. 239: 35–39.

Audley-Charles, M.G., A.M. Hurley & A.G. Smith. 1981. Continental movements in the Mesozoic and Cenozoic. In: T.C. Whitmore (ed.), Wallace's Line and Plate Tectonics: 9–23. Clarendon Press, Oxford.

Barlow, B.A. 1981. The Australian flora: its origin and evolution. In: A.S. George (ed.), Flora of Australia, Vol. 1: 25–75. Aust. Govt. Publ. Service, Canberra.

Barlow, B.A. 1983. Biogeography of Loranthaceae and Viscaceae. In: D.M. Calder & P. Bernhardt (eds.), The Biology of Mistletoes: 19–46. Academic Press, Sydney.

Barlow, B.A. 1984. Loranthaceae. In: A.S. George (ed.), Flora of Australia, Vol. 22: 68–131. Aust. Govt. Publ. Service, Canberra.

Barlow, B.A. 1987. Biogeography of Loranthaceae and Viscaceae and its bearing on vegetation history in the Malesian/Australasian region. In: H.C. Weber & W. Forstreuter (eds.), Parasitic Flowering Plants: 805–817. Philipps University, Marburg.

Barlow, B.A., & B.P.M. Hyland. 1988. The origins of the flora of Australia's wet tropics. Proc. Ecol. Soc. Aust. 15: 1–17.

Barlow, B.A., & D. Wiens. 1971. The cytogeography of the loranthaceous mistletoes. Taxon 20: 291–312.

Bowler, J.M. 1982. Age, origin and landform expression of aridity in Australia. In: W.R. Barker & P.J.M. Greenslade (eds.), Evolution of the Flora and Fauna of Arid Australia: 35–45. Peacock, Adelaide.

Burbidge, N.T. 1960. The phytogeography of the Australian region. Aust. J. Bot. 8: 75–212.

Chappell, J. 1974. Geology of coral terraces, Huon Peninsula, New Guinea: a study of Quaternary tectonic movements and sea-level changes. Bull. Geol. Soc. Amer. 85: 555–570.

Crook, K.A.W., & G.R. Taylor. 1985. Early opening history of the Southern Ocean between Australia and Antarctica. Otway 85: Resources of the Otway Basin: Summary papers and excursion guide. Geol. Soc. Aust., S. Aust. and Vic. Divs., No. 1.

Kemp, E.M. 1978. Tertiary climatic evolution and vegetation history in the southeast Indian Ocean region. Palaeogr., Palaeoclimatol., Palaeoecol. 24: 169–208.

Kershaw, A.P. 1981. Quaternary vegetation and environments. In: A. Keast (ed.), Ecological Biogeography of Australia: 81–101. W. Junk, The Hague.

Martin, P.G., & J.M. Dowd. 1989. A molecular evolutionary clock. In: Aust. Syst. Bot. Soc. Forum on Gondwanan elements in the Australian flora, Abstract.

McKenna, M.C. 1972. Possible biological consequences of plate tectonics. Bio-Science 22: 519–525.

Nelson, E.C. 1981. Phytogeography of southern Australia. In: A. Keast (ed.), Ecological Biogeography of Australia: 733–759. W. Junk, The Hague.

Nix, H.A., & J.D. Kalma. 1972. Climate as a dominant control in the biogeography of northern Australia and New Guinea. In: D. Walker (ed.), Bridge and Barrier: the natural and cultural history of Torres Strait: 61–91. Aust. Nat. Univ., Canberra.

Ollier, C.D. 1982. The Great Escarpment of eastern Australia: tectonic and geomorphic significance. J. Geol. Soc. Aust. 39: 431–435.

Ollier, C.D. 1986. The origin of alpine landforms in Australasia. In: B.A. Barlow (ed.), Flora and Fauna of Alpine Australasia: Ages and Origins: 3–25. CSIRO, Melbourne.

Pain, C.F., & C.D. Ollier. 1984. Drainage patterns and tectonics around Milne Bay, Papua New Guinea. Rev. Geomorph. Dyn. 32: 113–120.

Specht, R.L. 1981. Evolution of the Australian flora: some generalizations. In: A. Keast (ed.), Ecological Biogeography of Australia: 783–805. W. Junk, The Hague.

Takhtajan, A. 1969. Flowering Plants: Origin and Dispersal. Oliver & Boyd, Edinburgh.

Walker, D. 1972. Bridge and Barrier. In: D. Walker (ed.), Bridge and Barrier: the natural and cultural history of Torres Strait: 399–405. Aust. Nat. Univ., Canberra.

Webb, L.J., & J.G. Tracey. 1981. Australian rainforests: pattern and change. In: A. Keast (ed.), Ecological Biogeography of Australia: 605–694. W. Junk, The Hague.

Wellman, P., & I. McDougall. 1974. Cainozoic igneous activity in eastern Australia. Tectonophysics 23: 49–65.

Whiffin, T., & B.P.M. Hyland. 1986. Taxonomic and biogeographic evidence on the relationships of Australian rainforest plants. Telopea 2: 591–610.

Whitmore, T.C. 1981. Palaeoclimate and vegetation history. In: T.C. Whitmore (ed.), Wallace's Line and Plate Tectonics: 36–42. Clarendon Press, Oxford.

Wiens, D., & B.A. Barlow. 1971. The cytogeography and relationships of the viscaceous and eremolepidaceous mistletoes. Taxon 20: 313–332.

23 Elements of Pacific phytodiversity

S.H. SOHMER

Summary

The native flowering plants of the Hawaiian Archipelago have been derived principally from other Pacific islands. A comparison is made of endemism in the native flowering plants of Fiji, the Marquesas and Society Islands, New Caledonia, New Zealand, Samoa, Tonga, and the Juan Fernandez Islands, in ways that can be directly compared to the statistics now available for that in Hawai'i with the conclusion of the new comprehensive flora for that archipelago.

Introduction

What is presented here is the most recent knowledge concerning the abundance and diversity of the native flowering plant life of some of the more isolated Pacific archipelagos. Based upon the data presented by Wagner *et al.* (1990), in which statistics concerning the abundance and distribution of the native flora of the Hawaiian Archipelago are given, I have endeavoured, via communication with appropriate specialists, to obtain the same kind of data for the flowering plants for Fiji, Marquesas and Society Islands, New Caledonia, New Zealand, Samoa, Tonga, and the Juan Fernandez Islands. In addition, I had respondents for New Guinea and Vanuatu, but there were not enough data of the right kind for them available for this statistical overview. It is hoped that the data will provide new insights into the relationship of these floras. Certainly, it would be useful to have an update of our knowledge of these floras that could well serve as a catalyst for carrying this survey further in the future. Having more accurate numbers, as well as ideas of the largest families and genera, will certainly aid phytogeographers in their tasks. Furthermore, as the vast majority of the species of all these island groups (save the Juan Fernandez) are ultimately derived from what can be broadly defined as the Malesian region, we can think of that region as the ultimate source area for the native flora of the tropical Pacific islands.

P. Baas et al. (eds.), The Plant Diversity of Malesia, 293–304.

In order to aid comparisons I decided to ask for data in the same format as reported in Wagner *et al.* (1990). I also decided to concentrate strictly on the native taxa and not to consider the naturalised introduced taxa in order to more logically follow floristic patterns across the Pacific. The data will be provided below first by archipelago and then summarised.

Hawai'i

This archipelago is obviously the most isolated island group of any of the groups analysed. Its uniqueness has been chronicled many times (see Wagner *et al.*, 1990, for a summary of the history of our knowledge of the flora of this area).

There are a total of 956 native species in Hawai'i recognised under the species concept utilised in the *Manual of the Flowering Plants of Hawai'i* (Wagner *et al.* 1990). Of these species 89% are endemic. There are 13 families with 20 or more species, the largest being the Campanulaceae with 110 native species. The native species now make up only 53% of the total number of species native & naturalised in Hawai'i. See Table 23.1a–d.

Table 23.1a — Statistics summarising the native flowering plants of Hawai'i.

	Dicots	Monocots	Total
Families	73	14	87
Total genera	165 (76%)	51 (24%)	216
Endemic genera	31 (19%)	1 (2%)	32 (15%)
Total indigenous species	822 (86%)	134 (14%)	956
Endemic species	759 (92%)	91 (68%)	850 (89%)
Non-endemic species	63 (8%)	43 (32%)	106 (11%)

Table 23.1b — Largest families in Hawai'i.

Family	Total native species	Family	Total native species
Campanulaceae	110	Cyperaceae	45
Asteraceae	92	Caryophyllaceae	33
Rutaceae	55	Piperaceae	25
Rubiaceae	54	Malvaceae	24
Lamiaceae	54	Euphorbiaceae	21
Gesneriaceae	53	Fabaceae	20
Poaceae	47		

Table 23.1c — Twenty most speciose genera of Hawaiian flowering plants.

Genus	Number of species	Genus	Number of species
Cyrtandra	53	*Myrsine*	20
Cyanea	52	*Hedyotis*	20
Pelea	47	*Bidens*	19
Phyllostegia	27	*Pritchardia*	19
Peperomia	25	*Chamaesyce*	15
Clermontia	22	*Labordia*	15
Schiedea	22	*Sicyos*	14
Dubautia	21	*Coprosma*	13
Lipochaeta	20	*Lobelia*	13
Stenogyne	20	*Wikstroemia*	12

Table 23.1d — Numbers of species and total taxa of Hawaiian native and naturalised flowering plants treated in Wagner *et al.* (1990).

	Dicots	Monocots	Total
Native species	822	134	956 (53%)
Total native taxa	947	147	1,094 (56%)
Naturalised species	672	189	861 (47%)
Total naturalised taxa	679	190	869 (44%)
Total species	1,494 (82%)	323 (18%)	1,817
Total taxa	1,626 (83%)	337 (17%)	1,963

Fiji

According to A.C. Smith (pers. comm.), there are 132 native families, 467 native genera, and 1,302 native species (Table 23.2a–c). Compared to the flora of Hawai'i, this is a native flora nearly 36% larger. Only 2.6% of the genera and 61.4% of the species are endemic, however, compared with 15% and 89%, respectively, for Hawai'i. This obviously is directly related to the greater isolation of the Hawaiian Archipelago as compared to the Fijian one.

Table 23.2a — Statistics summarising the native flowering plants of Fiji.

	Dicots	Monocots	Total
Families	110	22	132
Total genera	337 (72%)	130 (28%)	467
Endemic genera	9 (3%)	3 (2%)	12 (2.6%)
Total indigenous species	1,014 (78%)	288 (22%)	1,302
Endemic species	686 (68%)	113 (39%)	799 (61%)
Non-endemic species	328 (32%)	175 (61%)	503 (39%)

Table 23.2b — Largest families in Fiji.

Family	Total species	Family	Total species
Rubiaceae	165	Urticaceae	28
Orchidaceae	162	Myrsinaceae	28
Euphorbiaceae	68	Arecaceae	26
Fabaceae (s.l.)	48	Poaceae	25
Myrtaceae	41	Cyperaceae	24
Gesneriaceae	37	Flacourtiaceae	23
Lauraceae	34	Elaeocarpaceae	21
Piperaceae (s.l.)	33	Sapotaceae	21
Melastomataceae	29	Sapindaceae	20
Meliaceae	29	Apocynaceae	20

Table 23.2c — Largest genera in Fiji.

Genus	Number of species	Genus	Number of species
Psychotria	76	*Glochidion*	18
Cyrtandra	37	*Astronidium*	15
Peperomia	23	*Ficus*	14
Ixora	22	*Elatostema*	14
Syzygium	22	*Aglaia*	14
Bulbophyllum	19		

French Polynesia - Marquesas and Society Islands

Jacques Florence (1987 and pers. comm.) has made available data concerning the island groups of French Polynesia. The archipelagoes he covers are the Austral Islands, Gambier, Marquesas, the Societies, and Tuamotus. I will report here on the Marquesas and Societies only and will treat these separately so as to allow a more meaningful comparison. One of the problems with the data as reported by Florence (1987) as far as these particular comparisons are concerned is that the endemism of the floras reported for each group of islands is reported relative to one another, and since all of the islands share species in common, the apparent percentage endemicity is lowered relative to outside regions. However, data providing absolute statistics for percentage endemism of the flowering plants of the Marquesas and Society Islands were provided by Florence late in 1989 (pers. comm.) and they are provided here (Table 23.3a–e).

There are some caveats in Florence's contribution that need to be put on record here. *Pandanus* species are not included in these statistics and in some cases the number of species includes infra-specific taxa.

There is a Flora of the Marquesas Project that will certainly eventually allow a better comparison with Hawai'i, and it is my own belief that the floras of these two island groups are more closely related to each other than to any other flora. Rubiaceae constitute the largest family in both the Society Islands and the Marquesas. There are 26 species of Rubiaceae in the Marquesas.

Table 23.3a — Statistics summarising the native flowering plants of the Society

	Dicots		Monocots		Total	
Families	68		9		77	
Total genera	141	(75%)	48	(25%)	189	
Endemic genera	1	(1%)	0		1	
Total indigenous species	345	(81%)	82	(19%)	427	
Endemic species	235	(68%)	34	(41%)	269	(63%)
Non-endemic species	110	(32%)	48	(59%)	158	(37%)

Table 23.3b — Largest families in the Society Islands.

Family	Total species
Rubiaceae	57
Orchidaceae	35
Euphorbiaceae	29
Peperomiaceae	20
Gesneriaceae	18
Poaceae	15

Table 23.3c — Largest genera in the Society Islands.

Genus	Number of species
Peperomia	20
Cyrtandra	18
Psychotria	16
Glochidion	15
Myrsine	13
Bidens	10

Table 23.3d — Statistics summarising the native flowering plants of the Marquesas.

	Dicots		Monocots		Total	
Families	59		6		65	
Total genera	101	(82%)	22	(18%)	123	
Endemic genera	2	(2%)	1	(5%)	3	
Total indigenous species	203	(84%)	39	(16%)	242	
Endemic species	142	(70%)	27	(69%)	169	(70%)
Non-endemic species	61	(30%)	12	(31%)	73	(30%)

Table 23.3e — Largest genera in the Marquesas.

Genus	Number of species
Psychotria	13
Bidens	13
Cyrtandra	12

New Caledonia

I have often been in awe of the relatively high diversity of life in New Caledonia. The data for the native flowering plants of the island and its outliers, obtained from Philippe Morat (pers. comm.), Morat *et al.* (1984), and Jaffré *et al.* (1987), demonstrate once again how rich the flora of the area is (Table 23.4a–d). According to Morat there are 2,973 species of native flowering plants with 2,363 of these, or 79.5%, endemic. What is interesting is that 87.7% of the dicots are endemic and 45.2% of the monocots. These are the highest figures after New Zealand and the Hawaiian Archipelago, yet New Caledonia is much closer, at least at this time in its geologic history, to source areas. The high endemicity rate is remarkable and may have a great deal to do with the significant amounts of ultrabasic rock found on the island. As Morat (pers. comm.) shows, 91% of the native angiosperm species found on ultrabasic substrates in New Caledonia are endemic to that substrate.

Table 23.4a — Statistics summarising the native flowering plants of New Caledonia.

	Dicots	Monocots	Total
Families			154
Total genera			719
Endemic genera			107 (15%)
Total indigenous species	2,398 (81%)	575 (19%)	2,973
Endemic species	2,103 (88%)	260 (45%)	2,363 (79.5%)
Non-endemic species	295 (12%)	315 (55%)	610 (20.5%)

Table 23.4b — Largest families in New Caledonia (from Jaffré *et al.* 1987).

Family	Total species	Family	Total species
Myrtaceae	223	Orchidaceae	197
Rubiaceae	218	Poaceae	120
Euphorbiaceae	207	Apocynaceae	104

(Table 23.4b continued)

Family	Total species	Family	Total species
Araliaceae	102	Celastraceae	31
Cyperaceae	92	Sterculiaceae	31
Cunoniaceae	90	Asteraceae	31
Rutaceae	86	Meliaceae	26
Sapotaceae	79	Asclepiadaceae	26
Sapindaceae	65	Dilleniaceae	25
Papilionaceae*	61	Clusiaceae	24
Myrsinaceae	55	Thymelaeaceae	21
Flacourtiaceae	55	Mimosaceae*	21
Lauraceae	46	Ebenaceae	20
Elaeocarpaceae	45	Epacridaceae	19
Verbenaceae	44	Caesalpiniaceae*	19
Proteaceae	43	Podocarpaceae	18
Pittosporaceae	43	Araucariaceae	18
Moraceae	39	Winteraceae	17
Pandanaceae	38	Piperaceae	16
Arecaceae	32	Oleaceae	15

* The Leguminosae (*sensu lato*) with 101 species would be the second largest family.

Table 23.4c — Largest genera in New Caledonia.

Genus	Number of species	Genus	Number of species
Phyllanthus	110	*Xylosma*	20
Psychotria	85	*Xanthostemon*	19
Pittosporum	43	*Cryptocarya*	19
Tapeinosperma	39	*Planchonella*	19
Eugenia	37	*Arthroclianthus*	19
Schefflera	37	*Homalium*	18
Oxera	34	*Caryophyllum*	18
Alyxia	33	*Diospyros*	18
Ficus	33	*Freycinetia*	17
Pancheria	30	*Zygogynum*	17
Elaeocarpus	30	*Cyclophyllum*	16
Acropogon	26	*Metrosideros*	16
Cupaniopsis	26	*Parsonsia*	16
Syzygium	25	*Ixora*	16
Hibbertia	24	*Austrobuxus*	15
Cunonia	23	*Guettarda*	15
Morinda	23	*Garcinia*	15
Polyscias	23	*Alstonia*	15
Pandanus	21	*Acianthus*	15

Table 23.4d — Occurrence of species in relation to substrate in New Caledonia (from Jaffré *et al.* 1987).

Substrate	Number of species	Number endemic to substrate	Percent
Ultrabasic	1,844	1,671	91
Non-ultrabasic	1,840	1,253	68

New Zealand

New Zealand is also a very interesting floristic area and made more so by a climatic cooling in recent geological time. Dr. John Dawson (pers. comm.), one of the world's leading authorities on the New Zealand flora, has supplied the data cited here (Table 23.5a–c). The specific endemism is the second highest percentage figure of all the groups presented here – second only to Hawai'i.

Table 23.5a — Statistics summarising the native flowering plants of New Zealand.

	Dicots		Monocots		Total	
Families	87		23		110	
Total genera	244	(69%)	108	(31%)	352	
Endemic genera	24	(10%)	4	(4%)	28	(8%)
Total indigenous species	1,476	(71%)	590	(29%)	2,066	
Endemic species	1,321	(89%)	372	(63%)	1,693	(82%)
Non-endemic species	155	(11%)	218	(37%)	373	(18%)

Table 23.5b — Largest families in New Zealand.

Family	Total species	Family	Total species
Asteraceae	338	Epacridaceae	41
Poaceae	198	Gentianaceae	37
Cyperaceae	180	Liliaceae	34
Scrophulariaceae	172	Rosaceae	31
Apiaceae	112	Caryophyllaceae	27
Orchidaceae	95	Fabaceae	26
Ranunculaceae	58	Thymelaeaceae	25
Rubiaceae	57	Pittosporaceae	19
Onagraceae	54	Araliaceae	18
Boraginaceae	52	Violaceae	18
Juncaceae	51	Myrtaceae	17
Brassicaceae	44		

Table 23.5c — Largest genera in New Zealand.

Genus	Number of species	Genus	Number of species
Hebe	108	*Raoulia*	30
Carex	85	*Brachyglottis*	30
Celmisia	57	*Dracophyllum*	27
Coprosma	51	*Luzula*	25
Myosotis	51	*Pimelea*	25
Epilobium	51	*Pterostylis*	24
Ranunculus	46	*Cardamine*	24
Aciphylla	41	*Senecio*	20
Olearia	36	*Astelia*	20
Gentiana	36	*Pittosporum*	19
Poa	35	*Gnaphalium*	18
Craspedia	34	*Rytidosperma*	18
Uncinia	33	*Euphrasia*	18
Leptinella (Cotula)	30	*Acaena*	18

Samoa

I have Dr. Arthur Whistler (pers. comm.) to thank for the data presented here for Samoa (Table 23.6a–c).

Table 23.6a — Statistics summarising the native flowering plants of Samoa.

	Dicots	Monocots	Total
Families	75	17	92
Total genera	203 (72%)	78 (28%)	281
Endemic genera	1 (0.5%)	0	1
Total indigenous species	390 (71%)	160 (29%)	550
Endemic species	139 (36%)	43 (27%)	182 (33%)
Non-endemic species	251 (64%)	117 (73%)	368 (67%)

Table 23.6b — Largest families in Samoa.

Family	Total species	Family	Total species
Orchidaceae	95	Euphorbiaceae	18
Rubiaceae	44	Cyperaceae	16
Urticaceae	24	Poaceae	14
Gesneriaceae	20	Sapindaceae	10
Fabaceae	19	Moraceae	9
Myrtaceae	19	Piperaceae	9

Table 23.6c — Largest genera in Samoa.

Genus	Number of species	Genus	Number of species
Cyrtandra	20	*Bulbophyllum*	11
Psychotria	20	*Dendrobium*	11
Syzygium	15	*Ficus*	9
Elatostema	14		

Tonga

Art Whistler (pers. comm.) also supplied the data for Tonga (Table 23.7a–c). The relatively small size, and low elevation of Tonga, as well as proximity to Fiji, makes the species endemism the lowest of all the floras dealt with here.

Table 23.7a — Statistics summarising the native flowering plants of Tonga.

	Dicots	Monocots	Total
Families	76	13	89
Total genera	189 (77%)	56 (23%)	245
Endemic genera	0	0	0
Total indigenous species	245 (77%)	75 (23%)	320
Endemic species	10 (4%)	0	10 (3%)
Non-endemic species	235 (96%)	75 (100%)	310 (97%)

Table 23.7b — Largest families in Tonga.

Family	Total species	Family	Total species
Orchidaceae	40	Apocynaceae	10
Rubiaceae	23	Poaceae	9
Fabaceae	22	Myrtaceae	8
Euphorbiaceae	14	Meliaceae	7
Cyperaceae	12	Sapindaceae	7

Table 23.7c — Largest genera in Tonga.

Genus	Number of species	Genus	Number of species
Syzygium	6	*Planchonella*	4
Ficus	4	*Dendrobium*	4
Ipomoea	4		

Juan Fernandez Islands

Of all the island floras provided vital statistics in this paper, these islands stand out. The native flora is the smallest and the floristic relationships are with South America, rather than with Malesia. According to Tod Stuessy (pers. comm.), there are 16 native families, 69 native genera (with 11 endemic), and 159 native species. The specific endemism is 65.4%, which itself attests to the islands' relative isolation from what would have been source areas (Table 23.8a & b). The largest family represented is the Asteraceae with 29 species.

Table 23.8a — Statistics summarising the native flowering plants of Juan Fernandez.

	Dicots		Monocots		Total	
Families	12		4		16	
Total genera	50	(72%)	19	(28%)	69	
Endemic genera	7	(14%)	4	(21%)	11	(16%)
Total indigenous species	120		39		159	
Endemic species	90	(75%)	14	(36%)	104	(65%)
Non-endemic species	30	(25%)	25	(64%)	55	(35%)

Table 23.8b — Largest genera in Juan Fernandez.

Genus	Number of species
Dendroseris	17
Robinsonia	7
Erigeron	6
Wahlenbergia	5
Eryngium	4
Solanum	4
Chenopodium	3
Gunnera	3
Peperomia	3

Conclusions

The individual statistics are interesting from a number of points of view, but the most interesting, in my opinion, is the summary table that follows. Whereas New Caledonia, with an area of 19,103 sq.km, has 2,973 species for a density per unit area of 0.156, New Zealand, over 14 times larger than New Caledonia, has a specific density almost 20 times less than that of New Caledonia, if these data are correct.

The most startling is the Juan Fernandez Islands where the species density is nearly 149 times that of New Zealand and nearly eight times that of New Caledonia. This is due mostly to the size of the island group. The smaller the islands, the higher the specific density. However, New Caledonia, which is not much bigger than Hawai'i, has a specific density that is nearly three times that of Hawai'i. In terms of its native flora, Hawai'i had quite a few more 'open' niches than New Caledonia did. The fully oceanic condition of Hawai'i and the isolation from source areas are probably the reasons. A contributing factor, however, must be the extinction of a significant proportion of the endemic plant life of the Hawaiian Islands since man arrived about 900 years ago.

Table 23.9 — Summary table for statistics regarding the native flowering plant species of Pacific island groups.

Country / Area	Area (sq. km)	Number of species	Species density	Percent endemic
New Zealand	269,057	2,066	0.008	81.9
New Caledonia	19,103	2,973	0.156	79.5
Fiji	18,274	1,302	0.071	61.4
Hawai'i	16,705	956	0.057	89.0
Samoa	2,849	c. 550	0.193	33.1
Societies	1,608	427	0.266	63
Marquesas	1,243	242	0.195	70
Tonga	699	320	0.458	3
Juan Fernandez	134	159	1.19	65.4

References

Florence, J. 1987. Endémism et évolution de la flore de la Polynésie Française. Bull. Soc. Zool. (France) 112: 369–380.

Jaffré, J., Ph. Morat, J.-M. Veillon & H.S. MacKee. 1987. Changements dans la végétation de la Nouvelle-Calédonie au cours du Tertiaire: la végétation et la flore de roches ultrabasique. Bull. Mus. Nat. Hist. Nat. (Paris) 9: 365–391.

Morat, Ph., J.-M. Veillon & H.S. MacKee. 1984. Floristic relationships of New Caledonian rain forest phanerogams. In: F.J. Radovsky, P.H. Raven & S.H. Sohmer (eds.), Biogeography of the tropical Pacific. Bernice P. Bishop Mus. Special Publ. 72: 72–128.

Wagner, W.L., D.R. Herbst & S.H. Sohmer. 1990. Manual of the flowering plants of Hawai'i. Univ. Hawai'i Press & Bishop Mus. Press, Honolulu.

24 Van Steenis' contributions to nature conservation and public education

WILLEM MEIJER

Summary

Van Steenis' life-long activities for nature conservation and public nature study were incubated in the Dutch Society for Nature Study and nurtured by the Netherlands Indies Society for Natural History with its Journal 'Tropische Natuur' and the Local Society for Nature Conservation. There he played an inspiring role as organiser, stimulator and adviser. In this role he had contacts with all layers of Society from high level government officials to jungle workers from who he learned much. On the basis of this experience he was able after the war to convince the new Indonesian goverment of the desirability of his ambitious Flora concept and to secure its funding. The high standards set, his intense interest in nature study, his enormous work power, his optimism and perseverance against all odds should inspire us to continue with his work. The new movements in Malesia for popular nature study and for nature conservation combined with new efforts to speed up the Flora Malesiana project might ultimately succeed to keep nature alive in this part of the tropical world.

Beginnings in the Netherlands

Van Steenis was already an ardent nature conservationist during his student years in Holland where he played a leading role in the Dutch youth society for nature study (Nederlandse Jeugdbond voor Natuurstudie-NJN, Ittmann 1927). He also was a junior member of the Dutch Botanical Society with its floristic club. Professor Frits Went told me that he entered with van Steenis a competition of bringing together a good herbarium collection of the Dutch flora. Apparently this was a recruiting method for the floristic club. Van Steenis lost, but he joined the club and Went became a famous plant physiologist with a great interest in floristics. Some members of the NJN criticised plant collecting as potentially dangerous, possibly leading to plant extinction, but van Steenis defended this method strongly (van Steenis 1924). The basic philosophy of the NJN was that with aesthetic awareness of nature, observation, exploration, conservation and education had to go together (founding father Professor J. Heimans as cited in Meijer 1980). This remained van Steenis' philosophy about his "Scientia Amabilis" for the rest of his life.

Initially he tried to make a study of plant distributions and vegetation history of the Netherlands the basis of a doctoral thesis (unpubl. letter to van Soest 12

P. Baas et al. (eds.), The Plant Diversity of Malesia, 305–311.

Oct. 1978), but at the advice of Pulle, his major professor, he choose the revision of Bignoniaceae of Malesia as his thesis subject since he aimed at a career in the tropics. Still while tropical floras started to lure him he did not lose interest in his local surroundings. April 22, 1925 he made a plea to the Town Council of Utrecht in the local newspaper to preserve a green zone of the forests of Rhynauwen. This was coordinated with actions by the local chapters of the Natural History Society and the Society for Protection of Birds. In this article van Steenis pointed out the close relation between public education and nature conservation, the need to educate the public to respect nature and to get people interested in outdoor nature study. When about 50 years later a four-lane highway was planned to intersect this forest great public protests against its original design were heard. Most of the forest is saved now. It is situated very close to the new Botanical Garden and Botanical Institute at the Uithof, Utrecht. Other areas like the fens near Oisterwijk, Plasmolen and Mook cropping up in van Steenis' early reports in Amoeba (the NJN Magazine) were only partly saved, and *Parietaria* – collected there by him – is still preserved in the Rijksherbarium. His beloved Soesterveen, where his friend Frans Verdoorn did some of his early bryological studies (Verdoorn 1926; Gradstein & Richards 1986) was not saved as a larger hydrological complex.

Activities in Indonesia

With self-confidence gained from the Dutch experience of successful actions van Steenis joined in the Dutch Indies the local Natural History Society and the Society for Nature Conservation and he kept joining forces with the bird watchers and wild life enthusiasts, supplying the needed botanical data and arguments for conservation and pointing out the needs for public education. He was from the start known as a very hard worker (de Wit, verbal comm.). A long time his activities were mainly of influence among the more affluent Dutch settlers or transients, the local goverment officials and some Dutch educated Indonesians, a public of around 1200 readers of 'Tropische Natuur' and of the magnificant album 'Natuur in Indie' (van Steenis 1937). Pioneering work had already been done in this field by Koorders, Backer, Docters van Leeuwen and others. In his new job as Plant Identifier at the Bogor Herbarium and his special interest in mountain floras, van Steenis established many contacts with foresters, mountain hikers, members of the Natural History Society and biologists of other research institutes besides many leading government officials (Rappard, pers. comm. 1989). Those foresters who received a solid taxonomic training at Wageningen, like Endert, de Voogd, Rappard and Steup sought his advice and assistance and let him and others share in their field experiences. His closest contacts with Indonesians with little or no formal Dutch education were with the technical staff, the so-called mantris of the Botanical Gardens and the Bogor

Herbarium, especially Nedi and Noerta. They shared with van Steenis a great ability of plant form memory. I could later profit from the wide knowledge gained by these people in association with van Steenis and other botanists who worked before me at Bogor. They still talked about him and especially when starting to go cross-country in the Gedeh forests above Tjibodas Noerta started to remember Tuan Steenis, tetapi dia selalu Hotpardon ("you now remind me of Mr. van Steenis, but he said all the time GVD", a Dutch curse, while he was sliding and falling over the slippery rocks).

Unfortunately most of the pre World War II work done on conservation was published in Dutch, like his many articles on botanical documentation of potential nature reserves and keys for identification of water plants, sea grasses, saprophytes and parasites and his 'Maleise Vegetatieschetsen', which were distributed among all forest offices and much read and used there (Rappard, verbal comm. 1989).

From 1935 up to 1942 he was the botanical co-editor of 'Tropische Natuur' with his old friend and NJN buddy entomologist Lieftinck as main editor (van Steenis-Kruseman 1972). According to Donk it took van Steenis three years to become confident with the overwhelming richness of the Malesian flora at the family and genus level. His first paper (van Steenis 1931) on nature conservation was written four years after his arrival: 'Bergsport en natuurbescherming' (Mountain sport and nature conservation).

This was a plea for close cooperation with the Society for Nature Conservation since there was still great scope for local tourism and nature study on the majestic volcanos and other mountains in Indonesia, especially in overpopulated Java where hydrological reserves were very essential for the lowland paddy fields.

The more substantial botanical reports on conservation by van Steenis are included in the 'Verslagen' (Annual Reports): 1929, Papandajan pages 77–78, with a reference to an article in 'Tropische Natuur' 1930: 51, 73–90. In 1932 van Steenis was the Secretary who edited the Annual Report. Forest botanist Endert, born in Indonesia, where he started his tropical career in 1915, wrote in this report about the sacred lowland forest Doengoes Iwul in West Java and also Udjungkulon, Harau canyon in West Sumatra and West Kalimantan received attention. In the report of 1933/34 we read on pages 24–25 about important new wild life reservations and nature reserves in Sumatra: Loeser 900,000 ha, mountains above 3000 m altitude, the South Kroe Wildlife Reservation of 214,500 ha with rhinos, elephants and tapirs in the Benkulu Residency. A road project to the summit region of Papandajan was prevented. The Society was a real watchdog organisation against wild development projects and served as an action and pressure group, preventing great technological mistakes. However, a road to the Tangkuban Perahu crater near Bandoeng could not be prevented. This is now the most famous drive-in volcano in Indonesia. In New Guinea mining inside nature reserves was forbidden by law and large reserves were set out, but they remained paper reserves for a long time. The album 'Natuur in Indië' edited by van Steenis (1937) was cited by Jacobs (1972) as a great event and a lasting monument for nature conservation and public education. Fortunately his contributions on the volcanos Gedeh-Pangerango and Papandajan in West Java were translated and published in English (van Steenis 1939b, c).

In 1933 van Steenis became the president of the Society. He gave together with ornithologist Hoogerwerf a report on the overall situation in 1935, the cooperation with the Botanical Gardens and the Forest Service, and the local Government authorities, other societies and the local press, and with journals in Holland. Padang Loewai and Loeser reservations were announced. Journalistic misinformation given by F. Koster in a Dutch newspaper was clearly rebutted. Some new wildlife reserves in Sumatra and Borneo were later devastated (Berbak, Padang Luwai) or simply forgotten. Van Steenis proposed a Nature Reserve at Indramaju (see also 'Tropische Natuur' and 'De Levende Natuur' 1935). In the last pre-war report of 1939 van Steenis contributed an extensive well documented and illustrated report on Danoe (Lake) near Serang in Bantam, West Java (pages 214–222). He never lost his early love for coastal sandy seashore botany and mangroves and the plants of fresh water lakes and swamps, peatbogs and heathlands and much later he was very glad to be able to recruit Kees Den Hartog, now professor in Nijmegen, for revisions of the bulk of waterplant families and seagrasses for Flora Malesiana. Two years before World War II his magnum opus about his botanical explorations during the 1937 Loesir expedition appeared in the Journal of the Dutch Geographic Society (van Steenis 1938).

Van Steenis also wrote about nature reserves he never visited: Kutei, New Guinea-Carstensz summits, Mount Kinabalu and others. Life was after all too short to see it all himself. He was always good in compiling data from various sources, but he had some trouble delegating part of the job of vegetation description, what he called his unfinished lifework, to us younger Malesian plant explorers. He published general papers on nature conservation in Malesia in Dutch (van Steenis 1939a), later in English (van Steenis 1961, 1971) and in 1980 he made with his very loyal and supportive wife Rietje his last nostalgic visit to Papandajan and North Sumatra; he and his hosts avoiding devastated logging areas and other places turned upside down (van Steenis-Kruseman 1988). This way he could live another four years in relative peace. His understudy Marius Jacobs gave masterly surveys of the aims and prospects of botanical exploration and conservation in Malesia from the great action centre of the Leiden headquarters of Flora Malesiana and in the Flora Malesiana Bulletin he always gave much attention to conservation matters. Ultimately the giant flower *Rafflesia,* much featured in 'De Tropische Natuur' while van Steenis was editor, became the symbol of Flora Malesiana and plant conservation. Thus Marius

Jacobs, for about ten years editor of Flora Malesiana Bulletin, and many others, inspired and supported by van Steenis were already carrying on the torch while the great master kept more in the background in the almost hopeless struggle of Malesian botanists to keep representative parts of the lowland forests intact. It was a great loss when Jacobs suddenly died 28 April 1983 (van Steenis 1983).

We all know that van Steenis could be very stubborn. However, after the war and the independence of Indonesia, in the interest of his flora project he soon adapted to the new political realities and he sang the praise of Soekarno while he recruited at a meeting of the Netherlands Botanical Society and in the 'Vakblad voor Biologen' me and others to become botanists at the famous Bogor Botanical Gardens (van Steenis 1951). Before I went in 1955 to West Sumatra as a lecturer and plant explorer he confessed to me that he and other old colonials had done darned little to give Indonesians a good chance for nature education. By stimulating me and others to go native as far as the language was concerned and to bring our ascetic NJN life style into practice and by causing the Flora Malesiana project to bud off to Australia, Africa and the USA he made school even outside Malesia. With his Schoolflora, with his moral support of the local crash course Akademi Biologi and with his stimulation of various workers to propagate nature study and conservation among local Malesian students and forestry workers and later attempts to open up the old literature van Steenis and his wife had a very positive though intangible influence on present-day love for nature movements among the students of Malesia. Hard work set a great example for younger workers. Bakhuizen van den Brink Jr, for example, was discouraged by the great master to read all the 80 French novels he inherited from his uncle (de Wit, verbal comm.).

One final point: to which extent is the Magnum Opus, the Flora Malesiana a contribution to public nature study? Vernon Heywood (1983), the Editor of Flora Europaea, argued in his provocative article entitled 'The Mythology of Taxonomy' that many 'colonial' floras are not enough user-friendly, thus causing a breakdown of communication between producers of those floras and potential users: "Nowhere has the divorce between herbarium and library and the world of nature been more acute and had such far-reaching consequences as in the study of tropical floras, especially trees ... the myth has been fostered that only in herbaria can proper, professional taxonomy be practised" (page 85, Heywood 1983). "Just as herbaria feed on themselves, so many of the publications of taxonomy are directed at taxonomists, rather than other consumers" (page 86). "Taxonomists must spend some time analysing consumer requirements" (page 87).

Maybe Heywood had a point here. As regards tropical trees much more use of field characters is needed. Moreover, few people can afford to buy the whole Flora Malesiana. However, herbaria and botanical gardens and their libraries remain the much needed documentation centres but it is also far more important to preserve plants *in situ* as living populations than merely as a kind of fossil remnants (herbarium sheets) of extinct species. We are now well aware that everywhere in Malesia Flora Malesiana will need offshoots of local field guides written in the native language and lots of nature reserves. As a whole the criticism of Heywood was not totally fair as regards Malesia and van Steenis shrugged it off with the following rejoinder "A curious example of the mythology of the author's state of mind, bristling with unwarranted criticisms on the state of scientia amabilis" (Flora Malesiana Bulletin 37: 98. 1984).

The philosophy of Flora Malesiana as founded by van Steenis remains that the Flora Work, Conservation, Exploration and Public Nature Education have to go together. All what van Steenis, the 'angry young man' of Malesian Botany, with his keen interest in nature, worked for his entire long life would go on the trash heap of history if we cannot communicate this any longer convincingly and persuasively to the political and public powers of the present and next century.

Acknowledgements

This paper owes much to the Bibliography of all the published work by van Steenis published by Kalkman (1987) and to his assistance and the excellent help given to me by Libarian Mr. C. Lut while I carried out my biographic studies at the Rijksherbarium, June–August 1989. Mrs. van Steenis supplied me with extra background and she critically read an early draft. Jonkheer Frederick Willem Rappard and Professor H.C.D. de Wit supplied me with data and their visions on the accomplishments of van Steenis. Thanks are also due to the organisers of the Symposium for their moral, financial and editorial assistance.

References

Bibliography of all the published work by van Steenis in Blumea 32 (1987) 5–37 and his Opuscula in the Library of the Rijksherbarium.

Gradstein, S.R., & P.W. Richards. 1986. Frans Verdoorn and his contribution to Bryology. J. Bryol. 14: 203–213.

Heywood, V.H. 1983. The mythology of taxonomy. Trans. Bot. Soc. Edinb. 44: 79–94.

Ittman, K. 1927. Dr. C.G.G.J. van Steenis. Amoeba 7: 41–42. (An article full of praise for the leadership and organising qualities shown by van Steenis in the NJN and for his enthusiasm and ability to lead fieldtrips.)

Jacobs, M. 1972. Man, Nature and Education in: Lines in the published work of C.G.G.J. van Steenis, tropical Botanist. Blumea 20: 7–24.

Kalkman, C. 1987. In memoriam C.G.G.J. van Steenis (1901–1986). Blumea 32: 1–37.

Meijer, W. 1980. Herinneringen aan Heimans. Natura 77 (10): 347–353. (Memories of J. Heimans, the instigator of the Netherlands Youth Society for Nature Study.)

Rappard, F.W. 1989. A Memo written for this Symposium on the influence of Van Steenis in pre-war Indonesia on nature study, public education and nature conservation. Deposited in the Library of the Rijksherbarium.

Steenis, C.G.G.J. van. Herbaria, plantenuitrukkerij, natuurbescherming en nog wat. Amoeba 3: 32–35.

Steenis, C.G.G.J. van. 1925. Het natuurschoon van Rhynauwen bedreigd. Utrechts Provinciaal & Stedelijk Dagblad, 22 April 1925.

Steenis, C.G.G.J. van. 1925. N.J.N.-Herinneringen. Love of nature is of lifelong value. Amoeba 4: 113–116.

Steenis, C.G.G.J. van. 1931. Bergsport en Natuurbescherming. Meded. Ned.-Ind. Ver. voor Bergsport, Maart 1931: 22–27.

Steenis, C.G.G.J. van. 1937. Editor. Natuur in Indië. 2e Albumserie, Ned. Ind. Ver. Natuurbescherming.

Steenis, C.G.G.J. van. 1938. Exploraties in de Gajo-landen. Algemeene resultaten der Loesir-Expeditie 1937. Tijdschr. Kon. Ned. Aardr. Genootschap 55: 728–801, 32 phot., 2 maps.

Steenis, C.G.G.J. van. 1939a. Natuurbescherming in Nederlandsch-Indië van botanisch standpunt. Natuurk. Tijdschr. Ned. Ind. 99: 148–163.

Steenis, C.G.G.J. van. 1939b. The nature reserve Pangrango-Gedeh. Nature Protection in the Netherlands Indies: 15–19 (transl. from Natuur in Indië, 1937).

Steenis, C.G.G.J. van. 1939c. The nature reserve of the Papandajan. Nature Protection in the Netherlands Indies: 25–26, 2 fig. (ibid.).

Steenis, C.G.G.J. van. 1951. De Plantentuin te Bogor (Buitenzorg), Java, Kebun Raya Indonesia. Heden en toekomst. Vakblad voor Biologen 31: 81–90. (Also in German: Naturw. Rundschau Heft 7: 312–316, 1 fig.).

Steenis, C.G.G.J. van. 1961. Preservation of tropical plants and vegetation, an essential to the welfare of people. An Outlook on the future. Nature conservation in West Malaysia. Mal. Nature J., Spec. Issue: 23–25, 4 phot.

Steenis, C.G.G.J. van. 1971. Plant conservation in Malesia. Bull. Jard. Bot. Nat. Belg. 41: 189–202. (More extensive unpublished text in the National Archives.)

Steenis, C.G.G.J. van. 1978. Unpublished letter to Van Soest. 3 pages enclosed in Opuscula. Library Rijksherbarium.

Steenis, C.G.G.J. van. 1983. Marius Jacobs (19 Dec. 1929–28 April 1983). Flora Malesiana Bulletin no. 36: 3869–3871.

Steenis-Kruseman, M.J. van. 1972. The Life of a Botanist. Blumea 20: 1–6.

Steenis-Kruseman, M.J. van. 1988. Verwerkt Indisch Verleden. 50 pp. (Memoirs of the Indonesian period.)

Verdoorn, F. 1926. De mosflora der Soestervenen (Bryologische aanteekeningen I). De Levende Natuur 31: 172–180. (The moss flora of the Soester fen bogs.)

25 Conservation of plants in Malaysia

RUTH KIEW

Summary

In spite of the diminishing area of forest due to large scale land clearance, widespread logging, tin-mining and quarrying of limestone and the 'development' of mountain resorts, and pressure on individual species from commercial plant collecting, very few species are certainly extinct (two examples are *Begonia eiromischa* and *Ranalisma rostrata*). For the few groups studied, the percentage of endangered species ranges from 10% (palms) to 14% (trees). Particularly vulnerable are rare species confined to a single locality, such as 70% of *Didymocarpus* species. Endemism among tree species averages 30%, while for herbaceous groups it may be as high as 90% (*Begonia*). At present only National Parks and Wildlife Reserves are protected in perpetuity. A more extensive network of protected sites is urgently needed to protect areas with high endemism or local populations of rare species, such as *Rafflesia hasseltii*, which is not protected in Peninsular Malaysia. Virgin Jungle Reserves protect small areas but présently suffer from disturbance and are easily degazetted. Commercial collecting should be controlled by an extension of the Wildlife Act to cover plants, and efforts should be made to artificially propagate valuable species thereby reducing pressure on wild populations. In spite of pressures on forest from logging and land clearance, there is a future for the conservation of plants in Peninsular Malaysia as states have begun to realise that pristine areas are economically valuable in attracting the tourist dollar, and as public awareness and pride in preserving their nature heritage are high.

Introduction

The conservation of plant species in Malaysia as compared with animals has been relatively neglected, in part because they lack the glamour and appeal of fluffy animals, particularly primates, or the gorgeousness of birds, such as pheasants. In addition, tradition in Peninsular Malaysia has separated the management of the exploitation of forests (under the Forestry Department) from the management of animal wildlife (under the Game Department, which has now become the Department of Wildlife and National Parks).

What do we know about Malaysian plants? Compared with many other tropical countries Peninsular Malaysia is well served with floras (Kiew 1988) so that we are in a position to begin to assess the conservation status of individual species. We know, for example, that for the 2,398 tree species studied by Ng and Low (1982) endemism runs at about 30%, while for many herbaceous

P. Baas et al. (eds.), The Plant Diversity of Malesia, 313–322.

families it is as high as 80% (Gesneriaceae) to 90% (Begoniaceae) (Kiew 1983). Ferns and mosses are also well covered by floristic accounts. The same cannot be said for liverworts, algae and fungi, which remain neglected groups.

The situation is very different when we consider the Bornean states of Sabah and Sarawak, where there are no local floras and for the vast majority of families there is no recent check list, nor keys for identification. Check lists of species in Borneo as a whole list 4,922 species (Merrill 1921) and 7,201 species (Masamune 1942). Merrill estimated his check list represented between 50–60% of the total flora and this has subsequently been borne out (Kiew 1984). The total number of species will probably be in the region of 11,000. Merrill put endemism at about 50%. Without a flora of any kind it is extremely difficult to know which species are rare or endangered or to begin to assess their conservation status. This paper will therefore refer mainly to the higher plants and will concentrate on Peninsular Malaysia.

Threats to plant survival

The threats to the continued survival of plants include land clearance, logging and forestry practices, tin-mining and limestone quarrying, the development of mountain tops and commercial collecting.

(a) Land clearance

In 1960 72% of Peninsular Malaysia was forested, by 1979 this was reduced to 41% (Davison 1982). Much (86%) of the state land in Peninsular Malaysia was cleared for land development schemes between 1950 and 1986. Many forest reserves on good agricultural soil in the lowlands were also cleared, with the result that logging shifted into hill forest.

One has to recognise that the clearing of forest, mostly for conversion to oil palm plantation, is justified as it has been carried out with a view to providing a livelihood for the rural landless. However, this will be a continuing trend as the government policy on population growth aims for an increase from the present 15 million to 70 million in the next thirty years.

The conversion of 20% of mangrove forests to other uses during this same period is less difficult to justify as it has aggravated the decline in estuarine and inshore fishing by the destruction of nursery grounds for fish, crustaceans and shellfish as well as resulting in coastal erosion.

Shifting (or swidden) cultivation is a traditional way of life for many indigenous people and has recently been blamed by politicians as a major cause of forest destruction. In Sarawak, it is estimated that about 23% of the total land area is under shifting cultivation. However, studies such as Chin's (1985) show that in fact old secondary forest is most usually used, the areas cleared are small and being surrounded by forest regenerate quickly. In addition, shifting cultivators recognise primary forest as a vital renewable resource, which it is in their interest to preserve.

(b) Logging

Commonly quoted figures for Malaysia are that Peninsular Malaysia will be logged out by 1990 and Sabah and Sarawak by the year 2000. While these figures may be overly pessimistic, the speed of logging operations in the last twenty years is alarming. In 1971 61% of Sabah's land area was covered by primary forest, by 1980 this was reduced to 27% (Davies & Payne 1982). Annual rates of logging range from 760 sq.km in Sabah to 900 sq.km in Peninsular Malaysia (Davis *et al.* 1986).

There is no doubt that logging is a destructive activity. One study showed that where 26 trees per ha were cut for timber, another 33 timber trees were broken or damaged and up to 40% of the exploited area was denuded of vegetation (FAO 1982). In many cases, these open areas do not regenerate as the soil has been severely compacted by heavy machinery. Shade-loving plants of the ground layer suffer particularly from the adverse effects of opening up the canopy and the erosion and silting of streams.

The forestry practice of cutting climbers to encourage the growth of economically important tree species has a detrimental effect on *Rafflesia hasseltii* in Peninsular Malaysia. There are frequent reports that its host vine, *Tetracera* spp., is cut even when the vine bears large buds of *Rafflesia.*

(c) Mining

Although tin-mining affects only 1% of Peninsular Malaysia's land area, its effects are significant as Ng and Low (1982) have shown that 66 tree species that grew in forest on tin-rich soil have not been recollected since the 1940's.

Quarrying has a particularly adverse effect on the limestone flora which includes a large number of endemic species about half of which (129 species) are confined to this substrate (Chin 1977). Many of these species are known from a single or a group of limestone outcrops. Particularly vulnerable are the plants that grow around the shaded base which is often cleared for agriculture, which may also result in the vegetation of the outcrop being burned (Kiew 1985a).

(d) Mountain tops

Mountain tops support a high level of endemic species, many of which are found on a single mountain. The tops are cleared for telecommunication stations, tea, vegetable and flower farms or for hill resorts. In addition, once these mountain tops become easily accessible to the public, their plants, particularly orchids, become prey to the ignorant tripper, who takes them home to the lowlands where they languish and die, or to the commercial plant collector who strips areas of their orchids, pitcher plants and rhododendrons in particular. This stealing of the native flora has a long term effect in reducing the attractiveness of the area to the tourist.

(e) Commercial collecting

Commercial collecting has become an increasing problem in the past few years. Sabah and Sarawak have been the target for pitcher plants,where unscrupulous collectors strip off all the pitcher plants they can find and carry them away by the sackfull (Briggs 1985). The situation has become so serious that all *Nepenthes* species are now included on CITES Appendix II, so that a special permit issued by the exporting country is needed. (Previously the only Malaysian plant listed on CITES Appendices was *Nepenthes rajah,* which was placed on Appendix I, where export of these plants or their products is not permitted.) Recently a professional plant collector was convicted in a British court of illegally trafficking slipper orchids. He had in his possession 20 plants of the extremely rare *Paphiopedilum sanderiana* from the Gunung Mulu National Park, Sarawak, as well as *P. rothschildianum* from Gunung Kinabalu National Park, Sabah, and *P. bougainvillianum* and *P. dayanum* from northern Borneo, both of which are on the verge of extinction. Recently endangered palms have also become a target for both the professional and amateur palm-hunters (Kiew 1989a).

The current situation

How much primary forest is there in Malaysia? And how much of this forest is permanently protected?

The only forest that is permanently protected is forest gazetted as National Parks or as Wildlife Reserves. In fact there is only one national, i.e. federal, park in Malaysia, the others in Sabah and Sarawak are state parks set up under their own state legislation. There is a reluctance by the individual states to turn over land in perpetuity to the federal government as the economic benefit from these areas accrues to the federal government not to the state. This is the reason why the Endau–Rompin area proposed as a National Park is being set up by the states of Johore and Pahang as a joint state park. However, there is then the worry that these state parks are easily degazetted as happened in the case of the Klias State Park in Sabah.

In Peninsular Malaysia the single National Park, called appropriately Taman Negara, which was set up in the 1930's, conserves about 4350 sq.km of prime primary lowland forest as well as the highest mountain in Peninsular Malaysia, Gunung Tahan, which together with Wildlife Reserves protects about 4% on a land area basis (Mohd. Khan 1988).

Accurate figures for the area of forested land are difficult to obtain, but current estimates of forested areas for Peninsular Malaysia are 41% for Peninsular Malaysia, 65% for Sabah and 72% for Sarawak. These figures include National and State Parks and Wildlife Reserves, lowland, hill, montane and swamp forests, as well as logged, unlogged and disturbed forests. It is therefore difficult to know exactly how much lowland primary forest, the most endangered forest type, there is outside National parks and Wildlife Areas.

In Peninsular Malaysia the rate of logging and land clearance was severely depleting the timber stocks of the country. To sustain forestry on the long term, 36% of land has been declared as Permanent Forest Estate and the annual cutting rates have been reduced to 637 sq.km to sustain a 30-year felling cycle. This area includes National parks and Wildlife Reserves.

Conservation status of species

Even for Peninsular Malaysia where the flora is well-known, relatively few species have been critically assessed for their conservation status. For example, of the 850 odd species of orchids in Peninsular Malaysia, 19 are included in the 1986 listing of IUCN Conservation Monitoring Unit of which six are considered endangered, seven as rare and two as vulnerable. The Malayan Nature Society is currently compiling a conservation data base on Malaysian plants and animals and publishes a series on portraits of threatened species (Malayan Naturalist, Vol. 37, 1983, onwards).

Are any plants extinct in Peninsular Malaysia? At present there appear to be very few strong candidates. Two species that are almost certainly extinct are *Begonia eiromischa* (Kiew 1989b) and *Ranalisma rostrata* (Kiew 1985a). Other species thought to possibly be extinct have either been rediscovered, such as *Begonia rajah* (Kiew 1989b), *Calospatha scortechinii* (Kiew & Dransfield 1987) and the fern *Botrychium daucifolium* (Kiew *et al.* 1985), or on further study were found to be synonymous with other species as were, e.g., *Iguanura arakudensis* (Kiew & Dransfield 1987) and *Didymocarpus perditus* (Kiew 1987a).

How many species are endangered? Ng and Low (1982) considered that 14% of the tree species they studied were endangered. (This figure, which is based on herbarium collections at KEP, is rather high and is likely to drop when field data are taken into consideration.) A more detailed study on palms showed that out of the 194 palm species in Peninsular Malaysia 20 species are endangered and 16 are rare (Kiew 1989c) and for Sarawak 20 of the 218 species are endangered and 26 are rare (Pearce 1989).

The concept of rarity in the tropics encompasses two very different types. One is the extremely local species, such as *Didymocarpus primulinus*, which is known from a single population of about 100 plants which grows on a single rock face; the other is species that are widely distributed but are nowhere common, such as *Ficus albipilus* which in Peninsular Malaysia is known from only a few trees at a single locality although its geographic distribution ranges from Thailand to Australia (Kiew *et al.* 1985). Many herbaceous plants fall into the first category, for example about 50% of the 50 species of *Begonia* and 70% of the 85 species of *Didymocarpus* are known from a single locality. These rare and local species are particularly vulnerable to habitat disturbance. In contrast, tree species tend to fall into the latter category.

However, in all cases conservation status based on herbarium study needs to be re-evaluated after field surveys. It is heartening to note that although figures for forest cover have declined drastically very few species are extinct and that field surveys have had good success in relocating endangered or rare species. An extreme example is *Musa gracilis*, which was thought to be very rare, but has in fact been able to expand its habitat along the edge of logging tracks where it is now quite common (Kiew 1987b).

There is, however, a general lack of check lists of species found in National parks. Our study on palms shows that 50% of Peninsular Malaysia's palm species are found within Taman Negara and the Endau–Rompin area. This at first sight appears satisfactory, however, a disproportionate number of these 99 species are common and widespread compared with only 1 endangered, 6 rare and 13 particularly vulnerable (V1) species (Kiew 1989c).

The problem of protecting the rare local species, which are scattered throughout the country, will continue to be a problem in the field of conservation, particularly as the need for setting up small nature reserves is not generally appreciated.

The role of Botanic Gardens

Tropical botanic gardens have two main roles to play in conservation. The first is to raise the level of the public's knowledge of and pride in their natural heritage. To this end, beautiful, interesting and bizarre local plants should be prominently displayed and labelled and information leaflets be made available. At present there are no botanic gardens in Sabah and Sarawak nor a highland botanic garden in Peninsular Malaysia. To see Malaysian rhododendrons in cultivation one has to go to Japan, New Zealand or the United Kingdom.

Secondly, they can serve an important role in propagating and distributing local plants to other institutes throughout the world, thereby reducing the need for repeated collection, sometimes done illegally, of plants from the wild. They can also maintain populations of plants as a permanent gene pool, though it should be pointed out that individual display plants do not meet the requirements of maintaining 500 plants to ensure genetic diversity.

However, having said that one cannot be complacent about the role of botanic gardens in saving species. Horticultural science has not yet advanced to the stage where parasites, such as *Rafflesia* species, can be cultivated. Nowhere is there authentic 'wild' material of *Begonia rajah* in cultivation, as the original collections have died out at Kew and Singapore. This is a common fate of forest plants which are not 'botanic garden' hardy and without conditions of exceptionally high and constant relative humidity die very quickly. Botanic gardens are also not free from economic cuts, which makes hothouse plants particularly vulnerable, or in the case of Singapore, from the effects of a world war. Ultimately, the best solution is protection in its natural habitat.

The future

What is required to conserve the maximum number of plant and animal species is a network of protected areas to safeguard a representative area of all forest types. From data base work it is possible to identify areas with a particularly high number of endemic, endangered or rare species, as for example Ng and Low's study identified Penang Hill as one such area. In addition to National Parks and Wildlife Areas, there needs to be a system of smaller reserves to safeguard pockets of forest where rare and endangered species with a very local distribution are found.

At first sight, Virgin Jungle Reserves (VJRs) in Peninsular Malaysia would appear to satisfy this need as they protect small areas (85% of them are less than 400 ha) designed to protect undisturbed representative areas of forest or areas which contain rare species. At present there are 85 VJRs covering a total of 19,028 ha making up 0.3% of the Permanent Forest Estate. However, in practice it has been easy to change their status (18 of the original total of 104 have been degazetted) and they are subjected to disturbance from poachers, illegal gatherers of forest products, illegal logging, shifting cultivation etc. (Borhan & Cheah 1987). In addition, because they come under the jurisdiction of the Forest Department, they do not include non-productive forest types, such as swamps or limestone vegetation.

In the absence of any other legislation VJRs should be invoked to protect populations of endangered plants with narrow distributions, such as *Begonia rajah, Ficus albipilus* and *Rafflesia hasseltii,* whose future is insecure as they grow in forest reserves.

In Peninsular Malaysia the reluctance to give permanent protection to any new areas is glaring. Taman Negara was set up in 1930s, the last VJR was gazetted in 1976. The Endau–Rompin State Park is only being gazetted after prolonged public pressure. Sabah and Sarawak with their smaller populations have been more ready to set up a National Parks system, though the monetary advantages of logging are never far from a politician's mind.

If forest is to be looked on as a renewable resource, the Forest Department will have to be stricter in enforcing good forestry practices to prevent wastage and damage to standing stock and will have to invest more in silvicultural methods that enhance the growth of timber species after logging. Without such methods, the amount of forest that can be relogged will dwindle and eyes will turn to the timber riches of National Parks.

With pressures on land use, it cannot be expected that protected areas remain inviolate and that the public is denied access to them. On the contrary, the public should be encouraged to visit these areas although their access should be controlled along well-marked trails. In fact with the revenue from tourism becoming increasingly important to the national economy, state governments are fast realising that scenic, unspoilt areas are an economic asset to attract tourists and that it is in their interest to conserve them. Some conflict remains as to how to develop these areas without destroying them.

There is a very heartening trend in the interest the general public in Malaysia shows towards their nature heritage. The Endau–Rompin expedition, which was reported regularly in newspapers and on television, elicited the response that the public wanted to know more. An important aspect of conservation should therefore be the production of well-illustrated books on Malaysia's natural history both for school children and the general public. For while the politician may respect the scientist's advice, it is ultimately public pressure that persuades him to act.

The problem of large scale commercial collecting of Malaysian plants needs to be approached from three angles:

1. At present, the market for these plants is overseas. Societies for orchid, palm and rhododendron specialists, as well as the horticulturalist in general, should encourage the ethic of buying only propagated rather than wild-collected plants.
2. Where there is a demand for rare species, these societies could support local concerns, particularly those linked with conservation, such as National Parks and botanic gardens, to set up local propagating units. A single pod of orchids, pitcher plants or rhododendrons produce many tiny seeds from which masses of progeny can be grown. This not only allows the country to benefit from its own natural resources, but by satisfying the market it makes illegal collecting unprofitable.
3. Local laws need to be gazetted to ensure that wildlife including plants, are protected from illegal export. Although Malaysia is a signatory of CITES, Peninsular Malaysia does not have any laws (apart from the Quarantine Act) for controlling the export of wild plants, even though they are listed on

CITES Appendices. The onus is therefore on the receiving country to demand certification. Sarawak has independently set up a State Legislative Assembly Special Select Committee on the Flora and Fauna 1986, which lists species that cannot be exported. For example seven species of palm are included on this list (Pearce 1989).

Conclusions

While the figures on forested area are bleak, I nevertheless remain optimistic about the future of Malaysian plants and vegetation. My reason for this is the surge in interest and pride by the Malaysian public for their nature heritage.

What can scientists and the public overseas do to help in the conservation of plants in Malaysia? They can, for example, support field work that is orientated to refind endangered species and to build up check lists of plants in areas of conservation importance; they can collaborate in funding local units to propagate threatened species of commercial value; they can be involved in the publication of inexpensive general interest books on local natural history; and they can support local conservation organisations.

References

Briggs, J.G. 1985. The current Nepenthes situation in Borneo. Malay. Nat. 38 (3): 46–48.

Borhan, M., & L.C. Cheah. 1987. The Virgin Jungle Reserves in Peninsular Malaysia: the need for a positive management strategy. In: Proc. Workshop on 'Impact of Man's Activities on Tropical Upland Forest Ecosystems': 81–90. Faculty Forestry, University Pertanian Malaysia.

Chin, S.C. 1977. The limestone hill flora of Malaya 1. Gards' Bull., Singapore 30: 165–219.

Chin, S.C. 1985. Agriculture and resource utilization in a lowland rainforest Kenyah community. Sarawak Mus. J. 35: 1–322.

Davies, G., & J. Payne. 1982. A Faunal Survey of Sabah. IUCN/WWF Project No. 1692.

Davis, S.D., S.J.M. Droop, P. Gregerson, L. Henson, C.J. Leon, J.L. Villa-Lobos, H. Synge & J. Zantovska. 1986. Plants in danger: What do we know? IUCN, Gland, Switzerland.

Davison, G.W.H. 1982. How much forest is there? Malay. Nat. 35 (1 & 2): 11-12.

FAO. 1982. Tropical Forest Resources. FAO Forestry Paper 30.

Kiew, B.H., R. Kiew, S.C. Chin, G. Davison & F.S.P Ng. 1985. Malaysia's 10 most threatened animals, plants and places. Malay. Nat. 38 (4): 2–6.

Kiew, R. 1983. Conservation of Malaysian plant species. Malay. Nat. 37 (1): 2–5.

Kiew, R. 1984. Towards a flora of Borneo. In: Ismail Sahid et al. (eds.), Research Priorities in Malaysian Biology: 73–80. Universiti Kebangsaan Malaysia.

Kiew, R. 1985a. The limestone flora of the Batu Luas area, Taman Negara. Malay. Nat. 38 (3): 30–36.

Kiew, R. 1985b. Commercial collecting of wild plants in Malaysia. Malay. Nat. 38 (3): 48–50.

Kiew, R. 1987a. The herbaceous flora of Ulu Endau, Johore-Pahang, Malaysia including taxonomic notes and descriptions of new species. Malay. Nat. J. 41: 201–231.

Kiew, R. 1987b. Notes on the natural history of the Johore banana, Musa gracilis Holttum. Malay. Nat. J. 41: 239–248.

Kiew, R. 1988. Herbaceous flowering plants. In: 'Malaysia': 56–76; ed. E. Cranbrook, Perga-mon Press, U.K.

Kiew, R. 1989a. Collecting endangered palms in Peninsular Malaysia. Principes 33: 94–95.

Kiew, R. 1989b. Lost and found – Begonia eiromischa and B. rajah. Nat. Malay. 14: 64–67.

Kiew, R. 1989c. The conservation of Palms in Peninsular Malaysia. Malay. Nat 43 (1 & 2): 3–15.

Kiew, R., & J. Dransfield. 1987. The conservation of palms. Malay. Nat. 41 (1): 24–31.

Masamune, G. 1942. Enumeratio Phanerogamarum Bornearum. Dept. Taxonomy and Ecology, Faculty Agric. Sci., Taihoku Imperial University, Japan.

Merrill, E.D. 1921. A bibliographic enumeration of Bornean plants. J. Str. Br. Roy. Asiatic Soc. Spec. No. 1–637.

Mohd. Khan, M.M. 1988. Animal conservation strategies. In: 'Malaysia': 251–272; ed. E. Cranbrook, Pergamon Press, U.K.

Ng, F.S.P., & C.M. Low. 1982. Check list of endemic trees of the Malay Peninsula. FRI Kepong Research Pamphlet No. 88.

Pearce, K. 1989. Conservation of palms in Sarawak. Malay. Nat. 43 (1 & 2): 20–36.

26 The impact of development on the fern flora in some areas in Malaysia and efforts toward conservation

A. A. BIDIN

Summary

The tropical rain forest of Malaysia occupies about 61% of the total land area. It is one of the greatest reservoirs of plant and animal diversity in the world. However, with the alarming rate of deforestation to accommodate the needs of the industrial, agricultural sectors and new settlements, the existence of certain plant species including ferns is threatened. The recent deforestation in some areas on offshore islands of Peninsular Malaysia such as Pulau Langkawi, Pulau Sembilan, Pulau Tioman and Pulau Aur for the development of tourist industry seriously threatened the survival of certain species of ferns which are endemic to these islands. *Platycerium wallichii, P. platylobum, Bolbitis malaccensis, Tectaria variolosa* and *Drynaria bonii* in Pulau Langkawi, *Brainea insignis* in Pulau Sembilan, and *Quercifilix zeylanica* in Pulau Tioman and Pulau Aur are a few of the examples. Quarrying and mining of marble and limestone especially in Pulau Langkawi together with the continuous extraction of timber from hill forests on mainland Peninsular Malaysia are posing a real danger to the fern flora, in terms of area depletion and extinction due to changes in edaphic conditions. Efforts to introduce these threatened taxa into a managed Fern Garden (Filicetum) are being undertaken.

Introduction

The primary forests in the lowlands of Peninsular Malaysia are rapidly being depleted to make way for the expansion of various economic activities such as agriculture (mainly oil palm, rubber and cocoa), logging and tin mining. Development of new settlements due to the rapid population growth also enhances the depletion.

Besides these factors that cause deforestation, the latest threat is from the tourist industry. Tourism is the third biggest revenue earner for the country after petroleum and agricultural commodities. In order to attract local and foreign visitors, a number of scenic places are opened up, and the existing ones upgraded. Most of these are islands around the country shorelines such as the Langkawi group on the west coast and Tioman on the east; other touristic resorts are developed in the highlands where the weather is temperate: Cameron Highlands, Fraser's Hill and Mount Kinabalu. In these places parts of the natural vegetation have been stripped off or modified to accommodate tourism-related infrastructures such as hotels, recreation facilities, roads, etc.

P. Baas et al. (eds.), The Plant Diversity of Malesia, 323–328.

Plant conservation

With an alarming rate of deforestation, which results in the loss of many plant species, and with some parts of the forest – especially in Sabah and Sarawak – not yet explored, it is difficult to give the total number of species in the flora of Malaysia. However, it is estimated there are about 14,500 species of flowering plants (Ng & Darus 1984) and 650 species of ferns and fern allies (Bidin 1989).

Of the 2,398 species of trees surveyed by Ng & Low (1982) for Peninsular Malaysia, 654 are endemic and 342 species of these are considered endangered. As for herbaceous species, the degree of endemism is much higher. For example, Kiew (1983) estimated that families such as Begoniaceae and Gesneriaceae are for 90% and 80% endemic respectively. The largest plant family in Malaysia, the Orchidaceae with about 800 species, has probably at least 50% endemism. Out of 500 species of ferns documented for Peninsular Malaysia by Holttum (1968), 59 are considered rare and endemic, having a restricted distribution but occurring at most in two localities.

The chances of survival of these endemic and endangered species has now reached an alarming level. For the purpose of this paper two areas have been selected to highlight human impact on the flora, especially the ferns.

Langkawi Islands

This group of islands, situated to the West of the Peninsula, consists of six medium-sized islands and 94 smaller ones. The highest point is Gunung Raya (880 m) on the main island.

The climate is monsoonal, with a marked dry period (December–March). Large areas of limestone rock are found on most of the islands and the vegetation is, consequently, slightly different from the rest of the country, with more Burmese and Thai elements in the flora (Henderson 1939).

Two types of vegetation are observed in Langkawi, on limestone and granite. Much of the limestone cliffs are honeycombed with caves, and the vegetation is often stunted and reduced. The granitic areas, on the other hand, have much softer appearances, the vegetation is markedly more luxuriant.

Recent studies (Bidin 1987) on the Langkawian ferns showed that of the 145 infrageneric taxa recorded, a good number were categorised as rare (*Drynaria bonii, Pseudodrynaria coronans, Bolbitis malaccensis, Tectaria brachiata, Tectaria variolosa*) and one of them is endemic to Langkawi (*Platycerium platylobum*).

Drynaria bonii (a new record for Peninsular Malaysia) is found in Langgun Island only living epiphytically or on limestone with direct exposure to sunlight and sea sprays. Herbarium records show that the plant has its centre of distribution in Indo-China and Thailand.

Pseudodrynaria coronans, again a mainland Asian fern, has its southernmost distribution in Langkawi near the summit of Gunung Raya. The fern is also found on Gunung Jerai in northern Kedah.

Bolbitis malaccensis is only recorded from two localities in Peninsular Malaysia, i.e. in Langkawi and in Tioman. *Microlepia strigosa,* though widely distributed on mainland Asia and quite common in Langkawi, so far has only been recorded once from mainland Peninsular Malaysia, i.e. in a limestone area of Perak at 3500 feet above sea level.

Tectaria brachiata is widely distributed on granite in light shade in Langkawi, especially in rubber plantations as well as along trails in forested area where the canopy is slightly open. Holttum (1981) reported that the species is adapted to a climate with a seasonal dry period, which Langkawi certainly has. The fern has a distributional area from Northeast India to South China, Thailand, Vietnam and northern Peninsular Malaysia (Langkawi).

Tectaria variolosa, a mainland Asian limestone fern, is found in Langkawi on limestone cliffs and in rock crevices by the sea on Timun Island. It is a new record for Peninsular Malaysia.

Platycerium platylobum is found in two localities on the main island. The ferns are absent from the other islands in the group, or from the interior of Perlis and northern Kedah, all of which share the same monsoonal climate.

The beauty of Langkawi has not only attracted visitors but also brought developments to the islands. New hotels and recreational facilities have been built to cater for the needs of visitors. Jungle treks and camping sites are also made available to tourists in the Gunung Raya Forest Reserve right to the summit of the mountain.

With these activities, the after-effects such as floods, erosion and landslips are inevitable and pose a real threat to the indigenous vegetation of Langkawi not only in terms of area depletion but also in degradation and invasion of exotic plants.

Tioman Islands

The group consists of 35 islands, which are sparsely populated and fairly rugged with numerous rocky outcrops and comparatively few flat lands. As in Langkawi, the climate is monsoonal.

Compared to Langkawi, the flora of which has been relatively little investigated, the Tioman group of islands has been fairly well botanised (e.g., Lee 1977; Latiff 1982; Ibrahim 1986; Bidin 1988).

A total number of 113 fern species has been recorded and there are some curious features found in Tioman, although in general the fern flora is characteristic of much of Malaysia. For example *Quercifilix zeylanica,* found in abundance growing on rocks near the seasides in Tioman and Aur, has never been collected from Peninsular Malaysia. *Bolbitis malaccensis,* which is found here, is not found on the Peninsula but has also been collected from Langkawi. Two species of small tree ferns, *Cyathea tripinnata* and *C. oosora,* are found here but not in Peninsular Malaysia. These species have also been collected on Mt Kinabalu in Borneo (Holttum, in Lee 1977).

With the increase in number of visitors to these islands, both Malaysians and foreigners, there is an intense pressure to open up more area for the development of hotels. Since flat land for building is limited in the inland, all development programs are focussed along the flat stretch of land near the beach. Unfortunately many of the fern species unique for these islands are found near the seasides and thus the survival of these species is threatened.

Conservation measures

Any intrusion to the delicate rain forest ecosystem, no matter how small, like the building of roads and recreational facilities for visitors as exemplified by Langkawi and Tioman, can have serious repercussions.

As for the ferns, to alleviate the problem of extinction of the rare and endangered species, two important measures must be considered:

1) Preservation of existing habitats – where more forest reserves must be gazetted and the existing reserves should be upgraded to national parks.
2) Collection of germ plasms especially of the rare and endangered species in a managed garden – a fern garden.

The first fern garden in Malaysia was set up in 1988 at the Universiti Kebangsaan Malaysia in Bangi in an area of about 22 acres. It is part of the Bangi Forest Reserve assigned to the University for teaching and research activities. The area selected is landscaped in order to create a variety of scenic features – lakes, streams, walking tracts, meadows and terraces. These various features provide the different ecological niches for various species of ferns.

Even though the main objective for the establishment of this fern garden is to conserve the rare and endangered ferns and fern allies of Malaysia, it is also intended to serve as a reservoir for the representatives of Malaysian as well as the tropical fern flora in general.

Since most of the ferns introduced into the garden are of wild origin, it will present an enormous opportunity to do research in the sowing, propagation and hybridisation of such plants. The garden will also serve as an outdoor laboratory of morphological, anatomical, cytological and phytochemical investigations.

The fern garden is divided into three main sections. The first section (c. 5 acres) is reserved for economic ferns. These are the medicinal, edible and vine ferns, the rachises of which are used in a variety of handicrafts.

As in most countries in Asia, traditional medicines are widely used in Malaysia, especially among the rural population. Among the ingredients used are parts of the ferns. To date at least 76 species belonging to 44 genera and 13 families of ferns in Malaysia are known to be used for medicinal purposes, either alone or as an ingredient in a mixture of herbs (Bidin 1986; Burkill 1935). A number of these species have been introduced into the garden: *Drynaria sparsisora, Asplenium nidus, Helminthostachys zeylanica, Platycerium coronarium, Cibotium barometz, Schizaea dichotoma, Angiopteris evecta,* and *A. angustifolia.*

The edible ferns are represented by four species, i.e. *Stenochlaena palustris, Diplazium esculentum, Ceratopteris thalictroides* and *Acrostichum aureum.* The fronds are eaten as vegetables after boiling or cooked with coconut milk.

A number of handicraft items such as handbags, trays, and strings produced by the rural people in Malaysia are made from the peelings of rachises of the vine fern *Lygodium.* The species that are commonly used for this purpose are *Lygodium circinnatum, L. flexuosum* and *L. polystachyum.* They are selected for their texture, colour and durability.

The nursery and the collections of the rare and threatened ferns are located in the second section of the Fern Garden (c. 10 acres). The nursery is equipped with 20 prothalli chambers for spore sowings and sections for propagation. The ferns introduced into this section are collected from various localities in the country where their existence is threatened or known to be threatened in the near future due to the opening of the forest. Most of the accessions are from Langkawi, Tioman, Cameron Highlands and Fraser's Hill, where massive developments are taking place for the tourist industry. Among the ferns already introduced here are *Platycerium wallichii, P. platylobum, Photinopteris speciosa, Osmunda javanica, O. vachellii, Angiopteris angustifolia, Cyathea tripinnata,* and *Drynaria bonii.*

The third section of the fern garden (c. 7 acres) is reserved for the ornamental as well as the introduced ferns. Among the many ferns gradually introduced are the many species and varieties of Maiden-hair ferns (*Adiantum*), *Pteris, Microsorum, Asplenium,* and *Nephrolepis.*

References

Bidin, A. 1985. Paku-pakis Ubatan Semenanjung Malaysia. Dewan Bahasa & Pustaka. Kuala Lumpur.

Bidin, A. 1987. A preliminary survey of the fern flora of Langkawi Islands. Gard. Bull. Sing. 40: 77–102.

Bidin, A. 1988. The coastal ferns of Pulau Aur. Nature Malaysiana 13 (3): 22–25.

Bidin, A. 1989. The role of the Fernarium as a sanctuary for the conservation of threatened and rare ferns, with particular reference to Malaysia. Proc. 2nd Intern. Bot. Gardens Congress, Réunion 24–28 April 1989.

Burkill, I.H. 1935. A dictionary of the economic products of the Malay Peninsula. Govt. Printing Press, Kuala Lumpur.

Henderson, M.R. 1939. The flora of the limestone hills of the Malay Peninsula. J. Roy. Asiatic Soc., Malayan Branch 17: 13–87.

Holttum, R.E. 1968. Flora of Malaya 2. Ferns of Malaya. Singapore Govt. Printer.

Holttum, R.E. 1981. The fern genus Tectaria in Malaya. Gard. Bull. Sing. 34: 132–147.

Ibrahim, H. 1986. Notes on Gingers of the Pulau Tioman. Nature Malaysiana 11 (4): 10–13.

Kiew, R. 1983. Conservation of Malaysian plant species. Malay. Nat. 37 (1): 2–5.

Latiff, A. 1982. Notes on the vegetation and flora of Pulau Pemanggil. Mal. Nat. J. 35: 217–224.

Lee, D.W. 1977. The natural history of Pulau Tioman. Universiti Malaya, Kuala Lumpur.

Ng, F.S.P., & M. Darus. 1984. Taman Kiara: Planning for Malaysia's National Arboretum. In: Proc. Intern. Symp. on Botanic Gardens of the Tropics. Penang, Malaysia.

Ng, F.S.P., & C. Low. 1982. Check list of endemic trees of the Malay Peninsula. FRI Research Pamphlet No. 88.

27 Conservation of mangrove formations in Java

SUKRISTIJONO SUKARDJO

Summary

The mangrove forests of Java cover about 50,000 ha and represent an important natural resource. They are managed by the Department of Forestry and/or the State Forestry Corporation (Perum Perhutani). Large areas are at present being converted into fish ponds or destroyed for other uses. However, as the mangrove forest ecosystem plays an important role in the conservation of commercially valuable marine species, in the stabilisation of coastal areas, and also provides valuable raw materials, their destruction will have serious economic consequences which will exceed immediate commercial gains from present development. The mangrove formations are also threatened by pollutants. In this paper structure and floristics of the mangrove forests in Java are described, and research programmes relating to their management and conservation are reviewed. Recommendations are made to allow optimal and long-term development of the important mangrove resource. Further research is needed to gain an understanding of the many factors, cultural, physical, and biological, that influence the future of the mangrove forests in Java.

Introduction

Five major forest types are recognised in Java, e.g., mangrove forest (c. 50,000 ha), teak forest (910,000 ha), pine forest (352,000 ha), natural mixed forest (1,337,000 ha), and forest plantations (354,000 ha) (Bina Program DirJen Kehutanan Departemen Pertanian 1981).

Luxuriant mangrove communities were previously extensive in Java, especially along the northern coast. Most of these mangrove forests have been destroyed, and mangrove vegetation is now much more limited, although areas of natural forest still occur, particularly in Ujung Kulon National Park. Due to the high cliffs no luxuriant mangrove forests can be found along most of the southern coast. The remaining mangroves are under threat of complete destruction due to various types of exploitation, and some natural reserves have been established in Java by the Directorate General of Forest Protection and Nature Conservation (PHPA), Department of Forestry, to preserve samples of these important ecosystems.

The long record of biological conservation in Indonesia has been reviewed by Basjarudin (1974), Jacobs & de Boo (1982), Kies (1935), Lamoureux (1974) and Pluygers (1952). Many nature reserves have been established for

P. Baas et al. (eds.), The Plant Diversity of Malesia, 329–340.

Table 27.1 — Mangrove and associated fish pond areas in Java.

Province	Total area (ha)*	Present fish pond area (ha)
DKI Jakarta	95	1,163
West Java	28,513	40,257
Central Java	13,576	23,166
East Java	7,750	49,917
Java	49,934	114,503

* In the five years planning of the Badan Penelitian dan Pengembangan Pertanian 10% has been set aside as potential fish ponds.

public recreation, water supply, maintaining climatic stability, protecting the sources of rivers, and various other reasons. However, conflicting interests of the population of Java complicate the management and protection of Java's coastal and estuarine resources (Sukardjo 1980a, 1982a, 1989a; Sukardjo & Akhmad 1982).

In this paper I will review the characteristics of natural and man-influenced mangrove forests and discuss their conservation and ecological management.

Extent and species composition of the mangrove forest

Mangrove forest is one of the most important vegetation types in the coastal zone of Java and occurs on muddy flats or fine textured soils in deltas, estuaries, coastal belts and along coastal islands. The total area in Java is about 50,000 ha (Table 27.1), i.e., 1.2% of the total mangrove forest vegetation of Indonesia or 0.4% of the total land surface of Java. Overexploitation, conversion and mismanagement in the past (Sukardjo 1980a, 1984a, 1989a) has resulted in the loss of substantial mangrove forest areas. Precise estimates of the extent of the loss are conflicting and must be looked into. Not only has the mangrove area diminished, also the quality of trees and vegetation has been adversely affected. In Central and East Java mangrove forests have been ruthlessly thinned, and their rehabilitation does not proceed well (Fernandes 1934; de Jong 1934; Schnepper 1933; Sukardjo 1980b, 1984b, 1989d). Immediate steps are required to conserve and rehabilitate the remaining mangrove forest. However, conflicts between local population and the State Forestry Corporation (Perum Perhutani) regarding forest boundaries, land tenure and land use complicate the problem.

In 1965, the Perum Perhutani successfully planted mangrove seedlings along the northern coast of West Java and Segara Anakan in order to reduce the rapid erosion there (Sukardjo 1984a, 1989a; Haditenojo & Abas Ts 1984). More research is required on mangrove establishment, and to develop planting techniques and provide information to coastal developers both public and private. Techniques should also be developed for the establishment of mangroves in the degraded areas along the coast of Java.

The floristic composition of the mangrove forest is geographically variable, depending on soil characteristics. Thirty-five mangrove species are currently recorded for Java (Table 27.2). The most common are *Avicennia marina, Avicennia alba, Rhizophora apiculata, Rhizophora stylosa,* and the weedy fern *Acrostichum aureum.*

Plant communities

Based on numerous studies (Azkab & Sukardjo 1987; Djaja *et al.* 1984; Buadi 1979; Harminto & Jusuf 1979; Indiarto *et al.* 1987; Kartawinata & Waluyo 1977; Sukardjo 1980b, 1982b, 1984b, 1987b, 1988, 1989b, c, d; Tim Ekologi Fakultas Perikanan IPB 1984) nine communities of mangrove forest have been recognised in Java. Their characteristics are summarised in Table 27.3. Most of them have a low canopy with profiles less than 15 m high. Distribution patterns of trees, saplings and seedlings do not differ significantly from one another in these communities. Zonation is discernible but the sequences are not constant or distinct (de Haan 1931; Watson 1928). Edaphic factors together with tidal phenomena and salinity, interacting in a complex way, are involved in determining zonation and community characteristics. In West Java *Avicennia* forest communities grow on soft mud along the seaward fringe. A pure community of *Avicennia* was found to extend along a transect of 200 m from the sea inland. Soil texture is known to modify the tolerance of *Avicennia* to soil salinity. *Rhizophora* communities are restricted to sites with more compact soils. In general mangrove soils can be classified as clay, clay/loam, loam, and sand. There is a decreasing trend for clay content in the landward direction (Sukardjo 1982b, 1987b, 1988; Soerianagara 1971).

Along the northern coast of West Java and Segara Anakan Cilacap, Central Java, a famous mangrove forest has been established of pure stands of *Rhizophora* spp. Other mangrove species may be occasionally present but are insignificant in density and are confined to the edges of this man-made forest.

Comprehensive knowledge of the structure of the mangrove forest in Java will help explain the intricate interrelationships and interdependence between the species, sites and environmental components. This information is required for management and optimal production of economic products on a sustained yield basis.

Table 27.2 — List of mangrove species in Java.

Apocynaceae	1.	*Cerbera manghas* L.
Bignoniaceae	2.	*Dolichandrone spathacea* (L. f.) K. Schum.
Combretaceae	3.	*Lumnitzera littorea* (Jack) Voigt
	4.	*Lumnitzera racemosa* Willd.
Euphorbiaceae	5.	*Excoecaria agallocha* L.
Meliaceae	6.	*Xylocarpus granatum* Koen.
	7.	*Xylocarpus moluccensis* (Lamk.) Roem.
Myrsinaceae	8.	*Aegiceras corniculatum* (L.) Blanco
Myrtaceae	9.	*Osbornea octodonta* F. v. M.
Palmae	10.	*Nypa fruticans* Wurmb.
Rhizophoraceae	11.	*Bruguiera cylindrica* (L.) Lamk.
	12.	*Bruguiera gymnorrhiza* (L.) Lamk.
	13.	*Bruguiera parviflora* (Roxb.) White & Arn. ex Griff.
	14.	*Bruguiera sexangula* (Lour.) Poir.
	15.	*Ceriops decandra* (Griff.) Ding Hou
	16.	*Ceriops tagal* (Perr.) C.B. Robinson
	17.	*Kandelia candel* (L.) Druce
	18.	*Rhizophora apiculata* Blume
	19.	*Rhizophora mucronata* Lamk.
	20.	*Rhizophora stylosa* Griff.
Lythraceae	21.	*Pemphis acidula* J.R. & G. Forster
Rubiaceae	22.	*Scyphiphora hydrophyllacea* Gaertn.
Rutaceae	23.	*Merope angulata* (Willd.) Swingle
Sonneratiaceae	24.	*Sonneratia alba* J.E. Smith
	25.	*Sonneratia caseolaris* (L.) Engl.
	26.	*Sonneratia ovata* Backer
Sterculiaceae	27.	*Heritiera littoralis* Dryand. ex W. Ait.
Verbenaceae	28.	*Avicennia alba* Blume
	29.	*Avicennia marina* (Forsk.) Vierh.
	30.	*Avicennia officinalis* L.
Acanthaceae	31.	*Acanthus ebracteatus* Vahl.
	32.	*Acanthus illicifolius* L.
	33.	*Acanthus volubilis* Wall.
Dennstaedtiaceae	34.	*Acrostichum aureum* L.
Leguminosae	35.	*Derris heterophylla* (Willd.) Backer

The need for mangrove management and conservation

Mangrove forests outside Java are little utilised in Indonesia. Java in contrast is a crowded island and utilisation and conversion of mangrove forests is a source of conflict (Sukardjo 1982a, 1987a). The population of Java has greatly expanded since about 1970, and the mangrove forest has dramatically declined concomitantly.

The northern coast of Java offers contrasting landscapes. The diversity in comparatively small areas is reflected in variation in the structure and floristic composition of the mangrove vegetation. Mangrove forest has always been subjected to exploitation and the forest resource has in many places gradually deteriorated into shrub forest. Residual large trees remain scattered between an increasing number of weeds such as *Panicum repens, Paspalum commersonii, P. vaginatum, Sporobolus* sp., as reported for the Cimanuk delta complex (Sukardjo 1984b; Sukardjo *et al.* 1985). Owing to the economic importance of the mangroves, the authorities will be required to rehabilitate mangrove forest by establishing nature reserves and planning their management.

There are seven large rivers running from the uplands to the northern coastal zone in West Java. The river Cimanuk constitutes the largest system among them, and has played an important role in the balanced agricultural development of the province. Although estuary mudflats in the mangrove ecosystem might look barren, they are in fact an important feeding ground for many commercial fish and crustacean species. Destruction of the mangroves means loss of habitat for these species. Lackey (1975) pointed out that fishery comprises the interaction of habitat, biota and man. Man is the key factor in the management and conservation of the mangrove forest, but man also poses conflicting demands on the mangrove resource.

The existence of rare or endangered estuarine communities, including migratory and non-migratory birds (Sajudin *et al.* 1984; Sukardjo 1987a; Allport & Ann-Wilson 1986; Hoogerwerf & Rengers Hora Siccama 1938) and other wildlife (Manullang *et al.* 1984; Supriatna 1984; Rusminarto *et al.* 1984) is of intrinsic value to the mangrove vegetation in Java. Rambut and Dua islands in the Jakarta Bay, Muara Angke and Cimanuk nature reserves on the northern coast of West Java are the largest and numerically richest nesting sites for birds among the mangrove areas in Java. On and near those areas over 190 bird species were recorded (Hoogerwerf & Rengers Hora Siccama 1938) of which 148 species were nesting. Among them 15 species that nest in the mangrove forest are classified as endangered.

The conservation of undisturbed mangroves is essential to the survival of stocks of marine organisms of commercial value such as Barramundi fish (*Lates cavifron*) and prawn. Present nature reserves are far from adequate for this purpose. Reversion of mangrove forest thickets to steppe formation as observed

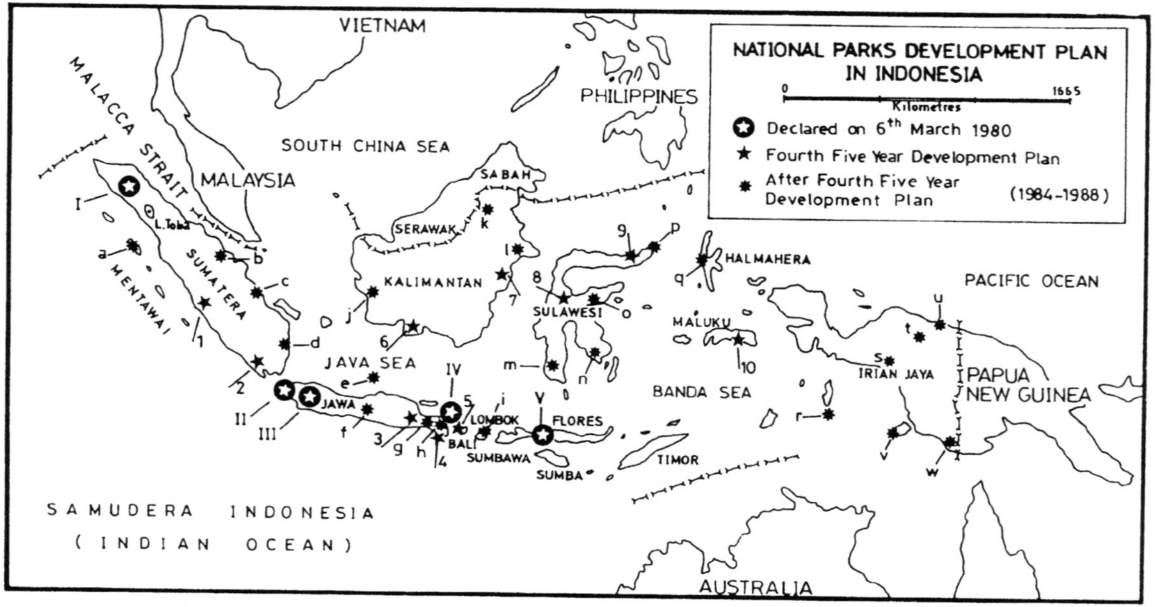

National Parks declared on 6 March 1980: **I** = Gunung Leuser; **II** = Ujung Kulon; **III** = Gunung Gede-Pangrango; **IV** = Baluran; **V** = Komodo. — *Fourth five-year development plan (1984–1988):* **1** = Kerinci-Seblat; **2** = Barisan Selatan; **3** = Bromo Tengger-Semeru; **4** = Meru Betiri; **5** = Bali Barat; **6** = Tanjung Puting; **7** = Kutai; **8** = Lore Lindu; **9** = Dumoga Bone; **10** = Manusela. — *National Parks after fourth five-year development plan:* **a** = Siberut; **b** = Siak Dua; **c** = Berbak; **d** = Way Kambas; **e** = Karimun Jawa; **f** = Merapi-Merbabu; **g** = Yang Plateau; **h** = Ijen; **i** = Rinjani; **j** = Gunung Palung; **k** = Khayan River; **l** = Sangkulirang; **m** = Bantimurung Goa Leang; **n** = Rawa Opa; **o** = Morowali; **p** = Tangkoko Batuangas; **q** = Halmahera; **r** = Aru; **s** = Lorenz; **t** = Poja/Memberamo; **u** = Cyclops; **v** = Wasur; **w** = Dolok.

Figure 27.1 — Conservation areas in Indonesia (source: PHPA Departemen Kehutanan 1987).

along the coast of Central Java and other forms of decline of mangrove forest clearly indicate that further conservation measures are required. The geographical zonation of new reserves must be based on the extent of each of the nine community types. For each zone the (future) appearance of natural features, procedures of conservation such as preservation, protection, restoration, and controls of various kinds of exploitation must be defined. Various kinds of planning for nature reserves have been attempted with input by ecologists and conservationists (Figure 27.1). The planning is based on readily identifiable features of the mangrove communities. Information on the ecological importance of mangroves to the fishery sector is required for management planning. Koorders' proposals (1912, 1916) can still be used in preparing plans of action.

Ecological management of mangrove conservation areas

In Java a number of nature reserves were established by the colonial government, e.g., Muara Angke, Pulau Dua, Pulau Rambut, Baluran, Banyuwangi Selatan, Ujung Kulon. They are permanently populated by inland birds (Hoogerwerf 1948; Kuroda 1933/1936; Kretschmer de Wilde 1939; Sody 1930, 1956). The conservation areas of Java vary in size and community, ranging from shrubby to luxuriant mangrove forest and abandoned wetland (e.g. Cimanuk). The mangrove forests must be managed intensively. In an earlier paper (Sukardjo 1987a) I proposed a number of steps for developing management systems for the Cimanuk nature reserve. These proposals are also valid for other mangrove vegetations in Java:

1. Inventory of mangrove communities in the area, preparation of a vegetation map, and a map of potential natural mangrove vegetation based on a phytosociological survey. A study of the dynamics of successional changes should be included in this inventory phase.
2. Evaluation of the mangrove communities in terms of disturbance, scarcity of species, and the degree to which they represent the mangrove ecosystem. Aspects of environmental protection, wildlife protection (e.g. nesting birds), conservation of natural resources (fish, crustaceans), and nature education should play a role in the evaluation.
3. Preparation of a management plan for each community based on 1 and 2.
4. Decision making about which mangrove forest community should be preserved, conserved or restored.
5. Implementation of the management plan in three stages: a) Zoning of the area concerned according to the conservation values of each community. Vegetation maps of the area are indispensable at this stage. b) Determination of how the communities should be preserved, conserved or restored. The basic principle here is to follow natural systems, to control natural succession, to ac-

Table 27.3 — Some characteristics of the nine mangrove community types in Java.

Locality	Predominant plant species in order of abundance by Importance Value	Species richness	Substrate	Reference
Pulau Dua	1. *Rhizophora stylosa – Rhizophora apiculata*	12	?	Buadi 1979
Pulau Rambut	2. *Scyphyhora hydrophyllacea – Lumnitzera racemosa*	6	sandy loam	Kartawinata & Waluyo 1977
	Rhizophora mucronata – Rhizophora stylosa	2	sand	
	3. *Rhizophora mucronata*	5	silty clay	
Muara Angke Kapuk	4. *Avicennia marina – Avicennia alba*	4	clay loam	Sukardjo 1989b
	Avicennia marina – Avicennia alba	3	clay loam	
	5. *Avicennia marina – Rhizophora apiculata*	7	clay loam	
	Rhizophora apiculata – Rhizophora mucronata	5	clay loam	
Ujung Karawangwang	6. *Avicennia marina – Aegiceras corniculatum*	9	?	Djaja *et al.* 1984
Indramayu	*Avicennia alba – Avicennia marina*	9	clay loam	Sukardjo 1980
Segara Anakan	7. *Bruguiera gymnorrhiza*	7	?	Sukardjo 1989d
	8. *Bruguiera gymnorrhiza – Rhizophora apiculata*	4	?	
	9. *Aegiceras corniculatum – Rhizophora mucronata*	7	?	Suroyo 1987
Baluran	*Rhizophora stylosa – Rhizophora apiculata*	16	?	Indiarto *et al.* 1987
Grajagan	*Rhizophora apiculata – Avicennia* spp.	14	?	Unpublished data

celerate succession, to create semi-natural communities (man-made forest) similar to wild communities for the use of natural resources. At this stage decisions also have to be made concerning protection against disturbance by wind and pollution (cf. Sumatra 1982; Sukardjo & Toro 1989), degradation of habitat, etc. c) Procedural steps to implement the management plan.

A flow chart for this fundamental approach, including all interacting steps and actions to be taken is available from the author on request.

Most of the mangrove afforestation and maintenance work in Java is carried out by Perum Perhutani (Sukardjo 1989a; Al Rasyid 1984) and by local inhabitants (Soeroso SD & Hadipurnomo 1984). Conservation of mangrove formations in Java is the responsibility of the government, which recognises the importance of ecological management of the mangrove resources as part of a general conservation ethic. Nevertheless conservation of mangrove communities in Java is of considerable complexity, and it has also been necessary to persuade provincial governments about the environmental and potential economic values of mangrove forests.

Ecologists have an important role to play in the implementation of a sound conservation policy. Their tasks can be summarised as follows:

1. Promoting an active and sound use of the country's natural resources, including: a) Co-operation in town planning and development. b) Co-operation in the planning of green belts and the conservation of historical parks. c) Nature conservation in general, i.e., of fauna, flora, water, air and soil.
2. Research: a) Of areas in need of conservation b) Biological studies of animals and plants requiring conservation. c) Studies of natural and man-induced changes. d) Study and documentation of sectors set aside for the building of 'masyarakat adil dan makmur' and also of future industrial regions as potentially interfering with the conservation of natural resources.
3. Cultural tasks: a) Propagation of principles, ideas and aims of nature conservation by means of lectures, exhibitions, films, and excursions. b) Publication of a journal, supported by the government agencies (Pusat, Daerah, KLH, Departemen Penerangan, Departemen Pendidikan dan Kebudayaan) and/or 'swasta', reflecting the efforts of the government in its conservation venture, and of articles on conservation in other media.

References

Al Rasyid, H. 1984. Program penghijauan di pantai teluk Jakarta (Greening programme along the coast of Jakarta Bay). In: I. Soerianagara *et al.* (eds.), Proc. Seminar II Ekosistem Mangrove: 144–146. LIPI Panitia Program MAB Indonesia, Jakarta.

Allport, G.A., & S. Ann-Wilson. 1986. Results of a census of the Milky Stork Mycteria cinerea in West Java. Study Report no 14, Intern. Counc. Bird Preservation.

Azkab, M.H., & S. Sukardjo. 1987. Komunitas semai hutan mangrove di Pulau Pari, Kepulauan Seribu (Mangrove seedling community in Pari Island, Thousand Islands). In: I. Soerianagara *et al.* (eds.), Proc. Seminar III Ekosistem Mangrove: 98–109. LIPI Panitia Program MAB Indonesia, Jakarta.

Basjarudin, H. 1974. Nature conservation and wildlife management in Indonesia. In: K. Kartawinata & R. Atmawidjaja (eds.), Coordinated study of lowland forests of Indonesia: 89–115. BIOTROP and IPB Bogor, Indonesia.

Bina Program DirJen Kehutanan Departemen Pertanian. 1981. Report on the forests of Indonesia. Department of Agriculture, Directorate General of Forestry Planning, Publ. no 18.

Buadi. 1979. Hutan bakau di Pulau Dua, teluk Banten Jawa Barat (Mangrove forest in Dua Island, Banten Bay, West Java). In: S. Soemodihardjo *et al.* (eds.), Proc. Seminar Ekosistem Hutan Mangrove: 69–71. LIPI Panitia Program MAB Indonesia, Jakarta.

Djaja, B., G. Sudargo & B. Indrasuseno. 1984. Hutan mangrove di Tanjung Karawang Bekasi Jawa Barat (The mangrove forest of Tanjung Karawan Bekasi, West Java). In: I. Soerianagara *et al.* (eds.), Proc. Seminar II Ekosistem Mangrove: 156–161. LIPI Panitia Program MAB Indonesia, Jakarta.

Fernandes, D.A. 1934. Over mangrove culturen. Naschrift. Tectona 27: 299–304.

Haan, J.H. de. 1931. Het een en ander over de Tjilatjapsche vloedbosschen. Tectona 24: 39–76.

Haditenojo, P.S., & Abas Ts. 1984. Pengalaman pengelolaan hutan mangrove di Cilacap (On the management of mangrove forest at Cilacap). In: I. Soerianagara *et al.* (eds.), Proc. Seminar II Ekosistem Mangrove: 65–73. LIPI Panitia Program MAB Indonesia, Jakarta.

Harminto, S., & J. Jusuf. 1979. Survai pendahuluan terhadap keadaan vegetasi hutan bakau di pesisir selebah barat teluk Jakarta (Preliminary survey on mangrove forest vegetation along the west coast of Jakarta Bay). In: S. Soemodihardjo *et al.* (eds.), Proc. Seminar Ekosistem Hutan Mangrove: 55–58. LIPI Panitia Program MAB Indonesia, Jakarta.

Hoogerwerf, A. 1948. Contribution to the knowledge of the distribution of birds on the island of Java. Treubia 19: 83–137.

Hoogerwerf, A., & G.F.H.W. Rengers Hora Siccama. 1937/38. De avifauna van Batavia en omstreken. Ardea 26: 1–51, 116–159; 27: 41–92, 179–246.

Indiarto, Y., S. Prawiroatmodjo, Mulyadi & I.N. Kabinawa. 1987. Analisis vegetasi hutan mangrove Baluran, Jawa Timur (Analysis of Baluran mangrove forest vegetation). In: I. Soerianagara *et al.* (eds.), Proc. Seminar III Ekosistem Mangrove: 92–97. LIPI Panitia Program MAB Indonesia, Jakarta.

Jacobs, M., & T.J.J. de Boo. 1982. Conservation literature on Indonesia: Selected annotated bibliography. Rijksherbarium Leiden.

Jong, B. de. 1934. Over mangrove culturen. Tectona 27: 288–298.

Kartawinata, K., & E.B. Waluyo. 1977. A preliminary study of the mangrove forest on Pulau Rambut. Marine Res. Indon. 18: 119–129.

Kies, C.H.M.H. 1935. Natuurbescherming in Nederlandsch Indië. Ned. Comm. Intern. Natuurbesch. Meded. 10, Suppl.: 12–24.

Koorders, S.H. 1912. Oprichting eener Nederlandsch-Indische Vereeniging tot Natuurbescherming. Nederlandsch-Indisch Landbouw-Syndicaat, Soerabaia.

Koorders, S.H. 1916. Beknopt overzicht van eenige Nederlandsch-Indische natuurmonumenten, die tot het landsdomein behooren en waarvoor staatsbescherming noodig geacht wordt. Meded. Ned.-Ind. Ver. Natuurbesch. 1: 1–64.

Kretschmer de Wilde, C.J.M. 1939. Vogelparadijs op Java's noordkust; van een bezoek aan het natuurmonument Poelau Doewa. In Jaren ...: 299–303.

Kuroda, N. 1933, 1936. Birds of the island of Java, I, II. Passares, Non-passares. Privately published by the author, Tokyo.

Lackey, R.T. 1975. Fisheries and ecological models in fisheries resource management. Dept. Fish. Wildl. Sci., Virginia Polytech. Inst., State Univ. Blacksburg, Virigina, U.S.A.

Lamoureux, C.H. 1974. Observations on conservation in Indonesia. In: K. Kartawinata & R. Atmawidjaja (eds.), Coordinated study of lowland forests of Indonesia: 116–121. BIOTROP and IPB Bogor, Indonesia.

Manullang, B.O., J. Supriatna & D.S. Hadi. 1984. Pengamatan lutung Presbytis cristatus di hutan mangrove Tanjung Karawang Bekasi Jawa Barat (Observation on the Presbytis cristatus in the mangrove forest of Tanjung Karawang, Bekasi West Java). In: I. Soerianagara *et al.* (eds.), Proc. Seminar II Ekosistem Mangrove: 238–242. LIPI Panitia Program MAB Indonesia, Jakarta.

Pluygers, L.A. 1952. Natuurbescherming en wildbeheer speciaal met betrekking tot Indonesië. Wolters, Djakarta, Groningen.

Rusminarto, S., A. Munif & B. Rijadi. 1984. Survei pendahuluan fauna nyamuk di sekitar hutan mangrove Tanjung Karawang Bekasi Jawa Barat (Preliminary survey on mosquitoes of Tanjung Karawang mangrove forest Bekasi, West Java). In: I. Soerianagara *et al.* (eds.), Proc. Seminar II Ekosistem Mangrove: 232–234. LIPI Panitia Program MAB Indonesia, Jakarta.

Sajudin, H.R., H. Rusmendro & Del Afradi. 1984. Inventarisasi avifauna di kawasan hutan mangrove Tanjung Karawang Bekasi Jawa Barat (Inventory of the avifauna of the Tanjung Karawang mangrove forest Bekasi, West Java). In: I. Soerianagara *et al.* (eds.), Proc. Seminar II Ekosistem Mangrove: 228–231. LIPI Panitia Program MAB Indonesia, Jakarta.

Schnepper, W.C.R. 1933. Vloedbosschen culturen. Tectona 26: 907–919.

Sody, R.J.V. 1930. Broedtijden der vogels in West en Oost Java. Tectona 23: 183–198.

Sody, R.J.V. 1956. De Javaanse bosvogels. Met enige indrukken over hun hoogteverspreiding en over het verband tussen deze en de plantengroei. Indon. J. Nat. Sc. 112: 153–170.

Soerianagara, I. 1971. Characteristics of mangrove soils of Java. Rimba Indonesia 16: 141–150.

Soeroso SD & Hadipurnomo. 1984. Rehabilitasi hutan mangrove di pantai Probolinggo, Jawa Timur: Kasus dasa curah Sawo (Socio-economic aspects of the management of mangrove forest on the coastal area of Probolinggo, East Java). In: I. Soerianagara *et al.* (eds.), Proc. Seminar II Ekosistem Mangrove: 78–81. LIPI Panitia Program MAB Indonesia, Jakarta.

Sukardjo, S. 1980a. The mangrove ecosystem of the northern coast of West Java. In: E.C.F. Bird & A. Soegiarto (eds.), Proc. Jakarta Workshop on Coastal Resource Management: 54–64. The United Nations University, Tokyo.

Sukardjo, S. 1980b. The mangroves in the new Cimanuk delta. Paper presented at the LIPI-UNU Joint Seminar on the coastal resource of the Cimanuk, West Java, Jakarta.

Sukardjo, S. 1982a. Mangroves of the Jakarta Bay need conservation. Duta Rimba III (51): 3–5.

Sukardjo, S. 1982b. Soils in the mangrove forest of the Cimanuk delta, West Java, Indonesia. BIOTROP Spec. Publ. no 17: 191–201.

Sukardjo, S. 1984a. The present status of the mangrove forest ecosystem of Segara Anakan Cilacap, Java. In: A.C. Chadwick & S.L. Sutton (eds.), Tropical rainforests: 53–70. The Leeds Symp., Spec. Publ. of the Leeds Philos. and Lit. Soc., U.K.

Sukardjo, S. 1984b. Ekologi permudaan alami hutan mangrove di delta Cimanuk (The ecology of natural rejuvenation of mangrove forest at the Cimanuk delta). In: I. Soerianagara *et al.* (eds.), Proc. Seminar II Ekosistem Mangrove: 329–339. LIPI Panitia Program MAB Indonesia, Jakarta.

Sukardjo, S. 1987a. Conservation of the marine life of mangrove forest, estuarine and wetland vegetation in the Cimanuk nature reserve. BIOTROP Spec. Publ. no 30: 35–52.

Sukardjo, S. 1987b. Tanah dan status hara di hutan mangrove Tiris Indramayu Jawa Barat (Soils and nutrient status in the mangrove forest in Tiris Indramayu, West Java). Rimba Indonesia 21: 12–23.

Sukardjo, S. 1988. Penelitian produktivitas hutan mangrove di cagar alam Cimanuk Indramayu: 1. Tanah-tanah mangrove di komplek delta Cimanuk: pendekatan 'multivariate' (The studies of mangrove forest productivity in the Cimanuk nature reserve: 1. Soils of the mangrove forest in the Cimanuk delta complex: multivariate approach). In: S. Anantawirya & S. Notosudarmo (eds.), Proc. Seminar Ekologi Tanah dan Ekotoksikologi: 15–41. The Satya Wacana Christian Univ. Press.

Sukardjo, S. 1989a. The present status of the mangrove forests in the northern coast of West Java with special reference to the recent utilization. Paper presented at the Intern. Conf. on Wetland: The people's role in wetland management, Leiden, June 1989.

Sukardjo, S. 1989b. Litter fall production and turnover rate in the mangrove forest in Muara Angke Kapuk Jakarta. BIOTROP Spec. Publ. no 37: 129–144.

Sukardjo, S. 1989c. Produksi primer neto hutan mangrove di Muara Angke Kapuk Jakarta (Net primary productivity of the mangrove forest in Muara Angke Kapuk Jakarta). Rimba Indonesia 23 (in press).

Sukardjo, S. 1989d. Struktur hutan-hutan mangrove di Cikeperan Cilacap (Structure of the mangrove forests in Cikeperan Cilacap). Rimba Indonesia 23 (in press).

Sukardjo, S., & S. Akhmad. 1982. The mangrove forests of Java and Bali. BIOTROP Spec. Publ. no 17: 113–126.

Sukardjo, S., M.H. Azkab & A.V. Toro. 1985. Weeds in natural regeneration area of mangrove forest in Cimanuk delta complex. Paper presented at the Workshop on the ecology and management of aquatic vegetation in the tropics, Jakarta, March 1985.

Sukardjo, S., & A.V. Toro. 1989. The tar balls pollution in the beach vegetation of Pari Island. Paper presented at the CAFEO VII, Bali, January 1989.

Sumatra, I.M. 1982. Insecticide residue monitoring in sediments, water, fish and mangroves at the Cimanuk delta. Majalah Batan 15 (4): 18–30.

Supriatna, J. 1984. Jenis-jenis ular di hutan mangrove dan makanan ular tambak Cerberus rynchop Schn. (Snakes of the mangrove forest and the food of Cerberus rynchop Schn., a brackish water snake). In: I. Soerianagara *et al.* (eds.), Proc. Seminar II Ekosistem Mangrove: 172–174. LIPI Panitia Program MAB Indonesia, Jakarta.

Tim Ekologi Fakultas Perikanan IPB. 1984. Ecological aspects of Segara Arakar in relation to its future management. Inst. of Hydraulic Engin. and Fac. of Fisheries, Bogor Agricultural Univ., unpubl. report.

Watson, J.G. 1928. Mangrove forests of the Malay Peninsula. Mal. For. Rec. 6: 1–276.

28 Gene banks and plant taxonomy

L. J.G. VAN DER MAESEN

Summary

Gene banks are usually specialised in certain crops or a group of crops. The first and preferably unambiguous entry of documentation for gene bank accessions is the Latin name of the plant(s) concerned, showing the importance of plant taxonomy in conservation of genetic resources. It is advisable to maintain a herbarium of key cultivars and wild species related to the crops as a permanent reference. In several collections, including living collections in botanic gardens, continuing efforts need to be made to update and verify proper naming. The role of botanic gardens as living depositories for germplasm should be increased and supported. For inventories of tropical ecosystems, e.g. rain forests, the need for taxonomy can hardly be disputed. A review is given of the gene banks in the Flora Malesiana region.

Introduction

Plant taxonomy or systematic botany is a key science for tropical research and documentation – for agriculture, forest ecology, biology other disciplines. A symposium on that topic was held in Uppsala in September 1987, and its proceedings (Hedberg 1988) highlight the need for taxonomy, even if the direct users are mainly other taxonomists (Heywood 1988). The need to produce both floras as well as practice-oriented publications was emphasised. In this paper I will consider the important role of taxonomy in establishing and maintaining genebanks for agriculture and conservation.

Conservation of natural diversity

Gene banks constitute a way to conserve genetic diversity, usually of crop plants, including their wild relatives. Ideally many flowering plants are stored as seed, the natural entity of the plant to survive bad times, sometimes encased in fruits or other diaspores. Orthodox seeds can be stored at low temperatures (–18 °C or lower) at a low relative moisture content (often as low as 5%) (Cromarty *et al.* 1982). When properly and slowly dried shelf-life of such seeds is at least 25 to 100 years. Storage in liquid nitrogen extends the life expectancy of seeds even more, and is already practised for some species. Recalcitrant seeds pose a problem. Seeds of cocoa (*Theobroma cacao*) and many tropical fruit trees show a rapidly diminishing viability when conserved at low temperatures and reduced moisture conditions. Living material of such species

P. Baas et al. (eds.), The Plant Diversity of Malesia, 341–349.

should therefore be maintained in the same way as heterozygous genotypes: by vegetative propagation in living collections, or in tissue culture or cryopreservation thereof. Although this is feasible, it is much costlier than seed storage. Both methods require optimum maintenance of embryo vigour to prevent genetic erosion in storage, and periodical rejuvenation is needed.

Another possibility is the maintenance of *in situ* genebanks, such as traditional cultivation of crop landraces, but very few of such genebanks exist as yet. Wildlife parks and nature reserves (Table 28.1) are *de facto in situ* gene banks: much of the natural diversity of an ecosystem or groups of ecosystems has the best chance to survive the onslaught by man in suitable nature reserves. A certain level of maintenance may be needed in some places, because a climax vegetation may not shelter all species present in preceding phases of the ecosystem, those occupying e.g. dry or savanna-type niches. The protection of animals in wildlife reserves also implies the protection of plants, soil and water. However, constant protection and monitoring is required, at considerable effort and cost. If protected animals breed prolifically, they may threaten the vegetation whereas in low numbers their intervention promotes distribution, pollination and rejuvenation. A climax vegetation is not static. Gene banks in the form of wildlife reserves require continuous expert supervision to conserve the vegetation in its diversity. The number of wildlife parks (IUCN 1982; van Lavieren *et al.* 1984) in South-East Asia is impressive, but not all areas are large or well-protected.

Hand-picked islands of diversity can be found in Botanic Gardens, of which only a few important ones are situated in or near South-East Asia, approximately the area covered by Flora Malesiana (Table 28.1). These obtained plantation and food crops from the region, and from other sources, such as the Americas. Species arrived frequently through Kew, Paris, Amsterdam or Leiden. These and botanic gardens in other regions, or even ones newly to be established, may play a role in preserving both endangered species (IUCN/WWF 1989) and research plants of future economic promise. The continued existence of such gardens should be safeguarded by public means. Their degradation to the status of public parks at the cost of reduction of the diversity (usually unperceptible to the layman's eyes) is (becoming) a reality. Therefore more emphasis on their important function in the future – that of a gene bank – is called for: they must become a recognised part of the world's genetic resources network. In 1981, Raven (in Smith 1985) estimated that the world's botanic gardens contain about 35,000 species, i.e. less than 15% of the world's flowering plant species. Often this pertains to a limited number of specimens, and few accessions per species, so that this still only constitutes a narrow genetic basis vulnerable to genetic erosion. It must be pointed out that in the last half century the diversity of species in botanic gardens decreased alarmingly. Before the second world war Bogor Botanic Gardens alone housed more than 12,000 species.

Table 28.1 — National parks, botanical gardens and arboreta, and living collections in South-East Asia.

Country	Wildlife parks*	Botanical gardens or arboreta	*In situ* (forest) conservation and living collection areas
Brunei	1 (6)		
Burma	13 (7)		working collections of rice, teak
Cambodia	6		–
Indonesia	15 (23)	Bogor	50 ha Durio zibethinus nr Bogor**
		Cibodas	1500 ha *Cocos nucifera*, Bone-Bone
		Pasuran	140 ha fruit species, Paseh
		Bedugul, Bali	153 ha economic crops, Serpong
			200 ha economic crops, Cibinong
			Saccharum, Pasuruan
Laos	0		
Malaysia	39 (5)	Kepong	oil palm, PORIM
		Rimba Ilmu	rubber, RRIM
		Penang	
Papua New Guinea	7 (17)	Lae	banana, Laloki
		Port Moresby	root crops, Aiyura, Laloki, Keravat
Philippines	60	Laguna	NPGRL coconut (PCA)
		Manila	sugar (PHILSUCOM)
		Quezon	tobacco (PTRTC)
		Siniloan	
Singapore	?	Singapore	orchid collections
Thailand	35 (7)	Saraburi	100 ha *Pinus merkusii*
		Trang	Dept. Agric., Katsetsart Univ., several collections of veg. propagated fruit and food crops
Vietnam	4 (3)	Ho Chi Minh City	

* Between brackets ( ) the number of proposed areas. Wildlife parks do not include forest reserves or other protected forests, of which many are established. Many such areas are under consideration to be proposed as national parks, nature reserves or wildlife sanctuaries. Some parks are very small.

** IBPGR (1983) mentions further living collections in Manado, Sukamandi, Lembang, Manokwari, Gambung, Yogyakarta, Pasar Minggu, Jember, Den Pasar and Solo. Sources: IBPGR (1984), Henderson (1983), van der Maesen & Sadikin (1989), Holmgren *et al.* (1981), van Lavieren *et al.* (1984).

Bogor introduced oilpalm (*Elaeis guineensis*) in 1848, Manila was instrumental in the transfer of cocoa from the Americas in the 17th century, Singapore obtained rubber (*Hevea brasiliensis*) for Malaya in 1877, Calcutta obtained tea (*Camellia sinensis*) for India, and through Calcutta Assam tea reached the world. Timber trees and many economic crops found their way through this venerated garden.

Peradenya was a nursery for *Cinchona* in Ceylon, now Sri Lanka (Burkill 1966; Smith 1985). The role in transferring and popularising the now common flowering plants and trees comes only second after the important function of transfer of germplasm of food and industrial crops. Ornamental plants offer increasing possibilities for sale and export. The success of orchid growing and multiplication in e.g. Singapore, Thailand and Indonesia is well known, while CITES regulations attempt to restrict wholesale removal of natural stands of orchids. Cultivation in the country of origin (or elsewhere) is a most commendable practice (see also Kiew 1990, this volume). Germplasm is needed for this growing branch of plant industry. In fact in the Netherlands, with its billion-dollar flower production, the Centre for Genetic Resources (CGN/CPO 1989) is directing attention to ornamental plants. Rapid turnover of flower cultivars, even faster than for food crops, induces genetic erosion, and the demand for new genotypes is growing.

Documentation for gene banks

The germplasm collections of the major crops and various cultivar groups within those crops are usually just labeled with an accession number. A documentation system is crop-wise, and infraspecific classification should form part of the passport data. It is in multi-species collections, such as collections of fodder species, vegetables, wild relatives, that the scientific name is the first descriptor. Genebank accessions arrive with a varying amount of data, the scientific name may be correct but quite often that is not the case. It is the task of genebank curators to verify the names during inventory or regeneration, a task especially difficult if numerous species and infraspecific taxa are stored. The job of a germplasm botanist is always taxonomic in part. The training of germplasm botanists necessitates inclusion of a schooling in both practical plant taxonomy and population genetics.

Specialised taxonomists should help in identification germplasm collections, solicited or unsolicited. The inventory of rain forests, from where useful plants and useful genes may be obtained, is far from complete yet, and particularly here the need for proper naming meets no dispute. For instance Booth (1988) reviewed procedures to apply appropriate systematics.

Intricate hierarchical classification is possible in many crop plants and their wild relatives, to express morphogenetic relationships. Of course breeders look for the closest relatives to transfer possible useful genes, but distant relatives are not overlooked. Therefore, sections and series serve a useful purpose in larger genera if these reflect true relationships. Often initial generations take a long time before the required characteristic is appropriately incorporated in an end-product, and fresh approaches tend to disregard previous attempts with negative results. In general simple crossing is still preferred over new techniques such as genetic engineering, but such methods will find wider application if gene transfer indeed proves to be speedier and more direct.

Hanelt (1986) reviewed opinions on formal and informal infraspecific classification. Formal classifications are based on some diagnostic characters, on overall variability or on ecogeographical principles; informal classification may follow economic assets with special guidelines. Database management to make taxonomy accessible can be useful; e.g. ILDIS (International Legume Database and Information System) is in full swing to provide a database system for all leguminous species in the world (Lockerman *et al.* 1988).

If for the Flora Malesiana area an initiative materialises to install a database to speed up identification of plants through standardised concentrated means, many users of systematic botany in and outside the region will profit. Gene banks may find this particularly useful as systematic verification of the holdings may become easier. An international agreement to use the nomenclature found in Flora Malesiana as generally accepted for gene bank specimens – unless stringent reasons arise to prefer other names – seems desirable. Of course, cultivar naming is not covered by Flora Malesiana. A series of books on useful plants and a database for these (c. 5,000 species) is in preparation by the PROSEA project. The present status is presented in the Proceedings volume edited by Siemonsma & Wulijarni-Soetjipto (1989).

Intricate classifications produced by (bio)systematic research tend to tire the practical users. It is somewhat cumbersome to place the long ternary names into documentation systems with a finite number of characters per column. Some systems do not have such restrictions. It is possible to indicate the infraspecific taxa with a group name without formal status, such as the *Sesquipedalis* group for the yard-long bean cultivars of *Vigna unguiculata,* the cowpea. Sometimes these cultivar groups are indeed named after formal varietal classifications. When treating the pulses for the PROSEA Handbook, a stand was taken to express cultigens in cultivars and cultivar groups, and maintain botanical varieties and subspecies for wild related taxa (van der Maesen & Sadikin 1989). This approach will be followed where possible for the other commodity groups to be treated in the PROSEA volumes.

Apart from the correct botanical name, which data can be provided by taxonomic botanists? Their monographs form the basis for floras, but also for the passport data and preliminary evaluation data, as is the parlance, of genetic resources documentation systems. Correct passport data, such as occurrence, flowering and fruiting periods, background information on ecology, pollination biology, perhaps practical experience with the growing of a wild species, all these aid in evaluation of germplasm. Soil and cultivation data, even limited ones, are valuable for plants not widely known. Maybe even more important, these data if well-described in notes, and not buried in lengthy descriptions, will indicate the germplasm collector when and where to collect as far as this is still possible.

There is a need for liberal assistance by plant taxonomists to germplasm curators in national and international institutions, while the employment of botanists trained in taxonomy is equally needed and carried out in practice. Most gene banks are located near a well-accessible herbarium, another documentation system required for all but the most common species. Available reports do not list herbaria attached to gene banks in South-East Asia, Index Herbariorum (Holmgren *et al.* 1981, being updated) mentions several herbaria in or near Bangkok, Bogor, Kuala Lumpur, Lae, Manila etc. A collection of key cultivars and wild species related to the crops needs to be available as permanent reference in a herbarium in or near the various gene banks.

Gene banks in the Flora Malesiana region

Gene banks operating in the Flora Malesiana region vary very much in size and targets. In a regional context maize, groundnut, soyabean, *Vigna,* sweet potato, *Citrus,* mango, other tropical fruits, forage crops and coconut are considered for priority action – collection and maintenance – in several countries (IBPGR 1983; Hanson *et al.* 1984). Other details on the regional activities of genebanks in South-East Asia are given by IBPGR (1984) and recent IBPGR Annual Reports. Indonesia, Malaysia, the Philippines and Thailand attracted most of the collection missions up to 1988, and from these countries most accessions were procured. A vast majority of the c. 13,500 accessions was identified at the species level. Further classification and verification is needed in some cases, and remains needed for verification with every rejuvenation cycle.

Gene banks outside the region are equally important in maintenance, conservation, documentation and supply of germplasm important for the region. To obtain germplasm from regional gene banks is easier because of Quarantine regulations, and the maintenance of germplasm collections constitutes a national and international duty. As far as recent information goes, Table 28.2 lists the major gene banks now in operation in South-East Asia. The International Board for Plant Genetic Resources (IBPGR) stimulates and assists the foundation and running of international and national genetic resources units. Technical handbooks such as by Cromarty *et al.* (1982), Ellis *et al.* (1985) provide excellent backup, but the explained terms deserve a vademecum in a number of languages.

In general genetic resources are freely exchanged and internationally freely accessible. Sometimes restrictions are necessary to avoid possible misuses. This is a politically loaded topic.

Table 28.2 — Genetic Resources Centres in South-East Asia.

Country	Centre or acronym	Major crops	Samples (c.)
Indonesia	BORIF	Indigenous trop. legumes, local landraces and advanced cvs of rice, soyabean, groundnut, mungbean, pigeonpea	20,000
	MARIF	soyabean, groundnut, mungbean, pigeonpea	70,000
Malaysia	MARDI	rice	7,000
	Bambong Lima	local landraces of rice	7,000
	Kuala Lumpur	winged bean (Univ. Kebangsaan)	
	PORIM	rubber	
Papua New Guinea	UPNG		
	Laloki		
	Aiyura		
	Bubia		
Philippines	IRRI	rice, local and advanced cvs, wild species, breeding lines	124,000
	NPGRL, Laguna	*Brassica, Capsicum, Vigna,* cucurbits, egg-plant, groundnut, mungbean, okra, pigeonpea, soyabean, tomato and winged bean	64,000
Thailand	TISTR	Asiatic maize, winged bean	2,000
	RRI	Local landraces, wild species, advanced rice cvs	8,000
Vietnam	Thanh Tri	soyabean (Hanoi)	450
	Hau Gang	soyabean (Univ.of Cantho)	400

* Almost all mentioned institutions have medium or long term cold storage facilities.

** Of much importance for the region are international institutes such as ICRISAT, Patancheru, India; IITA, Ibadan, Nigeria; CIAT, Cali, Colombia; and the regional institute AVRDC, Shanhua, Tainan, Taiwan.

Conclusions

Genetic diversity of both cultivated and wild plants keeps diminishing, but good advances have been made in safeguarding germplasm of major crops and some of their wild relatives. Interaction between or rather coinciding interests of taxonomy and economic botany will result in proper maintenance and documentation of the world's natural heritage in service of sustainable production of food and raw materials. Several countries in South-East Asia have good gene banks, but geographic and taxonomic coverage is far from complete.

For plants not yet recognised as having economical potential, the vast majority of plant species, the only hope is *in situ* conservation in the growing number of national parks, and to some extent preservation in botanic gardens is feasible.

The continued availability of well-trained taxonomic botanists and people conversant with botanical systematics is required. Those in charge of collections of natural diversity should call the help of taxonomic specialists, and strive to apply the most recent and correct classification and names, to monitor what is still a wealthy tropical flora.

Acknowledgements

For their critical reading of the text and useful suggestions I am grateful to Prof. Dr. H.C.D. de Wit and Dr. R.G. van den Berg. Ir. L.P. van Lavieren of Euroconsult assisted with data on protected areas in South-East Asia.

References

Booth, F.E.M. 1988. Economic botany, taxonomy and plant collectors. In: I. Hedberg (ed.), Systematic botany – a key science for tropical research and documentation: 29–39. Acta Univ. Upsal. Symbolae Bot. Upsal. 28.

Burkill, I.H. 1966. A dictionary of the economic products of the Malay Peninsula. 2nd ed. Ministry of Agriculture and Co-operatives, Kuala Lumpur. Vol. 1 (A–H). Vol. 2 (I–Z).

CGN/IVT. 1989. Genenbank voor siergewassen. Noodzaak en prioriteiten. Centre for Genetic Resources, and Institute for the Breeding of Horticultural Crops, Wageningen.

Cromarty, A.S., R.H. Ellis & E.H. Roberts. 1982. The design of seed storage facilities for genetic conservation. International Board for Plant Genetic Resources, Rome.

Ellis, R.H., T.D. Hong & E.H. Roberts. 1985. Handbook of seed technology for genebanks. Vol. 1. Principles and methodology. Vol. 2. Compendium of specific germination information and test recommendations. International Board for Plant Genetic Resources, Rome.

Hanelt, P. 1986. Formal and informal classification of the infraspecific variability of cultivated plants – advantages and limitations. In: B.T. Styles, Infraspecific classification of wild and cultivated plants: 139–156.

Hanson, J., J.T. Williams & R. Freud. 1984. Institutes conserving crop germplasm: the IBPGR global network of genebanks. International Board for Plant Genetic Resources, Rome.

Hedberg, I. (ed.). 1988. Systematic botany – a key science for tropical research and documentation. Acta Univ. Upsal. Symbolae Bot. Upsal. 28.

Henderson, D.M. 1983. International directory of botanical gardens IV. Koeltz Scient. Books, Koenigstein.

Heywood, V.H. 1988. Tropical taxonomy – who are the users? In: I. Hedberg (ed.), Systematic botany – a key science for tropical research and documentation: 21–28. Acta Univ. Upsal. Symbolae Bot. Upsal. 28.

Holmgren, P.K., W. Keuken & E.K. Schofield. 1981. Index Herbariorum Part I. The herbaria of the world. 7th ed. Antwerpen, The Hague, Boston.

IBPGR. 1983. A cooperative regional programme in Southeast Asia. Report of the fifth Regional Committee meeting. International Board for Plant Genetic Resources, Rome.

IBPGR. 1984. A cooperative regional programme in Southeast Asia. Five-year plan of action 1985–1989. International Board for Plant Genetic Resources, Rome.

IUCN. 1982. 1982 United Nations list of national parks and protected areas. International Convention for Conservation of Nature and Natural Resources. Gland, Switzerland.
IUCN/WWF. 1989. The botanic gardens conservation strategy. Final draft. Gland, Switzerland. 56 pp.
Lavieren, L.P. van, P.J.M. Hillegers & H.D. Rijksen. 1984. Assessment of trained manpower requirements and protected area management personnel for Southeast Asia. Report compiled by the School of Environmental Conservation Management, Bogor, Indonesia on behalf of the United Nations Food and Agriculture Organisation, Rome (Compilers not printed on title page).
Lockerman, R.H., F.A. Bisby & L.J.G. van der Maesen. 1988. Workshop: integration of information on plant diversity. In: R.J. Summerfield (ed.), World crops: cool season food legumes: 129–133. Kluwer Academic Publishers, Dordrecht, Boston, London.
Maesen, L.J.G. van der, & Sadikin Somaatmadja (eds.). 1989. Pulses. Plant Resources of South-East Asia No. 1. Pudoc, Wageningen.
Siemonsma, J.S., & N. Wulijarni-Soetjipto (eds.). 1989. Proceedings of the First PROSEA International Symposium, May 22–25, 1989, Jakarta, Indonesia. Pudoc, Wageningen.
Smith, N.J.H. 1985. Botanic gardens and germplasm conservation. Harold L. Lyon Arboretum Lecture No. 14. Univ. Hawaii Press, Honolulu.

29 Malesian origin for a domestic Cocos nucifera

HUGH C. HARRIES

Summary

At one time it was thought that the ancestors of modern *Cocos nucifera* reached the Western Pacific area by long distance dispersal along a southern route from America, with a fossil (*Cocos zeylandica*) in New Zealand as a remnant of such a pathway. The concept of a southern route is an unnecessary complication. An origin for the whole Cocoeae tribe in western Gondwanaland seems most compatible with the present day distribution. The tribe probably differentiated shortly before the break-up of that super-continent. Members radiated and became very diverse in the Americas; some rafted on the African and Madagascar Plates, where they survive to the present day; others rafted on the Indian Plate, where they are now extinct. With its ability to float, the coconut became independent of plate tectonics for its dispersal. The wild type evolved by floating between the volcanic islands and atolls where these fringed the continental plates and not on the land masses at all. Islands in the Tethys Sea could have been the ancestral home of the coconut, from where it dispersed by floating to other islands in the Pacific and Indian Oceans but not into the Atlantic. It would also have floated to continental coastlines but would have stood less chance of surviving competition from other plants or predation by animals until domesticated by early man. The continental coast and larger islands of Malesia were the sites for such domestication long before both wild and domestic types were taken into agricultural cultivation.

Introduction

The argument that *Cocos nucifera* reached the Western Pacific area by long distance dispersal along a southern route from America, with a fossil (*Cocos zeylandica*) in New Zealand as a remnant of such a pathway, was originally based on the need to explain the relationship between the coconut and cocosoid palms that occurred in South America. Even when these were assigned to other genera the apparent lack of a 'wild' relative in Asia or the Pacific kept this argument alive (Purseglove 1972). Similarly, the possibility that the modern distribution of the coconut was due mainly to human intervention also confused the issue. In effect, domestication was thought to have produced the coconut we know today and cultivation to have displaced the 'wild' relative (Fosberg 1960).

However, by considering and describing what the likely appearance and properties of a wild type coconut might be, it was possible not only to show

P. Baas et al. (eds.), The Plant Diversity of Malesia, 351–357.

these might be found as far apart as Palmyra Island in mid Pacific and the Seychelles in the Indian Ocean (Harries 1978), but also to locate previously unsuspected specimens in Australia (Buckley & Harries 1984) and in the Philippines (Gruezo & Harries 1984) and on the Malay peninsula in Thailand (Harries *et al.* 1982). This helps to confirm reports in the literature of coconuts found growing wild in Australia (Bentham 1863–1878) and Indonesia (Koorders 1911) and elucidate previously reported, but unexplained, different forms found in Papua New Guinea (Dwyer 1938) and Borneo (Harries 1981a).

The purpose of this paper is to make five points:

- that the coconut originated on the fringes of Gondwanaland;
- that it spread by floating comparatively short distances on the margins of oceans rather than long distances across them;
- that it had evolved to the recognisable coconut phenotype before man ever met it;
- that the process of domestication modified it without obliterating the original wild type;
- that the domestication probably took place in Malesia.

Evolutionary dispersal

The general consensus until fairly recently was that the coconut had its origins in Melanesia – an area extending southward (from the Equator) to the Tropic of Capricorn between 45° and 180° E – but that it could not have evolved there from any progenitors now alive or extinct. The postulated ancestor was assumed to have been carried by ocean currents from South America to Polynesia, or by a southernly migration via Antarctica at a time when the polar climate was very much warmer and when it was part of Gondwanaland (Child 1974; Purseglove 1972). This subtropical continent may have been vastly higher than anyone thought with the addition of the Kerguelan Plateau, which stretches for more than 3,000 miles but now lies about half a mile below the surface of the southern Indian Ocean. Fragments of dinosaur teeth and pieces of charcoal have been recovered by the ocean drilling vessel *Joides,* indicating the one-time presence of dense forests and animal life (Palmer 1988).

Indeed, it is possible to consider an origin for the whole Cocoeae tribe in western Gondwanaland, modern South America (Uhl & Dransfield 1987). It is suggested that the tribe probably differentiated shortly before the break-up of that super-continent. There would have been every ecological niche available, from montane to coastal, from temperate to tropical. It was from the coastal tropical group that the coconut emerged. Whilst the other members of the palm family radiated and became very diverse in the Americas, some rafted on the African and Madagascar Plates, where they survive to the present day (Drans-

Map locating living (Vg) and fossil (Cs, Cz, Ps) cocosoid palms and modern continental land masses relative to the Tethys Sea. Based on Smith & Briden (1977: map 7: 100 million years). Vg: *Voanioala gerardii*; Cs: *Cocos sahnii*; Cz: *Cocos zeylandica*; Ps: *Palmoxylon sundaram*.

field 1989), others rafted on the Indian Plate, where now only fossils are found (Sahni 1946; Kaul 1951). But it is suggested here that the coconut did not raft on any particular tectonic plate. Nor did it wait to be found and carried by man. With its ability to float the coconut became independent of plate tectonics for its dispersal, and for development of (almost) all of its characteristic features which were subsequently modified by domestication.

The wild type evolved by floating between the volcanic islands and atolls, where these fringed the coastal margins of the continental plates. The coral atoll is considered to be the world's most stable ecosystem and the coconut palm is its most successful plant form (Harries, in press). Although it is now seen as part of the typical strand vegetation of coral atolls and an early pioneer of newly emergent volcanic islands – its true location and origin is on coastlines (island or continental) which are *submerging*. It stands to reason that only when the land began to sink did the coconut learn to swim. It follows that islands in the Tethys Sea could have been the ancestral home of the coconut. The primordial Tethys Sea was once the size of the Indian Ocean but it vanished when the Earth's land masses closed in on it, forcing its sea bed upwards so that it turned into mountain ranges. Both the Alps and the Himalayas were originally part of the Tethys sea bed. Another such area has recently been located off the northwestern coast of Australia (Rabinowitz 1988). Unfortunately, the coconut has always grown on the narrow margin where the land meets the sea and good fossils will be hard to find. Nevertheless, the cocosoid fossil in the northwest Indian desert might represent the early dispersal in the Tethys Sea.

It seems reasonable to suggest that, as the size and location of the continents and oceans changed in time, the coconut merely dispersed by floating from islands in the Tethys Sea to those in the Pacific and Indian Oceans. The argument as to whether the coconut is able to float across the Pacific is misleading; in this respect the coconut merely had to keep pace with the newly emerging islands as they appeared. It is unlikely to have reached some Pacific island groups, such as Hawaii and the Marquesas, nor did it go into the Atlantic. The actual floating distances do not have to be great but, most importantly, the coconut would have repeatedly and continually floated to continental coastlines. Here, it would have stood less chance of surviving competition from other plants or predation by animals and it could not reach or survive inland on larger islands or on continents before it was domesticated (Harries 1978). The continental coast and larger islands of Malesia would be the obvious site for such domestication. And this must have happened long before the wild and the domesticated types were both taken independently into agricultural cultivation. In this respect it is important to realise that domestic coconuts are not necessarily the same as cultivated coconuts. Nor are wild coconuts primitive or small fruited (Harries 1981b).

Domestic selection

The ethnobotanical pathways of domestic selection detailed elsewhere (Harries 1978) are briefly restated here. Successful natural dissemination depends on a balance between fruit number and fruit size. Selection for one would be antagonistic to the other. If the palm produces a large number of fruit this would improve the chances of reaching more sites at which to become established. If it produces bigger fruit this would increase the distance over which it could travel and remain viable. Thus wild type coconuts found in the Laccadive Islands are small fruited, as are those in Vanuatu, since both these island groups consist of many small islands. The wild type fruit size in more remote islands such as the Seychelles, Palmyra and Christmas Island are bigger.

In contrast, the selection value of increased fruit size or number is irrelevant to domestication. The important selection to be emphasised is for increased

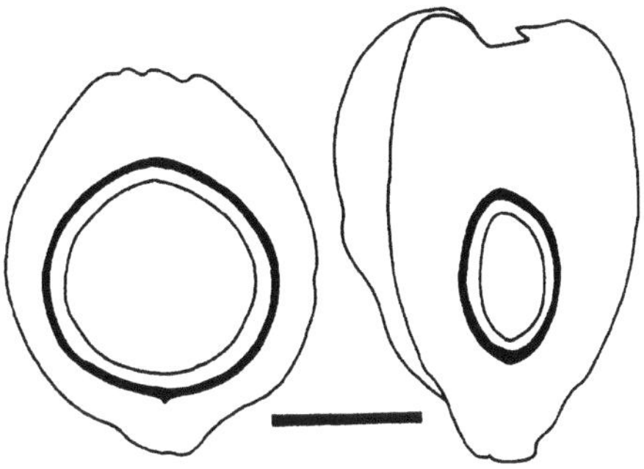

Vertical half sections of domestic type (left) and wild type (right). Scale bar = 1- cm. Taken from Dwyer (1938: plate 11).

endosperm (the drinkable part of the immature fruit or the edible part of the mature one). Without reducing either the total number or the individual fruit weight it is possible for selection to change the proportions of the fruit components. The husk is reduced from almost 70% to as little as 35% of the total fresh weight. Further reduction would eliminate the secondary function of the husk, which is to protect the nut inside when falling 30 m or more to the ground. Reduced ability to float is unimportant to domestication.

Selection that reduces the amount of husk would lead to a shorter, less angular and, eventually, almost spherical fruit. The facile explanation for the domestication of crop plants, that human selection increased the number or size of the edible parts, is inappropriate for an already large-fruited coconut. Although the weight of the fruit might be increased marginally by selection, the shape of the fruit is radically altered. The coconut fruit with a high proportion of husk is long and angular. These characters favour dissemination by floating as well as preventing it rolling in the surf so that it remains on the beach to root and does not easily wash away again. The shape of the fruit is reflected in the shape of the nut inside. The long fruit has a long, spindle-shaped nut, pointed at the end opposite the embryo, and with a thick shell.

In contrast the reduction in husk thickness and increase in water content lead to a spherical fruit which has a spherical or oblate nut and, because the proportion of shell does not change, the increased volume results in a thinner shell. A thin shell, more liable to crack, is a disadvantage to natural dissemination. A large nut with a flat base is useful as a bottle, a cup or a bowl, and the 'bottle' comes already filled with a drinkable liquid. Indeed, the coconut must come high on a list of plants first used by Palaeolithic man (Harries 1979).

The desirability of individual palms would become well known and they would be identified by name (as still occurs today). Visual selection for fruit shape is easy and would be applied automatically by adult or child, without conscious thought or effort, at every human and every coconut generation. Selection pressure under these circumstances would be high, despite cross-pollination. Any alternative explanation for coconut fruit variation would involve selection of the long-fruited type from a round or intermediate form. There are no compelling arguments for this.

A further unconscious selection readily made by primitive farmers is for speed of germination. Seedlings are taken from around the base of the palm, where they sprout after falling, or from nursery beds established for the purpose. In either case the slower germinating seedlings are over-shaded by early germinators, grow less well for that reason, and are not selected at planting time. Where there are differences in husk thickness, the sprout emerges sooner from the thin husk. As the fruit gets less elongated and more spherical, they would come to rest after falling, or they would be set, on their base rather than their side. This is common practice because the position gives quicker germination. It also gives lower total germination, which is a disadvantage to natural dissemination and to commercial production but is immaterial to primitive cultivators who often allow a natural understory of seedlings to develop at higher plant densities than are desirable (Wickremasuriya 1975). Ultimately, selection could result in germination on the bunch, as with *Nypa*. This has been observed under conditions of high humidity on the round-fruited varieties. It has never been reported under any condition from long-fruited varieties. The fruit of these fall from the palm when ripe. It need hardly be said that early germination is absolutely undesirable for dissemination by floating. Conversely, there is no reason to consider slow germination agriculturally desirable. The germination rate of the coconut, *Cocos nucifera* L., its long distance dissemination by floating and the fact that the speed of germination turned out to be a characteristic of taxonomic significance also showed how the genetic uniformity of the coconut populations on remote oceanic islands might be maintained (Harries 1981c).

Archeological evidence in support of the domestication theory was subsequently located in Borneo (Harries 1981a) but efforts to identify shell fragments in Papua New Guinea (Hossfeld 1948; Kirch, pers. comm. 1987) and Solomon Islands (Spriggs 1984) have been less successful. Until and unless more archaeological and palaeological coconut remains are unearthed in the Malesion region the probability that coconuts were domesticated there can remain only a speculation. In the meantime, taxonomists and other botanists should not assume that coconut is merely a cultivated species with no wild relatives. Closer examination of small groups of palms found outside obvious cultivations and particularly in isolated areas might reveal whether they resemble wild or domestic types. The presence of human populations, either now or in the past, is not necessarily a criterion for successful establishment of the coconut palm on the seashore but is essential to its survival inland.

References

Bentham, G. 1863–1878. Flora Australiensis: a description of the plants of the Australian Territory. Reeve, London.

Buckley, R., & H.C. Harries. 1984. Self-sown, wild type coconuts from Australia. Biotropica 16: 148–151.

Child, R. 1974. Coconuts. 2nd Ed. Longman, London.

Dransfield, J. 1989. Voanioala (Areioideae; Cocoeae; Butiinae) – a new palm genus from Madagascar. Kew Bull. 44: 191–198.

Dwyer, R.E.P. 1938. Coconut improvement by seed selection and plant breeding. New Guinea Agric. Gaz. 4: 24–102.

Fosberg, F.R. 1960. A theory on the origin of the coconut. In: Symposium on the impact of man on humid tropics vegetation, Goroka, Territory of Papua New Guinea: 73–75. Comm. Govt. Printers, Canberra.

Gruezo, W.Sm., & H.C. Harries. 1984. Self-sown, wild-type coconuts in the Philippines. Biotropica 16: 140–147.

Harries, H.C. 1978. The evolution, dissemination and classification of Cocos nucifera. Bot. Review 44: 265–320.

Harries, H.C. 1979. Nuts to the Garden of Eden. Principes 23: 143–148.

Harries, H.C. 1981a. The antiquity of the coconut palm in Western Borneo. Sarawak Mus. J. XXIX, no 50: 239–242.

Harries, H.C. 1981b. Practical identification of coconut varieties. Oleagineux 36: 63–72.

Harries, H.C. 1981c. Germination and taxonomy of the coconut palm. Ann. Bot. 48: 873–883.

Harries, H.C. (in press). The biogeography of the coconut palm. Principes.

Harries, H.C., A. Thirakul & V. Rattanapruk. 1982. Coconut genetic resources of Thailand. Thai J. Agric. Sci. 15: 141–156.

Hossfeld, P.S. 1948. The stratigraphy of the Aitape skull and its significance. Trans. Roy. Soc. S. Aust. 72: 201–207.

Kaul, K.N. 1951. A palm fruit from Kapurdi (Jodhpur, Rajasthan Desert), Cocos sahnii sp. nov. Current Science (India) 20: 138.

Koorders, S.H. 1911. Exkursionsflora von Java. Jena.

Palmer, A. 1988. Joides Resolution Expedition, Texas A & M U. In: A. Berry, Continent found under ocean. Daily Telegraph, 20th June 1988.

Purseglove, J.W. 1972. Tropical Crops: Monocotyledons. Longman, London.

Rabinowitz, P. 1988. Joides Resolution Expedition, Texas A & M U. In: A. Berry, Clue to ancient 'jigsaw puzzle'. Daily Telegraph, 10th October 1988.

Sahni, B. 1946. A silicified Cocos-like palm stem, Palmoxylon (Cocos) sundaram, from the Deccan Intertrappean beds. J. Indian Bot. Soc. Iyengar commem. vol.: 361–374.

Smith, A.G., & J.C. Briden. 1977. Mesozoic and Cenozoic paleocontinental maps. Cambridge Earth Science Series, Cambridge University Press.

Spriggs, M.J.T. 1984. Early coconut remains from the South Pacific. Polynesian Society J. 93: 71–77.

Uhl, N.W., & J. Dransfield. 1987. Genera Palmarum. Allen Press, Lawrence.

Wickremasuriya, C.A. 1975. Situation of the coconut industry in the Republic of the Maldives. FAO 4th Technical Working Party Coconut Production, Protection & Processing, Kingston, Jamaica.

30 On specific and infraspecific delimitation*

D.J. KORNET

Summary

This paper deals with the theoretical and philosophical inheritance of van Steenis as embodied in his essay 'Specific and Infraspecific Delimitation'. Only those topics are discussed which appear to have been problematic for van Steenis himself; the value of theory in taxonomy and the scientific demand of objectivity. The mystery why species seem not to be susceptible to strict definition is discussed in a general framework in which the origin of 'the species problem' is clarified. Some conditions for achieving a species concept with historical relevance are given. Finally Danser's commiscuum concept, highly valued by van Steenis, is discussed and compared with my own historical species concept and with the linneont concept of van Steenis.

Introduction

In his essay 'Specific and Infraspecific Delimitation', C.G.G.J. van Steenis dealt with a variety of theoretical and philosophical topics in order to reach a synthesis from which practical rules and recommendations for taxonomists can be inferred. I will discuss the issues of 'the value of theory for taxonomic practice' and of 'objectivity' because their relation seems to have puzzled van Steenis. I will then proceed to analyse the problem of defining species. Species appear to resist strict definition and van Steenis accepted this only reluctantly. To explain why the definition of species seems so difficult I will picture the species problem and clarify its origin. If we consider the historical aspect as more fundamental to a species than its morphological distinctness or its inborn interbreeding isolation, the defining criteria ought to secure spatio-temporal limitations of species. The historical species concept will return in a comparison with Danser's commiscuum concept of species with which this paper ends.[1]

---

*) The investigations were supported by the Foundation for Biological Research (BION), which is subsidised by the Netherlands Organization for Scientific Research (NWO).

1) Selected passages from van Steenis' essay will be cited in notes to follow.

P. Baas et al. (eds.), The Plant Diversity of Malesia, 359–370.

The ambivalence of van Steenis towards the value of theory for taxonomic practice

"Practical taxonomic work should confine itself to factual observation and not be hybridized with theoretical considerations ..." (p. ccxxiv)

This statement seems to imply that, according to van Steenis, in practical taxonomy we should stick to the facts and avoid theory. But his rejection of theory is far less absolute than is suggested here. The quotation is incomplete and refers actually only to particular theoretical considerations which van Steenis happened to mistrust.[2] He spoke quite differently, for instance, about Danser's theory of species which he appreciated.[3] But more interesting is that at other places van Steenis airs a different opinion about the value of theory.[4]

"There is therefore special reason for the tropical taxonomist to give a good deal of attention to the species concept in order that his work will be satisfactory and useful to others." (p. clxciii)

This contradiction shows that van Steenis was rather ambivalent in his appreciation of the value for taxonomic practice of theories about species. Though he occasionally doubted the merits of theory, van Steenis gave in his essay an extensive account of his own theory of species. He used this very theory to justify the seventy 'Rules and Recommendations' with which he concluded the essay.[5] It is this theory which prompted him to state that the morpho-

---

2) "Practical taxonomic work should confine itself to factual observation and not be hybridized with theoretical considerations on origin of species and genera or supposed phylogenetical relations. Such hypothetical deductions are not discouraged but should be kept clearly separate from the facts observed and not be introduced in their evaluation." (p. ccxxiv, Rules & Recommendations 6)

3) "Danser recommends identifying delimitation of Linnean species with that of the commiscua. This is in all probability the best approximation of the Linnean species ever proposed ..." "... It will be remembered that the Mendelejew natural system of elementary chemical elements seemed defeated or at least had to allow exceptions in the actual molecular weights which did not show the simple numbers to be expected. Those who stuck to the theory, ... appeared after all right ..." (p. ccvii)

4) "... and an allround background of theoretical taxonomy and its concepts and the digest of the papers on this subject will be useful and facilitate development of experience" (p. clxxv)

5) "... it appeared in most cases impossible to get a satisfactory identification with a key that worked or differential diagnoses that fitted. This often disgusting experience, about which I feel seriously concerned, has led me to a careful consideration of both the theoretical and practical side of specific delimitation and description ..." (p. clxviii)

logical distinctness of a group of organisms is not enough to grant it the rank of species.[6] From this theory he inferred what kinds of other data can offer relevant circumstantial evidence for species delimitation.[7]

Van Steenis should have relied on his theory of species, but did not, where he claimed that most species are real entities in nature and not convenient fictions of the human mind.[8] He attempted to justify this view by arguing that the observation of morphological gaps is replicable and therefore objective.[9] But this argument is not very satisfactory. As noticed by van Steenis himself, demarcations between likeness show in the same way the discontinuity between, for instance, atoms of different elements in chemistry.[10] But this objectivity of the pattern of likeness, as displayed by the properties of elements, does not imply at all that elements can exist in the same way as individual species can: as historical happenings.[11] Only a well-confirmed theory of species which implies that they are cohesive, spatio-temporally limited entities, with a beginning and a (possible) end in time, can justify a claim for the existence of species as historical individuals.

---

6) "... in the case of *Geum* the morphological differences between the two subspecies are very distinct; nobody will confuse them. ... Abundant material and detail observations are in such cases compulsory to come to the correct interpretation of the state of affairs." (p. clxxxix)
"... These few examples show how circumstantial evidence may lead to the evaluation of characters and establish their proper taxonomic value." (p. clxxi)

7) "For allied disjunct taxa which, by their morphological characters, could be distinguished as species, experimental taxonomy can provide useful data for the criteria of putative hybridization and miscibility, which may be important for the interpretation of their status." (p. ccxxviii, Rules & Recommendations 49)
"... to the weight which should be given to hybridization and miscibility of taxa, as this has an essential bearing on the concept of specific delimitation, ..." (p. cciv)

8) "Species are in the majority definable as distinct self-perpetuating units with objective existence in nature, ..." (p. clxvii)
(citing Julian Huxley) "In the great majority of cases species can be readily delimited, and appear as natural entities, not merely convenient fictions of the human intellect." (p. clxix)

9) "... the unpreoccupied eye of the rain-forest nomad tribes, who for their living are in closest contact with plants, frequently distinguishes the same genera and species (and sometimes the family) as the trained botanist."
"... most distinct in nature to human and animal observation" (p. clxvii)

10) "These demarcations between likeness show the discontinuity of living matter, similar to the discontinuity in the inanimate, the atoms in chemistry and the quanta in physics." (p. clxix)

11) "In these cases at least *the species is a historical happening:* a sudden event, followed by a period of existence, ending with extinction." (pp. clxix–clxx)

The desire of van Steenis for objectivity

It would not be fair to accuse van Steenis simply of being inconsistent in his statements. I think that his ambivalence towards theory reflects a more fundamental dilemma concerning 'objectivity'. With the influential positivists of his day van Steenis stressed the importance of objectivity in science:

> *"In science there is a general, reasonable tendency to analyze by objective values and avoid those subjective. Whatever charm the personal appreciation may have, methodologies containing it are considered with suspicion and conclusions based on them are found worthy of relative value only."* (p. clxxi)

Van Steenis' suspicion of theories could very well be rooted in the assumption that theories have a contaminating effect on the observed facts. He seemed to prefer the objectivity of observed facts over the supposed subjective character of a theoretical interpretation of those facts. But this rejection of theory forced van Steenis to introduce the notion of 'synthesis' instead, which he considered as particularly important for tropical botanists:[12]

> *"... two phases of scientific work in general, and taxonomic research in particular, ... the analysis being the observation and recording of all factual data prior to the synthesis in framing these data and assigning them their proper hierarchical place."* (p. clxxvi)
>
> *"A sense for synthesis is indispensable for good taxonomy ..."* (p. ccxxiv, Rules & Recommendations 10)

This 'sense for synthesis' comprises mental qualities like wisdom, discretion and common sense[13] (though it excludes intuition as a sixth sense[14]). Van Steenis concluded from this that the indispensable synthesis implies a per-

---

12) "The botanist engaged with tropical plants, ... is almost always compelled to confine his study to herbarium material, which requires far more thought and sense for synthesis to classify in a satisfactory and useful way." (p. clxviii)

13) "This synthesis according to the standard set by Linnaeus, requires a great measure of wisdom, discretion, and common sense in order to make it useful. ... Taxonomists, who among their psychical qualities show a deficiency of feeling for synthesis ... will naturally tend to keep the smallest units apart and their hierarchic synthesis will be poor." (p. clxxvi)

14) "... this romantic quality is a myth, unless intuition is generally defined as sublimated experience ... But intuition as a sixth sense is not included, and even not desirable." (p. clxxiv)

sonal valuation of facts. For taxonomy this means that judgements of the taxonomic value of particular morphological properties contain a personal element.[15] So, far from escaping from subjectivity in taxonomy, by introducing the notion of 'synthesis' van Steenis readmitted it in a disguised form. By 'synthesis' van Steenis means a body of interpreted facts, but such an intellectual construct is nothing other than a 'scientific theory' in the sense in which this expression is used in the philosophy of science. For the sake of ease of communication there is good reason to prefer the notion of 'theory' over that of 'synthesis'.

The worries of van Steenis about the subjectivity of scientific theories are exaggerated. Scientific theories are hypotheses about reality which can be tested independently and are, in that sense, objective. Scientific theories are fallible but not subjective because nature can say: no!

Van Steenis too thought that testing is important. He demanded that the pattern of distinctness as described by a taxonomist is intersubjectively observable. Testing is done in practice by those people who need to identify plants on the basis of that information. Unfortunately, van Steenis called this test-criterion 'practical usefulness'.[16] The term 'useful' extends also to arbitrary classifications which simply serve some practical needs of human beings, like a classification of food plants or medical plants. Scientifically interesting are those classifications which reflect discovered patterns of resemblance which demand an explanation, or classifications which constitute a hypothesis about nature, like those representing a historical reconstruction. These classifications are open to scientific testing.

The origin of 'the species problem'

Van Steenis noted with some regret that species cannot be defined strictly.[17] And he was, of course, not the only one who believed so. Biologists have long struggled with the definition of species. What, we might wonder, is the source of the so-called 'species problem'?

---

15) "However factual *observations* in the pure sciences may be, their *appreciation and evaluation* for the synthesis always contain a certain personal element." (p. clxxi)
"... a strongly personal element, specially as it means not only the *objective observation* of morphological characters but also the assigning of a *different degree of appreciation* of the value to the characters observed." (p. clxx)

16) "*The standard of quality of plant taxonomic work is found in its usefulness.* ... The usefulness is, I feel, an adequate control on the purely personal subjective element." (p. clxxii)

17) "*If it were possible to test all species experimentally, ... this would have the great advantage that we could delimit species according to one single principle, ...*" (p. ccvii)
"... the Linnean species has appeared the most essential, useful, and objective taxon. Though not susceptible of a strict definition, ..." (p. ccxxvi, Rules & Recommendations 31)

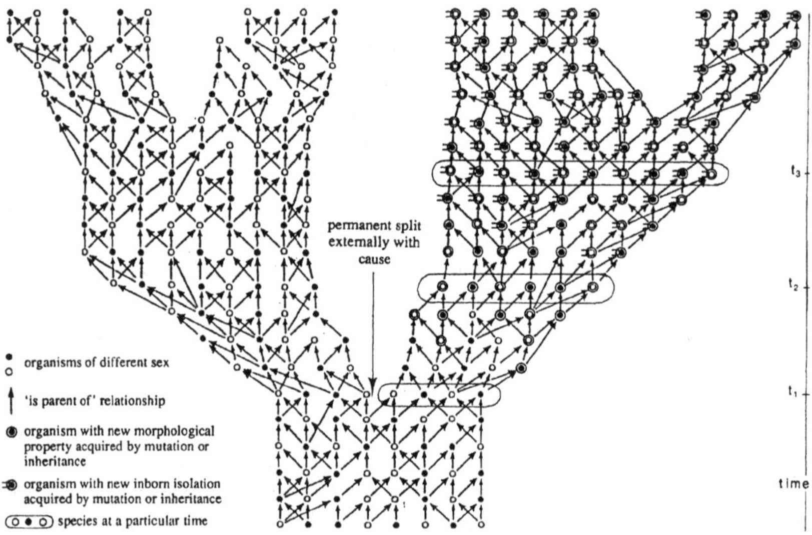

Figure 30.1 — Diagrammatical representation of historical species. The historical beginning of a species (t_1) need not coincide in time with the development of morphological distinctness (t_2) or inborn interbreeding isolation (t_3).

Species definitions often contain criteria which reflect more than one aspect of species. Important aspects of species are *morphological distinctness, inborn interbreeding isolation,* and their *historical character*. It has turned out that defining criteria which represent different aspects do not always jointly apply. Organisms belonging to groups which show an indisputable morphological distinctness, for instance, may very well have the capacity to interbreed. Every species definition which refers to more than one of the aspects mentioned above will have to allow exceptions. And any two species concepts, each of which refers to a different aspect, will overlap only partly in their extension. In other words: the species problem arises because defining criteria which reflect different aspects of species are mutually incompatible.

But why would this be the case? The answer is that the defining criteria show incompatibility because the different aspects of species which they reflect can originate at different times. This is illustrated in Figure 30.1, where a few related species are presented in their historical appearance. The picture shows when the organisms lived with respect to each other and their parental relations. At time t_1 the relational network of the ancestor species splits up for ever, giving rise to two daughter species. This moment, which marks the historical end of the ancestor and the historical beginnings of the daughter species, can usually be determined only afterwards. It may take some time before historical splits become irreversible and therefore permanent.

It is very important to realise that a split in a network can be initiated in more than one way. It can be caused by a genetic change which suddenly makes interbreeding impossible, or by ecological segregation. But a split can be caused also by external events like earthquakes or glaciation. When a permanent split is brought about by an external cause, the sister species need not at first be morphologically distinct at all and there is also no reason to doubt the inborn ability of their organisms to interbreed. Distinctness might develop later when new properties arise by mutation and become fixed in the species by natural selection (at t_2). Genetic incompatibility (inborn interbreeding isolation) develops in the same way and might very well evolve later (t_3). The sequence can, of course, be reversed as is the case with sibling species.

The species problem arises as soon as we try to apply mixed species concepts to a species in which the developments of the different aspects do not coincide in time, in the period between t_1 (the historical beginning of the species) and t_3 (before the two inborn aspects of distinctness and interbreeding isolation are developed). The aspects can indeed coincide in time. In speciation by allopolyploidy, for instance, the historical splitting is caused by sudden genetic change which makes interbreeding impossible and can be accompanied by abrupt morphological change. Only in a case like this will the species problem fail to arise, because the three aspects do indeed originate at the same time. In this light it is remarkable that van Steenis mentioned 'allopolyploidy' as an indisputable speciation mechanism which results in good species.[18]

So, to recapitulate, the actual and permanent split of the interbreeding network generated by either an external or an internal cause determines the historical beginnings of species. The development of genetically fixed morphological characters determines the morphological distinctness of species. And the development of genetically fixed sexual properties determines the genetic interbreeding isolation. The 'species problem' originated because the developments of these three aspects generally do not coincide in time.

Conditions for the definition of a historical species concept

If defining criteria which refer to different aspects of species are incompatible, a strict species definition is possible only if it is based on a single aspect. Which one should we prefer?

---

18) "It has further appeared unmistakable that at least in allopolyploidy one mechanism has been found for the origin of new Linnean species." (p. ccxxiii)

"... the discontinuous structure of the plant kingdom which can *a fortiori* hardly be explained otherwise than by jump-wise origin of the taxa (mutation in the wide sense), though theoretically origin by accumulation of small jumps cannot be excluded *a priori*. As far as we know at present, the sudden origin of good species by mutation (polyploids), both new and existing, has been proved without a shadow of doubt." (p. clxix)

Definitions based on any sort of property of individual organisms form classes or sets which can maximally claim the status of natural kind (Geesink & Kornet 1989). Morphological distinctness and inborn interbreeding isolation, being genetically determined properties of the individual, are properties suitable for the definition of natural kinds: on this conception of species any organism which possesses the required properties belongs to a natural kind constituting a species. But natural kinds are spatio-temporally unlimited and therefore inherently ahistorical; the notion of species defined on these two criteria thus fails to connect the life of an individual organism with its position within earth-bound evolution. A member of a species defined on these criteria might just as well have been created *in vitro* or introduced from elsewhere in the cosmos.

If we want a species concept with evolutionary meaning its definition must have historical relevance. Therefore the definition of the historical concept needs to secure its spatio-temporal limitation. We can manage to do so only if we define a species in terms of a relational network in time. In this case, the species is defined not by properties of individual organisms, but by actual relationships which connect them. On this analysis it turns out that for a strict definition of historical species we need to know only the relationships of parenthood which hold between organisms and the temporal ordering of their life-spans (Kornet, in prep.).

Organisms of a species may be connected to one another either directly or indirectly by a combination of relationships (a point already mentioned by van Steenis[19]). The distinction between direct and indirect connection is based on relationships all of which are actual, and therefore should not be confused with the distinction between actual and potential relationships. Imagine, for instance, two organisms at opposite ends of a race-chain ('Rassenkreise'). Their genetic constitution makes it impossible for them to interbreed but they may still be connected indirectly, by a chain of actual interbreeding relationships.

Though morphological distinctness and inborn interbreeding isolation are not fit as criteria for the definition of a historical species concept, they are valuable secondary properties. Although sometimes misleading, they are generally reliable indicators for the recognition of historical species, and as such invaluable for taxonomic practice. To specify that a certain degree of morphological distinctness will be considered necessary for us to recognise and officially name two species is a matter of convenience and does not determine the time at which the species came into existence.

---

19) "The fact that in populations there will be always be individual specimens which cannot be crossed with any random individual specimen is of only slight value to the commiscuum concept, as it is sufficient if they can be crossed with some other specimen, so that the commiscuum is then still a closed hybridization net, ..." (p. ccvii)

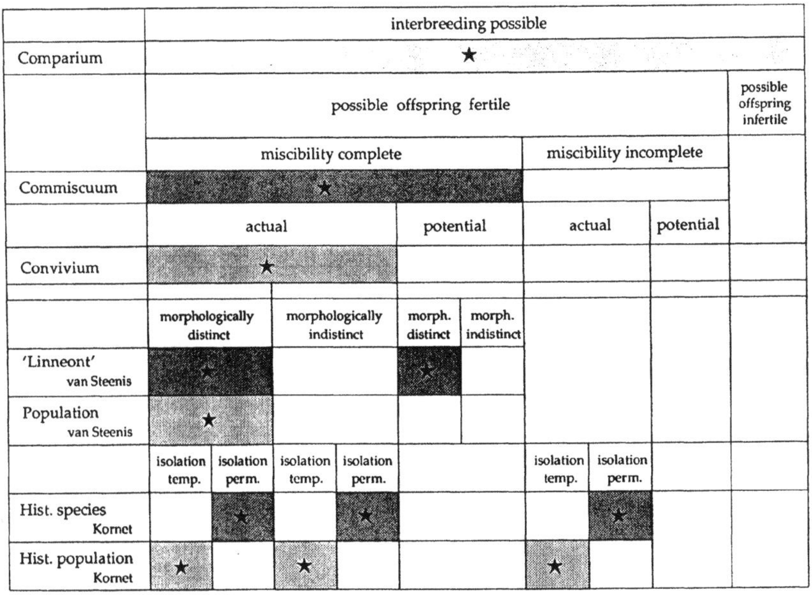

Figure 30.2 — Comparative table of the concepts of species and populations mentioned in the text.

Danser's commiscuum concept for species

According to van Steenis, the commiscuum concept of Danser is the best available species concept.[20] I think that van Steenis was quite right in this appraisal: it is interesting indeed, because it does not contain morphological distinctness as a defining criterion. I shall discuss Danser's concepts and compare them both with van Steenis' concepts of species and populations and with my historical view of them. In Figure 30.2 this comparison is represented in abstract form.

---

20) "Reversely I am fully convinced that through these three clear concepts Danser has furnished a critical content to a hitherto badly defined field of theoretical taxonomy, and feel that his concepts are giving sense, emphasis, and direction to professional taxonomy against a clear background.
Although Danser's concepts have for reasons unknown not come into the general use they deserve, I am perfectly convinced that they approach the real biological situation of specific delimitation in nature as closely as can be expected in this matter." (p. ccviii)

The concept of commiscuum occupies in Danser's view the middle level of a hierarchy of three concepts (Danser 1929). The widest of his concepts is that of the 'comparium': all organisms between which interbreeding is possible belong to a single comparium, regardless of whether the potential hybrids are fertile.[21]

Danser's species concept is that of 'commiscuum'. It is defined by two criteria. First, organisms belong to a single commiscuum (species) if they are able in principle to produce fertile offspring. Secondly, conspecific organisms should be susceptible to complete miscibility, i.e., it should be possible for their morphological differences to merge within a few generations.[22]

It does not matter for the definition of commiscua whether a possible interbreeding network is actualised or only potential. This does matter for the definition of a 'convivium' (population), the members of which must be connected by a combination of actual interbreeding relations. Convivia of a single commiscuum are as a matter of fact isolated but are still miscible and able to interbreed.[23]

Even though Danser's definition of the commiscuum concept does not include the criterion of morphological distinctness, this concept is still ahistorical. Danser's definition still refers in part to potential interbreeding networks. The ability to interbreed is an inborn (genetic) property of individual organisms, so the definition of commiscuum still hinges on properties of individuals, which, as explained above, threatens the historical relevance of the concept.

For all van Steenis' appreciation of Danser's notion of commiscuum, he added 'morphological distinctness' among the criteria for the definition of his

---

21) "The *comparium* (hybridization community) comprises all individuals which are connected by the possibility of hybridization.. The hybrids must be viable, but it is irrelevant whether they are capable of producing sexual organs." (p. ccvi)

22) "The *commiscuum* (miscibility community) is the total number of individuals which are connected genetically through miscibility. ... It is not necessary that *all* individuals are *directly* connected by miscibility." (p. ccvi)
"*A sharp distinction should be made between hybridization possibility and miscibility; under the latter concept is understood the possibility, either in nature or in cultivation, of obtaining by hybridization, fertile specimens showing an almost complete number of various combinations of the differential characters between the taxa, blending them into one whole without demarcations.*" (p. ccxii)
"... are not only capable of fertile hybridization but are miscible (producing in the F2-x a complete series of intermediates blending them) ..." (p. ccxxviii, Rules & Recommendations 50)

23) "The *convivium*, defined as each partial population which is, by geographical causes or otherwise, genetically isolated from the other part(s) of the population, with which it is, however, completely miscible." (p. ccvi)

concept of 'linneont'.[24] This move makes his linneont concept even more remote from historical relevance than the commiscuum concept is. Inspired by his ever-present concern for taxonomic practice, van Steenis reintroduced morphological distinctness in the strict definition of species and ended up with a concept referring to a mix of three aspects again. The fruitfulness of Danser's reduction of his commiscuum concept to only two aspects is lost and van Steenis had to conclude once more that his species criteria are incompatible with one another.[24]

For population and species concepts with historical relevance the criterion of permanent or temporary isolation will have to be added to the scheme of Figure 30.2. The criterion of actual networks applies, but the criterion of miscibility does not.

Conclusions

Despite his occasional rejection of theory as pernicious for taxonomic practice, van Steenis made ample use of theory for the evaluation of morphological characters and for the selection of relevant circumstantial evidence for species delimitation. He had to rely on his own species theory also to justify the claim that species exist since an objective observable pattern misses that aim. His fears about the subjectivity of scientific theories were groundless because theories are testable. Scientific theories are fallible, but not subjective. The same cannot be said of the 'sense of synthesis' which therefore is not a suitable substitute for scientific theorising.

The 'species problem' originated because the historical beginnings of species need not coincide at all with the development of a characteristic property of the species or of the genetically-determined interbreeding isolation. A historically-relevant species concept cannot support any genetically-determined property as a defining criterion since it will lose the decisive spatio-temporal limitation of its reference.

Danser's commiscuum concept for species was promising since it contained no morphological criterion. The inclusion of potential interbreeding networks impairs the spatio-temporal limitation so that it is still denied a historical relevance. Van Steenis could not resist adding morphological distinctness in his

24) *"Though they represent commiscua from the genetical standpoint, they should taxonomically not be arranged on the specific level along with Linnean species as the factor responsible for the incompatibility is not bound with a sufficient clear and large set of morphological characters. Their genetic incompatibility prohibits unfortunately to accept a strict application of the principle to equalize Linnean species with commiscua, though this would be – and still is theoretically – the best way of introducing a biological agency responsible for specific demarcation in nature."* (p. ccxiii)

concept of linneont. In so doing he again enlarged the number of incompatible criteria supposedly to be used in the definition of species.

References

Danser, B.H. 1929. Über die Begriffe Komparium, Kommiskuum und Konvivium und über die Entstehungsweise der Konvivien. Genetica 11: 399–450.

Geesink, R., & D.J. Kornet. 1989. Speciation and Malesian Leguminosae. In: L.B. Holm-Nielsen, I.C. Nielsen & H. Balslev (eds.), Tropical Forests: 135–151. Academic Press, London.

Steenis, C.G.G.J. van. 1957. Specific and infraspecific delimitation. Flora Malesiana I, 5^3: clxvii–ccxxxiv.

31 Species and monophyletic taxa as individual substantial systems

R. LÖTHER

Summary

The philosophical notion that species are to be regarded as individuals, not as universals, clarified that the biological species problem does not belong to the philosophical dispute about universals. The relation 'species–organism' accords to the relation 'system–element', not with the relation 'universal–individual'. The species category denotes a class of biosystems. These systems constitute one of the organisational levels of the biosphere. Typological as well as nominalistic species concepts miss their subject. There are different kinds of species in nature. Evolved species are elementary units of the phylogeny. Monophyletic higher taxa are "closed phyletic communities of nature" (*sensu* Ax) or "historical groups" (*sensu* Wiley).

Introduction

The species problem is fundamental for the entire field of biology, not exclusively for taxonomy and evolutionary biology. Essential philosophical (ontological and epistemological) aspects are connected with this problem. The terms 'species' and 'genus' were introduced by Aristoteles as terms of logic which he used in his zoological writings as well. As terms of logic they are applied with a relative meaning. They express a subordination of concepts. The term 'genus' is related to a superior, more general concept, and the term 'species' to the subordinate, less general one. In other words, a relation of concepts is stated in a hierarchical system independent from the absolute position of the concepts in the system.

Ray and Tournefort were the first to distinguish the logical and the biological-taxonomic meanings of the terms. Ray reserved the term 'species' for the lowest unit, the logical 'infima species', of the classification of plants and animals. Simultaneously he interpreted the species as reproductive units with constant essential characters. Tournefort coupled the fixed assignation of the more logical terms to the levels of classification he distinguished as taxonomical categories. In his system of classification of the plants he applied the hierarchy 'species–genus–sectio–classis'. The terms 'species', 'genus', etc., designate, as terms of taxonomy, definite levels of the classificatory system of living

P. Baas et al. (eds.), The Plant Diversity of Malesia, 371–378.

beings. The terms are fixed to them and have an absolute meaning. Ever since, the absolute hierarchy of taxonomic categories has rested on the species category, and the species problem in biology was created.

Formerly there was much controversy between the views of creation and constancy of species and descent from other species with modification, as well as on the objective or subjective character of the species concept. Today the situation is quite different, although remainders from earlier times tend to persist. Modern biology taught us: "Nothing in biology makes sense except in the light of evolution, *sub specie evolutionis*" (Dobzhansky 1965). The species concepts from the context of evolutionary theory (the 'biological species concept' and the 'evolutionary species concept') leave no doubt about the objective existence of species in nature. Phylogenetic systematics (cladistics) is elaborating the phylogenetic system as the true natural system of species. The species concept is the immediate conceptual connecting link between phylogenetics and taxonomy. The species concept, introduced by Ray as the basic unit of classification of living beings in the framework of a static world view, means nowadays also the elementary evolutionary unit of phylogenesis. A modern debate on the species problem focusses on the following questions:

a) What is the place of species as a form of organisation of life in the structure and evolution of living nature?
b) What is the meaning of infraspecific structure and what exactly is its relation to the dynamics of infraspecific and transspecific evolution?
c) Which are the modes of origin of new species?
d) Is it possible to develop a species concept which is applicable to all living beings?

Species as material systems

Species can only sensibly be understood if a particular species concept is presupposed. This concept has developed long after the visionary step of Ray, and particularly after the establishment of population genetics. It received particular expressions in the biological and in the evolutionary species concept. The presupposed general concept is that a species is *not* a logical class of organismic individuals with features that distinguish it from other logical classes of organismic individuals, but a *material, supraorganismic system*, spatiotemporally organised, forming an integrated whole. Species come into existence, change, and go, because new species arise from them or because they become extinct. These systems form one of the organisational levels of living nature in the biosphere of the planet earth. Other levels are that of organisms, the level of populations, the biocoenotic level, and the biostromatic level (Zavadskij 1968).

In other words, the species problem is *not* a problem of the dispute about the existence of universals, as derived from Plato and Aristoteles. Disputes on the species problem (e.g. on the question whether there are species in nature or whether they are constructions of the human mind, on the basis of morphological-typological and/or nominalistic species concepts) *are* held in the scope of the problems around universals, and such discussions simply miss their subject in spite of the position taken in the discourse. The biological term 'species' does not refer to a totality of logical classes but to a totality of substantial systems, of biosystems (Löther 1972, 1974, 1983). If species are not logical classes (or sets), then they must be individuals. One issue of 'Biology and Philosophy' (Vol. 2, 1987: 127–225) was entirely devoted to this question, and my views agree with those of Ghiselin and Hull expressed therein.

Consequently, the realistic basis for reference of taxonomy, of its comparing, classifying, and name-giving activities is the totality of individual biosystems called 'species', and not the totality of single living organisms. The relation 'species–single organism' conforms with the relation 'system–element' and with 'whole–part', not with the relation 'universal–individual'. Local populations are subsystems of the species system. The occupation of taxonomy with single organisms (specimens) belongs to the process of indirect recognition. In this process, characters of organisms are *not defining properties* of class- or set-membership, but signs or *indicators of their partship* of the particular species they belong to.

In case of the particular species, the elements of the system are mostly more similar among each other than with the elements of other systems. For this reason it is possible to deduce general features of the particular species from the organisms by comparing their characteristics. Because taxonomists did so for a long time, they recognised species correctly without realising that species are material systems. But the reliability of the deductive activities is limited by the phenomena of infraspecific polymorphism and other kinds of infraspecific variability on the one hand and on the other hand by the phenomena of extreme similarity among specimens of different species (in that case they are called sibling-species). Both these kinds of phenomena cause incompatibility with morphological-typological species concepts.

As mentioned above, I am of the opinion that the biological and the evolutionary species concepts are special versions of the more general, real system concept of species. The biological species concept stresses reproductive isolation and genetic unity. This concept is an unmistakable denial of nominalistic and typological species concepts. Mayr (1969: 26) defined: "Species are groups of interbreeding natural populations that are reproductively isolated from other such groups" or, in its latest version (Mayr 1982: 273): "A species is a reproducing community of populations (reproductively isolated from others) that occupies a specific niche in nature." The background of Mayr's definitions (and other versions of them, e.g. in 1963) is that he looks upon species as a trinity of 1) reproductive unity, 2) ecological unity interacting with other species sharing a common environment, 3) genetic unity consisting of an intercommunicating gene pool.

There are many followers of Mayr, but also critical objections have arisen. The main objections are that Mayr's definitions neglect the spatial and temporal dimensions of species and that they refer to species of biparentally reproducing organisms only. According to Mayr the word 'species' in biology is a relational term: "A is a species in relation to B and C because it is reproductively isolated from them. It has its primary significance with respect to sympatric and synchronic populations, and these are precisely the situations where the application of the concept faces the fewest difficulties ('the non-dimensional species'). The more distant two populations are in space and time, the more difficult it becomes to test their species status in relation to each other, but the more irrelevant biologically this also becomes" (Mayr 1969: 26, 27). Mayr confuses, in this text, the actual species problem and the practical problem of the delimitation of species in relation to sympatric and synchronic populations of biparental organisms. This is of great interest to field naturalists confronted with populations of such organisms in such 'non-dimensional' situations and the concept is undoubtedly useful in this respect. But delimitation is *not* the major conceptual problem.

Species are not the limits of species, but instead what is within these limits: the biosystems called 'species'. The relations of species to other species are the products of their internal conditions. To summarise, the term 'species' is not a relational term but denotes a class of biosystems which exist in space and time. Mayr's species concept is not simply false, but it is too narrow to define species as real units of nature. This is also true with regard to the excluded uniparental organisms.

The evolutionary species concept of Simpson (1961) is an extension of the biological species concept, in order to take into consideration the spatial and temporal dimensions of actual species, and to include uniparental organisms. Referring to Mayr's species definition and the evolutionary process, Simpson (1961: 152) states: "It is the fact of evolution that has made genetical species separate and it keeps them from always being sharply, clearly separate. It is also evident that the general definition of species has evolutionary significance. Still it is striking that the definition does not actually involve any evolutionary criterion, or say anything about evolution. It would apply equally well, or in fact a great deal better, to species that did not evolve." In order to remove limitations of Mayr's definition, Simpson proposed another definition which does not contradict Mayr's definition, but includes it as a more restricted version which is correct for a definite part of the totality of species only in one particular time-slice, in his definition of the 'evolutionary' species. Simpson's definition reads as follows: "An evolutionary species is a lineage (an ancestral–descendant sequence of populations) evolving separately from others and with its own evolutionary role and tendencies" (Simpson 1961: 153). A later, modified formulation of the evolutionary species definition was proposed by Wiley (1981: 25): "An evolutionary species is a single lineage of ancestor–descendant populations which maintains its identity from other such lineages and which has its own evolutionary tendencies and historical fate." It is apparent to me that the evolutionary species concept must be complemented with a definition of the term 'population'. In the biological species concept populations are Mendelian populations. Definitions of populations in the literature refer, as a rule, also to Mendelian populations, or include the term 'species', or both. Such definitions are of course not logically compatible with the evolutionary species concept. It presupposes a general definition of populations which includes Mendelian populations, clones, and pure lines alike by their common genetic, ecological, and evolutionary characterisation.

It is important that Simpson clarified that groups of uniparental organisms act in evolution in essential aspects like populations and species of biparental organisms. Simpson (1961: 162) summarises: "Evolution as unit, or the retention of a unitary role is maintained by a community of inheritance, by the capacity for genes to spread throughout the population (which therefore has a gene pool) and by the inhibition of their spread to other populations. All three factors occur in both biparental and uniparental populations."

This demonstrates the existence of natural units, supra-organismic systems, also of uniparental organisms. Both kinds of supra-organismic systems belong to one type of systems which were called "discrete or corpuscular systems" by Malinowski (1969). In their pure form, these systems consist of elements which in practice are not mutually related to each other. The elements are united by the same kind of relation to the environment which forces them into similar behaviour, also without direct mutual relations. A dynamic example of such systems in the non-living nature is the motion of sand grains in a river. The grains are certainly not mutually connected, but they become deposited according to discrete classes of similar size, forming mighty sand banks and spits of land, as a consequence of their similarity in shape and weight. Species of uniparental organisms come quite near the pure kind of this type of systems whereas species of biparental organisms have a higher grade of internal organisation by recombination of genes and, as a consequence, the possession of a shared, common gene pool.

The reflections of Zawadskij (1968) aim at a species concept broader than the biological species concept and not contradictory to the evolutionary species concept. He emphasises two principles: 1) the principle of the universality of species, i.e., the integration of all organisms in species, and 2) the principle of nonequivalence of species, i.e., that there are different kinds of species, one among them is the biological species *sensu* Mayr. Zavadskij does not formulate a short definition, but characterises species by 10 attributes common to all different kinds of species: multitude of organisms, type of organisation, reproduction, discreteness, ecological definiteness, diversity of forms, historicity, endurance, and wholeness (cf. Geesink 1987: 93, 94, and Sershantow 1978: 159–162). The concept of Zavadskij raises questions, for instance: What is the meaning of evolution of the species as one of the fundamental forms of existence of life since its origin on the earth over 3.5 billions years ago? What are the different kinds or types of species as a result of an evolution on supra-organismic level?

Monophyletic higher taxa as individuals

If we have the task to master a multitude of things, we can look upon them as sets and then investigate whether they have definite characteristics, by which they are classes, and definite characteristics by which they are systems. We can define the terms set, class, and system as follows: 1) a set is a totality of elements which is established only by enumeration of the elements which establish the set; 2) a class is a totality of elements which have at least one characteristic in common; 3) a system is a totality of elements if there are relations between the elements from which characteristics of the totality emerge, that are not present in any isolated element. Sets may also be classes. In a way it may be useful for analytical purposes to look upon elements of systems as elements of sets or of classes, but ultimately it will always be an inadequate approach. Systems are essentially not sets or classes, they are individuals. But what about higher taxa, more precisely, about monophyletic (in the sense of Hennig) higher taxa of phylogenetic systematics?

Monophyletic higher taxa are supra-specific taxa, units of two or more evolutionary species including the ancestral species which is exclusively common to them. They come into existence by the process of the splitting up of an ancestral species. This ancestral species is exclusively shared by all these products of the splitting process, and the resulting higher taxon includes all resulting new species of this ancestral species. The successive speciation events give rise to a hierarchically structured pattern of relationships between the species of a monophyletic higher taxon.

Ax (1984, 1988) reflects on monophyletic higher taxa as "closed phyletic communities of nature" ("geschlossene Abstammungsgemeinschaften der Natur"). According to him they are spatiotemporal units. Closed phyletic communities exist in space; the totality of their species has a definite geographical and ecological distribution. Closed phyletic communities are temporally characterised by a definite origin in time, which is the splitting of one evolutionary species into at least two daughter species. These closed phyletic communities have an historical continuity by the connection of species through the subsequent speciation events. Simultaneously there are *neither* vertical or horizontal relations between these species forming a closed phyletic community, *nor* any relations to their environments common to all of them, *nor* shared evolutionary mechanisms by which the community evolves as a unit. Closed phyletic communities are units of *phylogenesis,* not units of evolution.

Closed phyletic communities are not systems because there are no actual relations between their species (their elements) or between the totality of the species and the totality of their environments. But they differ also essentially from sets and classes because they are spatiotemporally definite and temporally continuous entities. They are entities *sui generis*. Following the proposal by Wiley (1981) we can call these entities 'historical groups'. Historical groups are substantial units by virtue of an existential (not logical) relation between their elements. Lewin (1922) recognised this relation already and called it 'Genidentität' (gene identity), a relation exclusively caused by the subsequent origin of one element from another one. According to Lewin 'Stammgenidentität' (phyletic gene identity) is a special case of gene identity. I feel that Lewin's concepts, which are completely faded into oblivion, need revival and further elaboration.

Phyletic gene identity implies that historical groups of species, closed phyletic communities of nature, are individual units. They are individuals by their common, inherent past, their inherited history, which is written down in the gene pools of their species: the successful expressions of the struggle for life in branching sequences of species during the course of evolution. The problem of the objectivity of higher taxa is the problem of the identification of historical groups, of the closed phyletic communities, of nature, which forms a large field for taxonomic work (Kreisel 1988).

References

Ax, P. 1984. Das phylogenetische System. Systematisierung der lebenden Natur auf grund ihrer Phylogenese. Fischer, Stuttgart.

Ax, P. 1988. Systematik in der Biologie. Darstellung der stammesgeschichtlichen Ordnung in der lebenden Natur. Fischer, Stuttgart.

Biology and Philosophy, Volume 2, 2. 1987: 127–225 (theme issue on the species-as-individuals concept).

Dobzhansky, T. 1965. Biology, molecular and organismic. The Graduate Journ. 7: 11–25.

Geesink, R. 1987. Theory of classification of organisms. In: E.F. de Vogel (ed.), Manual of herbarium taxonomy, theory and practice: 91–126. UNESCO/MAB, Jakarta.

Kreisel, H. 1988. Abstammung und systematische Einordnung der Pilze. Biol. Rundsch. 26: 65–77.

Lewin, K. 1922. Der Begriff der Genese in Physik, Biologie und Entwicklungsgeschichte. Eine Untersuchung zur vergleichenden Wissenschaftslehre. Springer, Berlin.

Löther, R. 1972. Die Beherrschung der Mannigfaltigkeit. Philosophische Grundlagen der Taxonomie. Fischer, Jena.

Löther, R. 1974. Zur Auffassung der Art als materielle System. In: W. Vent (ed.), Widerspiegelung der Binnenstruktur und Dynamik der Art in der Botanik: 41–51. Akademie-Verlag, Berlin (G.D.R.).

Löther, R. 1983. Das Werden des Lebendigen. Wie die Evolution erkannt wird. Urania-Verlag, Leipzig.

Malinowski, A.A. 1969. Einige Fragen der Organisation biologischer Systeme. In: A.I. Berg (ed.), Organisation und Leitung. Fragen der Theorie und Praxis: 135–163. Akademie-Verlag, Berlin (G.D.R.).

Mayr, E. 1963. Animal species and evolution. Belknap Press of Harvard Univ. Press, Cambridge, Mass.

Mayr, E. 1969. Principles of systematic zoology. McGraw-Hill, New York.

Mayr, E. 1982. The growth of biological thought. Diversity, evolution, and inheritance. Belknap Press of Harvard Univ. Press, Cambridge, Mass.

Sershantow, W.F. 1978. Einführung in die Methodologie der modernen Biologie. Fischer, Jena.

Simpson, G.G. 1961. Principles of animal taxonomy. Columbia Univ. Press, Oxford Univ. Press, New York, London.

Wiley, E.O. 1981. Phylogenetics. The theory and practice of phylogenetic systematics. John Wiley & Sons, New York.

Zavadskij, K.M. 1968. Vid i vidoobrazovanie (Species and speciation). Nauka, Leningrad.

32 The unified theory, macroevolution, and historical ecology

DANIEL R. BROOKS

Summary

The Darwinian revolution was based on the proposition that biological diversity is evolved diversity, and that evolution is a combination of genealogical (phylogenetic), developmental (ontogenetic), and environmental (selective) effects, but current theory concerns only the latter. The unified theory of evolution attempts to restore the macroevolutionary components of phylogeny and ontogeny to evolutionary explanation, along with selection. The unified theory asks new kinds of questions in evolutionary ecology. Phylogenetic constraints may limit the ways in which and the extent to which species adapt to different and changing environments, so it is possible that closely related species may have the same ecological or behavioural traits despite living in different environments. Two or more closely related species may live in similar habitats and yet exhibit divergent ecological or behavioural characteristics. Historical ecology uses phylogenetic trees to produce direct estimates of the origin and persistence of various aspects of ecological diversity and associations. It invokes two evolutionary processes, speciation and adaptation, to explain the evolution of diversity within clades, and invokes complementary processes of co-speciation and co-adaptation to explain the evolution of diversity in ecological associations.

Introduction

The Darwinian revolution was based on the proposition that biological diversity is evolved diversity, and that evolution is a combination of genealogical, developmental, and environmental effects. Paradoxically, developments during the past 50 years have seen a narrowing of evolutionary theory to the extent that much of *neo*-Darwinian explanation is based on attempts to reduce all explanations in evolution to changes in gene frequencies in populations under different environmental conditions. The unified theory of evolution (Brooks & Wiley 1988) attempts to restore phylogenetic and ontogenetic components to evolutionary explanation, along with selection, and to postulate general expectations about their relative contributions to the evolutionary process; hence, while not non-Darwinian it might be viewed as an expansion of neo-Darwinian theory. The unified theory attempts to integrate a variety of influences operating at different rates, and on different temporal and spatial scales, in providing evolutionary explanations.

P. Baas et al. (eds.), The Plant Diversity of Malesia, 379–386.

A widespread view of evolution asserts that the evolutionary 'play' takes place on a 'stage' organised by the environment. The unified theory implies a *wider* rather than an *opposed* view since the major component of the 'stage' consists of organisms, demes, populations, and species, all of them being 'players' as well as 'stage'. Environmental selection is also a major part of the 'play' that takes place on a 'stage' whose organisation is provided by 'phylogenetic constraints' and 'developmental constraints'. *Phylogenetic constraints* is a synonym for persistent ancestral traits that have not evolved rapidly enough to be affected by environmental selection during any given episode of microevolutionary change. *Developmental constraints* is a synonym for the necessary integration of any new trait with the rest of the developmental program in order to produce a viable organism which is then potentially acted upon by environmental selection.

According to the unified theory, just as with standard Darwinian theory, evolution results from an interaction between genealogical and ecological processes. Salthe (1985) and Eldredge (1985, 1986) have termed these the *genealogical hierarchy* and the *ecological hierarchy*. Ecological processes tend to have homeostatic effects, forcing populations into equilibrium conditions. By contrast, the genealogical processes are viewed as having developmental, nonequilibrium or diversifying effects. The impact of phylogenetic and developmental constraints is to slow the natural entropic accumulation of genealogical diversity, providing an organised but dynamic 'stage' upon which the environment can be seen as acting out the 'play' of natural selection. Natural selection increases the degree of organisation even further. Biological systems obey rules of self-organisation carried genealogically (historically) and expressed ontogenetically within environmentally defined boundaries. The environment is not inherently organised as the ecological hierarchy; its organisation into an ecological hierarchy is largely a consequence of organisation intrinsic to the genealogical hierarchy. Or, in other words, species do not fill empty niches, they create their own niches. At the same time, the environment provides an important constraining influence on biology, and the (self-generated) ecological hierarchy plays an important feedback role in evolution. The ecological hierarchy is the means by which two different genealogies, or two different generations in one genealogy, can causally influence one another.

The predominant physical manifestations of the interaction between genealogical and ecological processes differ depending on the time scale chosen for observation (Brooks & Wiley 1988; Brooks 1989; Maurer & Brooks, submitted). For example, on extremely short time scales the primary manifestation is physiological loss, or the dissipation of heat due to metabolic activities. On intermediate time scales the primary manifestation is in the accumulation and maintenance of biomass, evidenced by ontogenetic, reproductive, and successional phenomena. And on the longest time scales, the primary manifestation is the accumulation of genetic diversity. The dynamics of genealogical processes determines a 'phase space' that is filled to the extent that different possible genetic combinations are realised. It grows over time in proportion to the accumulation of novelties arising through changes in genetic systems. Reproduction and recombination tend to fill that space. That the phase space is only partially filled tells us that there are constraints on entropy increases, either inherent constraints, such as those arising from the 'rules' of the genetic system (e.g., those discussed by Lima-de-Faria 1983), constructional constraints (e.g., those discussed by Lauder 1982), or natural selection. The bulk of constraints on evolution originate in the genealogical flow of information, transmitted through reproduction. It is at the phylogenetic, or macroevolutionary level, that the effects of these constraints can be seen most clearly.

Macroevolution and the unified theory

Microevolutionary research comprises two fields, *population genetics* and *population ecology*. Population genetics determines the micro-phylogenetic patterns of variation upon which processes studied by population ecologists operate. Microevolutionary processes are those that occur on relatively short time scales and over relatively small spatial scales. They are ***within-species phenomena***, and correspond to the level of local adaptive phenomena. Macroevolutionary processes, by contrast, are those that occur over relatively long time periods and relatively large spatial scales. They are ***among-species phenomena*** whose effects are found in patterns of phylogenetic constraints on any system being studied. Macroevolution, in parallel with microevolution, comprises two fields, *historical ecology* and *macroecology*. Figure 32.1 depicts an heuristic view of the relative relationships among various areas of research in evolutionary biology as represented herein.

Macroecology may be characterised as being concerned primarily with the identification of large-scale *statistical patterns* of diversity (both diversification and association) that involve more than local populations of a single species, or more than the species in a localised habitat (see Brown & Maurer 1987, 1989). It is also concerned with the study of size, spatial and temporal *scaling effects* in explaining patterns of diversity (see Brooks & Wiley 1988; Brooks 1988; Maurer & Brooks, submitted; Brown & Maurer 1989).

Historical ecology is concerned with documenting the patterns of diversity produced by genealogical processes.The unified theory predicts three macroevolutionary aspects of genealogical processes. First, the most informative evolutionary summary of data about similarities among organisms will result from the use of analytical methods that maximise the degree of phylogenetic constraints for a given data set. Second, the necessity for developmental integration of all evolutionary innovations means that phylogenetic systematic

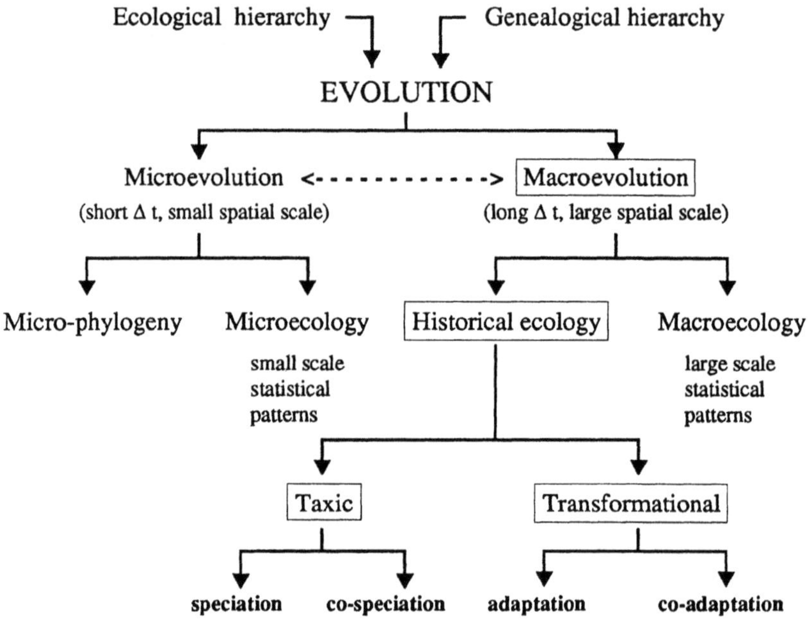

Figure 32.1. Heuristic diagram depicting the major research programs within historical ecology viewed by taking a hierarchical perspective on evolutionary biology. Historical ecology is shown nested within macroevolution, which is nested within the unified theory of evolution. There are many interconnections between the research areas depicted as separate entities.

analysis of data from different portions of the developmental program (such as larvae and adults) will result in highly concordant phylogenetic trees. And third, application of phylogenetic systematic methods to data derived from relatively independent sources, such as ecological, behavioural, anatomical, and biochemical characters, will result in highly concordant phylogenetic trees.

Historical ecology and the unified theory

The unified theory also suggests new kinds of questions in evolutionary ecology. To paraphrase Darwin (1872: 219), if there are phylogenetic constraints on ecological and behavioral diversification, it is possible that closely related species may have the same ecological or behavioural traits despite living in different environments. Further, contemporaneous species may exhibit ecological and behavioral traits that originated in environments quite different from those in which the species find themselves today. If this is true, it must be possible for phylogenetic constraints to limit the ways in which, and the extent to which, species adapt to different and changing environments. Alternatively, two closely related species may live in similar habitats and yet exhibit divergent ecological or behavioural characteristics; this indicates a certain degree of independence of the rules characterising the genealogical and the ecological hierarchies.

The use of phylogenetic trees to produce direct estimates of the origin and persistence of various aspects of ecological diversity and associations has been advocated recently among evolutionary ecologists (Ridley 1983; Clutton-Brock & Harvey 1984; Pagel & Harvey 1988), but is currently little heeded, even by its advocates. Few evolutionary ecologists who express concern about the significance of phylogenetic constraints actually incorporate phylogenetic explanations into their own work. Those that do usually equate taxonomy with phylogeny, which can lead to undesired ambiguity or circularity.

Historical ecology extends phylogenetic studies of the origin of key innovations in ecological and behavioural traits, of ecological and behavioural correlates of speciation, and of patterns of biogeography and ecological associations, in an attempt to incorporate different causal agents in evolutionary explanations (Brooks & McLennan 1991). Ross (1972a, b) may be regarded as the modern father of historical ecology (Brooks 1985). He suggested that for a variety of insect groups, ecological diversification was consistent with, but much less frequent than, phylogenetic diversification. He found that approximately one out of every thirty speciation events in a variety of insect groups was correlated with some form of ecological diversification. Furthermore, he felt that there were no predictable patterns explaining those shifts that did occur and suggested that ecological diversification in evolution comprised a biological 'uncertainty principle'. This interpretation of the relationship between ecological and functional diversification and phylogenetic diversification was certainly at odds with extreme ecological determinism intepretations of phylogenetic diversification.

Historical ecology comprises research involving the evolution of particular clades as well as the evolution of ecological associations (Brooks & McLennan 1991). The study of the evolution of individual clades invokes two evolutionary processes, ***speciation*** and ***adaptation***. Persistent ancestral traits represent instances in which the production rules of the genealogical hierarchy have constrained the structure of the ecological hierarchy. Adaptive changes in such traits represent instances in which the genealogical hierarchy has changed in response to the effects of the ecological hierarchy. The study of the evolution of multi-species ecological associations invokes ***co-speciation*** and ***co-adaptation***. Macroevolutionary processes associated with speciation (and extinction) represent the 'taxic' aspect of macroevolution; by contrast, macroevolutionary processes associated with adaptation represent the 'transformational' aspect of macroevolution (Cracraft 1985). This is an important interface between the ecological and genealogical hierarchies, because the range of energy-flow pathways in ecological associations such as ecosystems is limited by the attributes of the component species as well as abiotic conditions. Macroevolutionary questions about the relationship between ecology and evolution can be examined by considering how closely phylogenesis in any clade is associated with evolutionary changes in ecological or behavioural traits relevant to diversification or to particular theories about causal mechanisms in evolution.

Integrating phylogenetic and historical ecological patterns

If the ecological hierarchy exerts an organising influence by acting as a homeostatic rather than developmental force on biological systems, the unified theory predicts that the ecological and behavioural (functional) correlates of phylogeny should be conservative relative to the morphological and developmental correlates of phylogeny. Hence, it is expected that most closely related species will be morphologically distinct but ecologically and behaviourally similar. In addition, we would expect to find evidence of phylogenetic constraints on suites of ecological and behavioural traits. This has also been found to be true for studies performed to date (Brooks & Wiley 1988; McLennan *et al.* 1988; Brooks & McLennan 1991).

The genealogical hierarchy exerts an organising influence on biological systems through phylogenetic and developmental constraints. However, because increasing diversity and complexity is an entropic phenomenon, evolution will occur despite the various constraints on its expression. Hence, because the developmental 'rules' of the genealogical hierarchy are relatively independent of, and can supersede, the homeostatic 'rules' of the ecological hierarchy, ecological and behavioural diversification should lag behind developmental and morphological (including macromolecular) diversification on a phylogenetic scale (see Brooks 1989 for an empirical example).

Conclusions

The unified theory of evolution proposes an expanded agenda in evolutionary research. It is based on a common leitmotif, the use of energy in maintaining and transforming ordered states of matter, linking research efforts. Using information and entropy concepts as an accounting system, a common phenomenology for processes occurring on different levels of organisation in biological systems has been discovered (Brooks & Wiley 1988). Processes relevant to evolution are generative (originating or diversifying) or conservative (maintaining or selective) in their effects. The interplay of these processes through time produces historically-constrained order. Many processes affect biological systems, at all levels of organisation and at all times, but their effects are often manifested on different time scales. Changes occurring on time scales shorter than speciation rates are microevolutionary; those occurring on time scales longer than speciation rates are macroevolutionary. Macroevolution is neither autonomous from, nor reducible to, microevolutionary theory.

Acknowledgements

I wish to thank Robert Geesink and Deborah McLennan for reading this manuscript and many helpful suggestions. This study was supported by operating grant No. A7696 from the Natural Sciences and Engineering Council of Canada.

References

Brooks, D.R. 1985. Historical ecology: A new approach to studying the evolution of ecological associations. Ann. Missouri Bot. Garden 72: 660–680.

Brooks, D.R. 1988. Scaling effects in historical biogeography: A new view of space, time and form. Syst. Zool. 38: 237–244.

Brooks, D.R. 1989. The phylogeny of the Cercomeria (Platyhelminthes: Rhabdocoela) and general evolutionary principles. J. Parasitol. 75: 606–616.

Brooks, D.R., & D.A. McLennan. 1991. Phylogeny, ecology, and behavior: A research program in comparative biology. Univ. Chicago Press, Chicago.

Brooks, D.R., & E.O. Wiley. 1988. Evolution as entropy: Toward a unified theory of biology. 2nd Ed. Univ. Chicago Press, Chicago.

Brown, J.H., & B.A. Maurer. 1987. Evolution of species assemblages: Effects of energetic constraints and species dynamics on the diversification of the North American avifauna. Amer. Nat. 130: 1–17.

Brown, J.H., & B.A. Maurer. 1989. Macroecology: The division of food and space among species on continents. Science 243: 1145–1150.

Clutton-Brock, T., & P. Harvey. 1984. Comparative approaches to investigating adaptation. In: J. Krebs & N. Davies (eds.), Behavioral ecology: An evolutionary approach: 7–29. 2nd Ed. Sinauer, New York.

Cracraft, J. 1985. Species selection, macroevolutionary analysis, and the 'Hierarchical theory of evolution'. Syst. Zool. 34: 222–229.

Darwin, C. 1872. The origin of species. 6th Ed. J. Murray, London.

Eldredge, N. 1985. Unfinished synthesis. Columbia Univ. Press, New York.

Eldredge, N. 1986. Information, economics and evolution. Ann. Rev. Ecol. Syst. 17: 351–369.

Lauder, G.V. 1982. Historical biology and the problem of design. J. Theor. Biol. 97: 57–67.

Lima-de-Faria, A. 1983. Molecular order and organization of the chromosome. Elsevier, Amsterdam.

McLennan, D.A., D.R. Brooks & J.D. McPhail. 1988. The benefits of communication between comparative ethology and phylogenetic systematics: A case study using gasterosteid fishes. Can. J. Zool. 66: 2177–2190.

Maurer, B.A., & D.R. Brooks. Submitted. Scaling of entropy production in biological systems. Biosystems.

Pagel, M.D., & P.H. Harvey. 1988. Recent developments in the analysis of comparative data. Quart. Rev. Biol. 63: 413–440.

Ridley, M. 1983. The explanation of organic diversity: The comparative method and adaptations for mating. Clarendon Press, Oxford.

Ross, H.H. 1972a. The origin of species diversity in ecological communities.Taxon 21: 253–259.

Ross, H.H. 1972b. An uncertainty principle in ecological evolution. In: R.T. Allen & F.C. James (eds.), A Symposium on Ecosystematics. Occ. Pap. No. 4: 133–160. Univ. Arkansas Mus. Press, Fayetteville, Arkansas.

Salthe, S.N. 1985. Evolving hierarchical systems: Their structure and representation. Columbia Univ. Press, New York.

33 Nomenclatural stability, taxonomic instinct, and flora writing – a recipe for disaster?

P.F. STEVENS

Summary

Systematics might seem to be in danger of fission if recent authors are taken literally. Some suggest that only monographs or floras are needed, with funding for phylogenetic studies being discontinued. Stability of nomenclature for the users of classifications (largely other than biologists) is considered to be a paramount need; species (and other taxa) should be readily recognisable, and a 'broad view' taken of variation, with all characters being carefully evaluated. Taxonomic intuition is seen as being of great importance, but an understanding of function and ecology is also central for classification. Academic argumentation is to be isolated as far as possible from the public perception of taxonomy and systematics. Is this what herbaria are for? Others see phylogeny reconstruction as central to comparative biology and ultimately to a stable classification. The probable difference between many taxonomic species and the units involved in, or even proximally produced by, evolution seems clear, as is the concomitant need for a much finer analysis of variation than is common. Taxonomic 'intuition' is seen as the result of the taxonomic apprenticeship system; the human mind is capable, if inconsistent, when recognising gestalt; it is quite inefficient when simultaneously comparing numerous variables. Is this what academia is for? I attempt to understand these positions, drawing both on my work on the Malesian flora and as an academic. One important point is simply that of relating the limited goals and expertise of individuals to a broader vision of systematics. However, we have but a single system of nomenclature and, for many groups, a single accepted classification, and therein lies much of the current tension. Systematics is a discipline that deals with the history of life, and we forget this at peril of our professional integrity.

Introduction

Recent discussion has emphasised the connection between taxonomic classifications and the way human minds naturally classify. Here I focus on taxonomic instinct, or more generally, on neural (mental) analysis in classification, and attempt to evaluate the claims made for it; taxonomic instinct is analysed as it works at higher taxonomic levels, and also at the level of species. I also explore the relationship between neural classifications and nomenclatural stability, since the relationship between names and our perception of the world is the main topic of this paper. It seems particularly appropriate to raise these issues here since many of them are particularly common on the eastern side of

P. Baas et al. (eds.), The Plant Diversity of Malesia, 387–410.

the Atlantic, and commonly surface in discussions on the applications of classifications, such as in floras.

This paper is aimed at practical taxonomists, those of us who will write hundreds or thousands of descriptions of plant species during our working lives, although my observations may offer us little comfort. What I have to say is hardly new, but we seem to run into the same problems in taxonomy time and time again, in large part because the mental operations that occur when we form groups are so poorly understood. The next step in the analysis presented here is the development of a deeper knowledge of cognition and perception and how they relate to the taxonomic process.

Two issues obscure the discussion on taxonomic instinct. One is a feeling that classifications will be most easy to use if they are constructed by the same mental processes as occur in the users of classifications (Heywood 1988a); such classifications will be most readily understood by these users. There have been a number of recent publications emphasising how important it is for taxonomists to think about who uses classifications; these users include lawyers, architects, landscape gardeners, and civil engineers (e.g. Bisby 1984; Hawksworth & Bisby 1988; Heywood 1984, 1988a; Morin *et al.* 1989). This is very appropriate, especially in the context of writing floras and other 'applied' taxonomic works; we have to know for whom we are writing if our products are to be most helpful for their particular purposes. The relationship between neural classifications and users suggested above is, however, believed to be at a much deeper level than simply producing keys and descriptions that can be understood by non-taxonomists.

A second complicating factor is the relationship between Gilmourian philosophy and taxonomy; Gilmour's ideas on classification and its purposes are still cited, and with approval. For Gilmour, classifications which served a large number of purposes (i.e. were of greatest general utility) were called natural. The most natural classification was that about which most inductive generalisations could be made (Gilmour 1940, 1951). The conclusion that a multipurpose classification was best and most natural was inevitable, and the needs of all users were legitimately to be taken into account in deciding on the shape of this 'most natural' classification. Evolutionary classifications were necessarily special purpose classifications. Gilmour distinguished sharply between sense data, given once and for all and unchangeable, and the 'clips', classifications of one sort or another, by which reason built up the logically coherent pattern called the outside world (Gilmour 1940).

To Walters (1988: 18) it seemed that "... when, however, we find areas of controversy persisting through generations – as we do with regard to the evolutionary purposes of systematic botany – we ought to look to the philosopher and the historian to see what is wrong. This, I feel, was Gilmour's great contribution." It is not my purpose to analyse Gilmour's work in detail, but I would suggest that Gilmour's work was flawed in two areas, at least. In particular, the 'natural method' in systematics was developed by A.-L. de Jussieu who believed that there were no sharply bounded groups in nature, with higher taxonomic groups in particular being circumscribed very largely for matters of convenience since they did not exist in nature. The taxonomic procedures associated with the 'natural method' reflect that particular view of the world. That taxonomists continued to use a modification of those methods after the acceptance of evolution, which, with extinction, should lead to the production of distinct groups, is a major reason why the 'evolutionary purposes' of systematics have been so controversial (Stevens 1984, 1986). Further, Gilmour adopted an extreme view of the world, since it is now difficult to follow the idea that sense data are 'givens' (see especially Sattler 1986) since it is recognised that theory and observation are intimately entwined. It is interesting to note that Gilmour inadvertently rendered his position rather vulnerable when he noted that a library classification based on subject matter had a stronger influence on the total characteristics of books than did a classification based on binding design (Gilmour 1951). If indeed characters of organisms allow us to recognise phylogenies, as in the ciliated sperm of some gymnospermous groups, the pollen tetrads of most Ericaceae, the xanthones and lignified exotegmen of many Clusiaceae plus Bonnetiaceae (the latter includes our own *Archytaea*), and the remarkable foliar laticiferous system of *Tripetalum, Pentaphalangium,* and some species of *Garcinia,* then surely there is a biological equivalent of a library classification based on subject matter. The 'subject' matter of biological classification is phylogeny, a point to which I shall return.

A recent statement about the method and purpose of tropical taxonomy very much along Gilmourian lines is that of Heywood (1988a: 27):

> "Nearly all the taxonomic work on tropical organisms is undertaken using phenetic principles and procedures, whether numerical or neural. The classifications are judged by their ability to store large amounts of information, by their predictivity, by their simplicity and stability. Whether or not they may or may not correspond to some estimate of a phylogenetic or cladistic set of relationships is simply not relevant. In simple terms the cladistic relationships of the groups in a phenetic classification are simply irrelevant in judging the usefulness of the classification. Classifications have to be related to the way we as human beings perceive nature, and we know from our own experience and from cognitive psychology that our perception is based on overall assessment of similarities and certainly not on dichotomous branching patterns."

Stability of nomenclature may seem to be a desirable feature of such classifications for many taxonomists (e.g. Walters 1988; Prance & White 1988), the debate leading to change in circumscription and name of groups on phylogenetic grounds being characterised as simply arcane (Heywood 1988b; Cronquist 1987). However, if classifications were adopted that best fit the needs of the heterogeneous groups of users mentioned above, stability would not necessarily ensue; if a new group of users was identified, their needs might differ from those of others and cause a change in the classification. Gilbert (1986) adopted a rather 'splitty' (sic) approach in his classification of the *Vernonia galamensis* group because it is a potential oilseed crop – enter users, exit stability.

Others, however, will want classifications to reflect the best current estimate of phylogeny. Change of name will reflect changing ideas of relationships and will be welcomed, rather than avoided, since for many evolutionary questions an understanding of phylogeny is essential (see also Brooks [this symposium]; Donoghue 1989). Thus Patterson and Smith (1989) showed that groups of likely monophyletic origin fared differently through periods of mass extinction than did paraphyletic or polyphyletic groups; this clarifies the effect of these extinctions on the biota. Closer to home, the recent studies of Nixon and Crepet (1989) and Crepet and Nixon (1989) have recently suggested that the three species placed in *Trigonobalanus* formed a paraphyletic group and should be placed in three genera. These new ideas of relationship, in conjunction with fossil evidence, suggested particular notions of the biogeography of this part of the Fagaceae. Melville (1982) working under the assumption that *Trigonobalanus* was monophyletic and lacking fossil data, thought that its distribution resulted from the breakup of a Pacific continent, but there is now little evidence that this is so.

The major problem is that both workers who are interested in phylogeny and those who could not care less about it usually use the same system of classification. However, although most biologists interpret classifications, however constructed, in an evolutionary context, a naive equation of a 'natural' with an evolutionary classification, let alone with a phylogenetic classification, is justified neither logically nor historically (Walters 1988; see also Stevens 1986, and references).

So the issues revolve around the particular needs of users of classifications, the relationship between classifications and names, and how neural classifications in particular are constructed. In what follows I will first discuss taxonomic 'instinct' ('mental' or 'neural' classification) as it is believed to operate when taxonomists work on higher level groups. I then switch to the species level and focus on some neural operations that affect the circumscription of species. I emphasise that an understanding of local discontinuities is essential in any broader-scale survey. In the discussion I emphasise the fallability of neural methods and the inevitability of tension if a single system of names serves a variety of conflicting purposes, and that we cannot hope to resolve these conflicts if we do not understand better how we construct taxonomic groups, whether using the computer or the unaided mind.

Taxonomic instinct and intuition

One of the problems when talking to taxonomists about what they do is their feeling that the way in which the human mind reaches taxonomic conclusions is best described by words such as 'instinct', 'intuition', 'taxonomic genius' (or 'heuristic', Cronquist 1987). This is associated with the feeling that taxonomy is the art of describing what 'is' in nature, a process in which theory does not intrude. As an eminent European taxonomist recently wrote to me, "You are very brave to investigate the conceptual history of species concepts. Do they normally have such a basis? In practical taxonomy there remains little else but discontinuity ..." Gilmour's philosophy is certainly very compatible with this feeling, since Gilmour did not suggest that there was any particular relationship between theory and observation.

Heywood (1988b) noted that the human mind was very good at recognising the general appearance, or facies, of organisms. This is indeed true, but we do not form groups simply on their gestalt. The taxonomist is also commonly believed to be able to integrate effectively the large bodies of taxonomic data that have accumulated over the last two centuries. In so doing, the human mind may be seen to function like a computer (Kalkman 1987). Estes and Tyrl (1987) distinguished between these two methods of neural analysis, one, gestalt groupings formulated on the 'totality' of the features observed, and the other, groupings resulting from a linear examination of selected characters, the specific comparison of similarities and differences.

In this latter operation, which I discuss first, Estes and Tyrl (1987) suggested that the human mind was prone to seize on one or two variables which were considered more important than the others, other variation being relegated to second place. This apparently describes Ashton's 'intuitively weighted' characters in the Dipterocarpaceae (Ashton *et al.* 1984), or taxonomic weighting generally, but the weighting is based on an evaluation of the general pattern of variation (see below). Appreciation of the variation depicted in Edgar Anderson's pictograms is constrained by an overweighting of the parameters of the two main axes (Estes & Tyrl 1987; McNeil 1984). Indeed, the accuracy with which expert witnesses evaluate complex questions when giving evidence in trials has been questioned (Dawes *et al.* 1989 and references; Faust *et al.* 1990; cf. Kleinmuntz 1990). It has been found that a simple probabilistic approach is likely to lead to the evidence being evaluated as effectively, or more effectively, than many experts can. As Dawes *et al.* (1989: 1671) note, "a unique capacity to observe is not the same as a unique capacity to predict on the basis of integration of observations." The problems involved in taxonomic decision-making can usefully be compared with the problems these expert witnesses face, since both taxonomists and expert witnesses look for pattern among several only poorly-correlated variables.

Two examples must suffice to indicate how difficult it is to understand the workings of taxonomists' minds. George Bentham is one of the greatest plant taxonomists that has ever lived, and Cantino and Sanders (1986) talked about Bentham's taxonomic 'genius' because the major taxa that Bentham recognised in the Labiatae agreed rather well with Erdtman's subfamilial classification, based largely on pollen. They found that several other features, not simply pollen morphology alone, separated Erdtman's two subfamilies. Thus variation in both endosperm and plant odour is largely congruent with Erdtman's two subfamilies, and it is clear that Bentham was aware of the taxonomic importance of these features, even if he did not describe them in all groups (Bentham 1832–36). Hence in this rather simple case 'genius' can perhaps be related to the observation and taxonomic evaluation of particular characters. It is also probably connected with Bentham's remarkably sophisticated understanding of the problems encountered in the use of inductive logic (Bentham 1827).

A more complex example is provided by Prance and White (1988). They suggested that their intuitive classification of the Chrysobalanaceae, which summarises their very substantial endeavours in the family, accorded best with some numerical taxonomic algorithms, notably single-linkage clustering. This, they thought, simulated most closely how taxonomists actually constructed classifications (cf. also Cain & Harrison 1958). However, it proves very difficult to retrieve their classification from the data they present in their generic descriptions. These data were analysed using a single-linkage clustering algorithm, as well as unweighted pair-group arithmetic average (UPGMA) clustering and cladistic analyses. Interestingly, none of the topologies of the trees resulting from these analyses, and that from a phylogenetic analysis using only the data in the tribal descriptions, agreed much either with each other or with the topology implicit in Prance and White's formal classification; the tribes were largely dismembered (Chappill 1990). [Single linkage clustering has an unfortunate propensity to chain groups, that is, to add groups singly when they are all rather similar to one another (Sneath & Sokal 1973). UPGMA is currently one of the algorithms favoured for use in phenetic analyses at higher taxonomic levels.]

Of course, when Prance and White (1988) circumscribed the taxa that they recognised, they may have used data that were not included in their descriptions, or weighted particular characters, or, in general, used their taxonomic experience; unfortunately, details of weighting possibly used cannot readily be obtained from the detailed discussion of the characters presented. [An example of the effect of weighting in the construction of phylogenies using mental ('neural' in the sense used here) analyses is provided by Ball and Maddison (1987), who teased apart some of the reasons for the differences between computer-assisted and neural analyses of the 'same' data.] Along the same lines, it is not too helpful to be told that a taxonomist uses "a workable number of characters that are evidently correlated with many unmentioned ones" ('omnispective' classification: Blackwelder 1967: 191). Furthermore, given the great diversity of variation patterns that exist, 'taxonomic experience' can be used to justify almost any course of action. There is infraspecific variation in whether the ovary is superior or inferior in *Peliosanthes tata* (Jessop 1978), but of course this does not render ovary position *a priori* valueless when recognising higher level taxa. In all cases where 'experience' is invoked, justification ultimately has to be in terms of overall parsimony unless circularity is to result.

Let us grant Heywood's supposition (1988a, see above) that it is not relevant for some people whether or not classifications correspond with a phylogeny, but only on how well they store large amounts of information. This suggests that there may be conflict between different kinds of classification in these regards, although this is a difficult position to defend (see Farris 1979a, 1979b, 1983), at least when phenetic and cladistic classifications are compared. Neural classifications are very difficult to analyse from this point of view, since as we have seen, it may be difficult to establish just what data are being analysed. But if data in the descriptions cannot be related to a classification nominally based on those data, then the ability of neural classifications to store and retrieve information is questionable. We are a long way from being able to evaluate the claims for neural classifications advanced above. Only if characters and classifications are clearly related can we begin to talk about information retrieval, and then it will also be evident how new data support (or do not) the existing classification.

Other studies comparing numerical and neural classifications of taxonomists and non-taxonomists found general agreement (e.g. Sokal & Rohlf 1970; Moss 1971), enough to encourage a belief that taxonomic experience was not so important as had previously been thought. An 'intelligent ignoramus' could serve as the backbone of the taxonomic labour force of the future (Sneath & Sokal 1973 and references).

Still others have suggested that neural classifications are hierarchical, or even phylogenetic: "The hierarchical nature of the system is a consequence of the fact that the order of the living is itself hierarchically structured (Riedl, 1978), and that selection has therefore endowed our unconsciously operating cognitive apparatus with the most adequate system [i.e. hierarchical] for processing its patterns" (Jeffrey 1987: 27). This of course runs directly counter to Heywood's assertions mentioned earlier (see also Stevens 1986, 1987). But such a neural system is unlikely to be the only one we use; the hierarchical relationships of a wolf-like animal with pointy teeth would seem to be a matter of the utmost inconsequentiality if that animal were chasing you. Furthermore, as noted by

Maddison *et al.* (1984), in sequential mental analyses decisions early in the sequence may result in subideal phylogenies, or hierarchies of any sort (see also Jardine & Sibson 1971).

The use of facies to help circumscribe taxa is probably particularly important at lower levels of the taxonomic hierarchy, and is akin to the ability to recognise an individual in a crowd. Here data are integrated so that it is some totality of the appearance of the organism that is being recognised, not individual features. Facies can, however, be partly decomposed into its elements, and this may provide the difference between simply asserting that one plant is clearly different from another and providing evidence that the two differ.

How the plant grows contributes to this facies, and a remarkable, albeit severely constrained, amount of valuable taxonomic information about plant growth can be obtained even from herbarium specimens. With care and a knowledge of the principles of plant construction it is possible to communicate elements of what go to make up this aspect of the facies of the plant. In the Clusiaceae there is much variation in shoot growth that is valuable at several taxonomic levels. It is perhaps appropriate for my theme that the functional scaly terminal buds of *Cratoxylum* section *Isopterygium* that characterise the section rarely (never?) feature in taxonomic descriptions, but were clearly depicted by the artist Ruth van Crevel (see Robson 1974: figure 8a). Even the number of leaves per innovation varies in a species-specific way in genera such as *Calophyllum* and *Mammea*. *Mammea* in particular is a difficult genus, and any help in delimiting species is welcome. Although I do not pretend to understand the variation pattern in *M. novoguineense,* that some specimens have innovations with two pairs of leaves, and others with six, makes one pause and think when these same collections also differ in other rather subtle characters.

Interestingly, the facies of adult trees of *Garcinia* and *Kayea* in Malesia, both of which are described as being monopodial with spreading branches, is similar, but this similarity belies their fundamentally different construction. In this case, I did not have an inkling of how *Kayea* puts together its trunk from examining herbarium specimens. The main trunk of those species of *Kayea* that I have seen is a sympodium produced from axillary buds; the terminal buds, although functional, grow laterally and produce branches. The trunk of *Garcinia,* on the other hand, is a genuine monopodium.

Note, however, that the classical meaning of facies is "*external appearance,* as opposed to reality, *a pretence, pretext*" (Lewis & Short 1897, italics the authors'). Rather ironically, 'facies' and 'species' had connected meanings, the latter being characterised by their facies, thus representing appearance rather than reality. Although we might not subscribe to such definitions today, facies does mean general external appearance, and as we have just seen, external appearance is not a reliable guide to an understanding of the plant.

Species concepts

Some theory

Van Steenis's writings on specific and infraspecific delimitation in the Malesian region (van Steenis 1957, see also 1949) are rightly considered to be essential reading for any student of tropical floras, not simply those of Malesia. Here I focus on one aspect of van Steenis's ideas, the broad species concept, and how such species are formed by taxonomists. One point should be made clear immediately. Suggestions that the numbers of species be reduced may have two components. One is that it is simply inconvenient to have too many species. This is clearly an extra-biological and completely irrelevant reason if species are believed to be the building blocks of evolution or the fundamental unit of taxonomy, as van Steenis himself believed. The other component may be the desire to have species that are not connected to other species by intermediates; if there is some connection, infraspecific rank may be more appropriate. This is a more biologically-based reason, so long as we are aware of how we are analysing variation.

Van Steenis's support of broadly-circumscribed species is clear, the epigraph to the long section 'General considerations' in which variation at around the species level is exhaustively discussed, being "We should endeavour to determine how few, not how many species are comprised in the Malaysian flora." (van Steenis 1949: XII; 1957: CCXXVII). Although van Steenis emphasised the value of field work and field observations, he prized species that had been defined in the broader context usually not available to the field worker, but which was more easily obtained with access to large collections. Furthermore, he noted, "for binomiums the Linnean canon has *priority of conception*" (van Steenis 1949: XL, his italics). Of course, Linnaeus is also generally believed to have had a broad species concept, witness the opposition between 'Linneons' (broadly circumscribed species) and 'Jordanons' (narrowly circumscribed species), although this probably significantly misrepresents Linnaeus's species concept when he was dealing with plants he knew in the field.

Van Steenis (1949) also quotes extensively from Hooker and Thomson's preface to the abortive 'Flora Indica' (Hooker & Thomson 1855), again, required reading for all taxonomists, in which a very similar position is elaborated. Hooker articulated criteria for circumscribing species taxa in his lengthy introduction to the 'Flora Novae-Zelandicae': "... with all [Linnaeus, Bentham, the de Candolles, Decaisne, Gray, de Jussieu, Lindley, and the Richards] the tendency would be to regard dubious species as varieties, to take enlarged views as to the range and variation of species, and to weigh characters not only *per se*, but with reference to those which prevail in the Order to which the species under consideration belong." (Hooker 1853: vi).

There are two elements to Hooker's position. One is an explicit appeal to authority, both living and dead, and the other an equally explicit claim that taxonomists with access to large collections were necessarily best placed to make decisions about species limits because they could better understand the pattern of variation throughout the ranges of plants, which was the important thing to understand. In this introduction, Hooker attempted to train untutored colonial botanists in the right ways of seeing and thinking, to school them in the habits of observation and the ability to reason on what they read in the book of nature. This was a recurring theme in Hooker's writings: in the 'Flora Indica' he thought that narrow local species recognised because of bad example should be exposed so that local botanists could recognise the error of their ways (Hooker & Thomson 1855). As is clear from those parts of Hooker's correspondence that have been published (e.g. Huxley 1918, especially chapter XXIII) the issue of species limits greatly exercised him, and he felt that he was making headway in converting his colleagues, including George Bentham, into adopting his broad concept. Indeed, Hooker's position is by no means unrepresentative of a substantial portion of the botanical community then and since; justification of species limits by making recourse to authority and established tradition remained common (Stevens 1990a). Taxonomic species represent primarily a particular level of morphological discontinuity that has historically been accepted as evidence of specific distinctness; taxonomists may hope these species are in some way the units of evolution, but biologists in general are less sanguine about this (e.g. Davis & Heywood 1963; Levin 1979; Kay 1984, and references). In practical terms the result of both Hooker's and van Steenis's position is what Kalkman (1987) describes as an operationalist 'pseudodefinition' of a species: a good species is what a good taxonomist calls a species, and *that* is generally an easily recognisable portion of the biological world.

Asa Gray, in his review of the 'Flora Novae-Zelandicae', made some perceptive comments about Hooker's position. After remarking that the ablest and most experienced botanists were indeed adopting more enlarged circumscriptions of species, he suggested possible problems with Hooker's approach to variation. Gray was not sure that taxonomists working with large amounts of material examined it all as carefully as they might. The 'intuitive judgment' (sic) of the taxonomist was involved when a taxonomist was confronted with the vast amount of material, and it was through exercise of this judgment that the true genius of a taxonomist was revealed (Gray 1854).

So we are back to intuitive judgment again. However, it is unclear that 'intuition' represents anything more than the biologically suspect exigencies of a formal taxonomic system interacting with the unanalysed mental algorithms being used in the analysis of the variation (Stevens 1990a). Van Steenis (1957) characterised instinct somewhat more charitably as 'sublimated experience', and, as he emphasised was the result of 'synthesis'.

Some practice

Taxonomic species are to be readily recognisable, but how are descriptions to be interpreted or data analysed in deciding that species fulfil this requirement? Hooker himself thought that distinctions that appeared to separate species locally would break down if the study had a broader geographical base, with different combinations of characters occurring in different countries. We can see how this might happen if we skip a century. Leenhouts (1967: 309, italics added) frankly admitted: "But if one combines species this means not only that specimens are brought together under one and the same name, but also that the description, and that means the range of variability of many characters, is widened. *And, even if some local form is not directly connected by intermediates with any race of the complex species, it may fall completely within the description of the complex as a whole.*"

Thus the pattern of variation that is taken from the description may have little to do with the pattern of variation in nature. How can this be? The answer is that a combinatorial approach to the description of variation is linked to the formal requirement that species be readily separable morphologically. The variation encompassed in a species is represented by all possible combinations of all the variation described for the species. To take an oversimplified example, some members of a taxon may have both small fruits and large leaves when compared with other representatives of the species. A variant with small fruits and small leaves will be linked on to the species even if there is a discontinuity in leaf size between it and the other small-fruited specimens. Taking leaf size in isolation there is continuity; this continuity is transposed from an imaginary to a real situation. Similarly, two subspecies of a species may have anthers 1–2 and 2–3 mm long respectively, with the extreme measurements uncommon and the distribution of measurements bimodal, with means at, say, 1.5 and 2.5 mm long. A description of the species as a whole will record the anthers as being 1–3 mm long with an implied mean at around 2 mm long (this is why composite descriptions of species with infraspecific taxa are of little use). It is the total range of variation that represents the species, and the taxonomist searches for clear gaps between these grand combinatorial complexes, each of which represents a species.

A real example comes from *Oreomyrrhis* (Apiaceae). If one looks at variation in the *O. linearis* complex as a whole, when specimens are ordered by increasing leaf width regardless of geography (Figure 33.1A) variation appears continuous. On the other hand, variation in the northwestern part of the Owen Stanley Range is such that there appear to be three discrete taxa (Figure 33.1B). One has narrow, longitudinally folded leaves that are glabrous on the adaxial surface and usually have one or two pairs of teeth (*O. linearis s. str.*); only the odd leaf lacks teeth. Another has equally narrow and glabrous but flat leaves that nearly always lack teeth (*'planifolia'*); only the odd leaf has a single tooth,

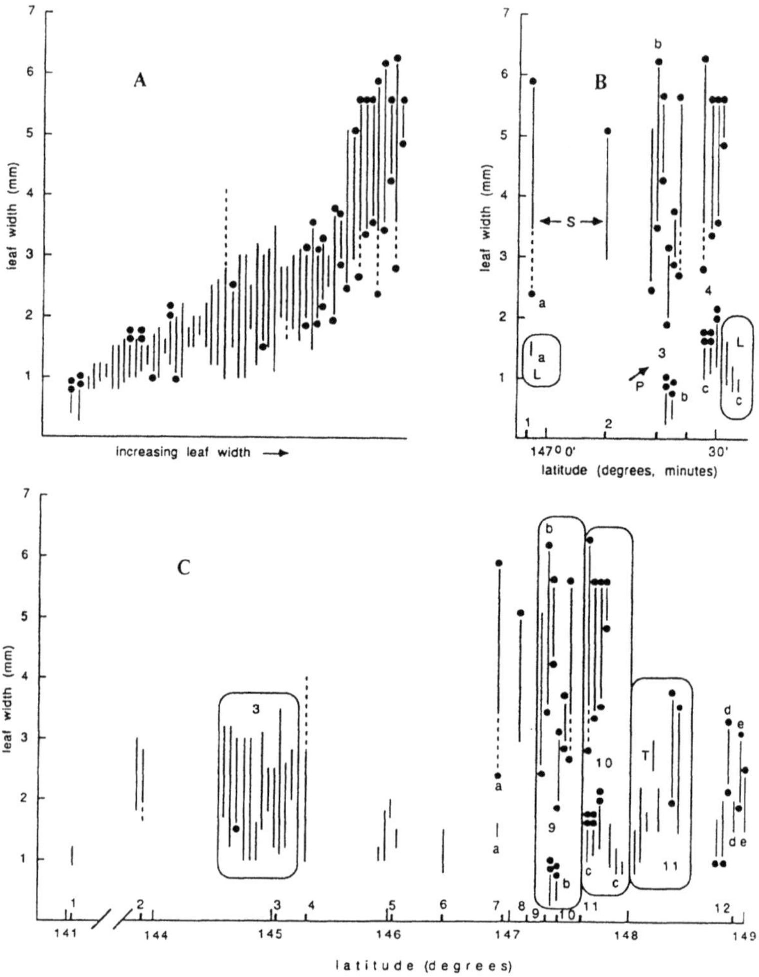

Figure 33.1 — Variation within *Oreomyrrhis linearis*. Bars denote variation within an individual collection, dotted lines extreme values. Above bar, •, leaves drying plane, ••, leaves drying plane, also, apex at most with a single tooth, usually none; below bar, •, leaf with hairs on adaxial surface. — A: Total variation analysed without regard for provenance of collections. — B: Variation in the northwestern Owen Stanley Range: L, *Oreomyrrhis linearis s. str.*; P, *planifolia* variant; S, *spathulata* variant; a–c, mixed collections; 1: Mt Strong; 2: Mt Dickson; 3: Giuni River, Wharton Range, Neon Basin, etc.; 4: Mt Albert Edward. — C: Total variation in a geographic context. In each mountain or mountain complex, specimens are arranged from lowest to highest altitude at which they were collected. T, type specimen of *O. linearis*; a–e, mixed collections; 1: Star Mountains; 2: Mt Giluwe; 3: Mt Wilhelm; 4: Mt Michael; 5: Mt Piora; 6: Mt Saruwaged; 7: Mt Strong; 8: Mt Dickson; 9: Giuni River, Wharton Range, Neon Basin, etc; 10: Mt Albert Edward; 11: Mt Victoria, Mt Scratchley, Isuani Swamp, etc; 12: Mt Suckling.

none has two. The third has broad, flat leaves with hairs on the adaxial surface and often more than two pairs of teeth (*'spathulata'*). All taxa appear to consist of populations, that is, several individuals with the same morphologies grow together, but apparently at most rarely in mixtures. Finally, there is no simple correlation of ecology with morphology. However, when the variation pattern in the species as a whole is superimposed on geography (Figure 33.1C), the character combinations mentioned above break down in the southeastern Owen Stanleys (see Stevens 1990b for more details).

This emphasises what is so easily forgotten: pattern is not something that is inherent in data, but only appears when the data are analysed in a particular way. Kellogg and Weitzman (1986) made this clear in their analysis of the variation pattern in Oceanic taxa of *Melicytus* (Violaceae). The same phenomenon occurs in Papuasian taxa of *Drimys* (Winteraceae) where what appears to be locally distinct variation in petal number (as between entities *'heteromera'* and *'polymera'* on Mt Kaindi and elsewhere) seems to be correlated with flower type, pistillate individuals having fewer petals than staminate individuals (pers. obs.; cf. Vink 1970). Although Vink may have been mistaken in this case, the whole point of his careful study was to understand the context in which taxonomic pattern in *Drimys* became evident. One more example must suffice. In the Oceanic *Garcinia pseudoguttifera*, variation in carpel number, fruit shape, and fasciclode morphology in the pistillate plant, and of pistillode in the staminate plant, as well as petal number and leaf shape, is all correlated with geography. Specimens from the southern Solomon Islands plus the New Hebrides, Fiji, and Tonga represent three distinct taxa (cf. Smith & Darwin 1974). Remove the geographic context, and dismiss some of the variation as the sort of variation that is common in the genus, and the pattern goes away.

This is one reason why experience, 'deductive analogy' (van Steenis 1949), 'synthesis' (van Steenis 1957), or the 'instinct' of the worker, can be such a mixed blessing. Any taxonomic generalisation must be built up using individual examples which do not assume the truth of the generalisation. Plants do not show law-like behaviour at the level of variation patterns; this is abundantly evident from van Steenis's work, and has been something of a theme in taxonomic literature for one hundred and fifty years, at least (see also Stevens 1986). Certainly, simply acquiring more data will not necessarily improve our understanding of a particular taxonomic situation, but a different way of analysing the data we already have may, as may thinking about the questions we are attempting to answer before acquiring the data, a point to which I shall return.

Let us turn again to the process by which we group specimens. In the herbarium, it can be difficult to recognise slight, but consistent differences between two taxa occurring sympatrically. It is somewhat easier to 'see' the same differences if there is a geographic component to the variation (Stevens 1980). Slight variation occurring in plants from the same locality is all too easy to dismiss as the sort of variation as might be expected within a population. But even in many 'well-known' species, it is the exception rather than the rule to have a clear knowledge of the morphology of the plant at any one locality, that is, material from all the stages of the growth cycle. All too often, what one is effectively doing when describing a species is describing a kind of individual, one in which the flowers are known from localities other than those from which fruits are known, and different again from the localities at which the bulk of the vegetative variation comes from – this latter being taken from specimens in bud or with immature fruits. When linking together specimens with ripe fruits from one locality with those with flowers from another locality, we are often forced to downplay differences between the specimens (see also van Steenis 1957 for an excellent discussion). We hope that these differences do not translate to differences in those parts of the life cycle that are not represented in the specimens. In such circumstances, the general process by which one forms taxa may be described as chaining, with each specimen differing only a little from others, but *in toto* possibly producing a highly heterogeneous taxon (for the operation of this process, which has been recognised for two centuries, at higher taxonomic levels, see Stevens 1984, 1986, and references). Taxa that are locally distinct may nevertheless be placed together because they are apparently joined by intermediates, but these intermediates grow elsewhere.

A final point which also confuses the delimitation of species is the distinction that is often drawn between qualitative and quantitative variation. It is sometimes suggested that species should be separated using the former kind of variation if at all possible (e.g. Ashton 1978, 1982), since qualitative variation is more useful taxonomically than quantitative. Thus Ashton (1984) recognised sympatric subspecies, rather than sympatric species, in *Hopea* and *Shorea* when the differences between the subspecies was quantitative (size of parts), rather than qualitative. Unfortunately, it is *especially* at the species level that many, perhaps even the majority, of qualitative differences are in fact only semantically qualitative, with the distinction between qualitative and quantitative characters existing in the mind, rather than in nature. For example, the texture of a plant part, a 'qualitative' difference, may convert to the amount of lignification of the tissue in it, a quantitative difference (Stevens 1990c).

Species as biological units versus species as taxonomic units

To the extent that such processes affect the circumscription of species taxa, one may legitimately wonder whether a widespread, very variable taxon is a biological phenomenon or simply the automatic result of the neural analysis carried out. *Calophyllum biflorum, C. blancoi* and *C. canum* are species that possibly suffer from this problem, and it would also be very easy to let heterogeneous taxa develop in *Vaccinium* section *Neojunghuhnia* and in *Mammea*. In *Kayea* the problem is especially acute in the *Kayea paniculata* / *K. philippinense* complex. The *K. racemosa* / *K. grandis* species pair in Malaya was in danger of becoming a widespread variable species: Whitmore (1973) thought that monographic studies might lead to the merging of the two, and *K. grandis* has been reported from Sarawak, as well (Ashton 1988b). However, with the description of two new species in Malaya and the identification of some specimens of the complex there as *K. rosea,* and the description of three new species from Borneo, the situation will be somewhat in hand. In *K. beccariana,* on the other hand, I have managed to link specimens in peninsular Malaya that have sepals free from the fruit, with others in Borneo in which the sepals and pericarp are completely fused in the ripe fruit. The biological significance of such a remarkable species is unclear, and before being used as an example of anything other than bad taxonomy, it needs more study.

The problems discussed above are particularly acute in *Garcinia* and *Mammea,* both largely dioecious genera (the latter is apparently androdioecious), in which there is taxonomically important variation in all parts of the plant. In the Ericaceae there is less variation in the fruiting stage; there is, however, some, but appropriate material is all too infrequently collected. Genera like *Oreomyrrhis* present the other extreme, where individuals with both flowers and fruits are common, and collections represent a number of individuals. Here one has no guarantee that the collector, let alone the taxonomist, appreciates local discontinuities in variation as large numbers of small specimens are assembled during a busy day's collecting; mixed collections result.

If we are more critical about how we evaluate variation, the importance of the field botanist in particular and students of a local flora in general is going to be increased. An understanding of local discontinuities is crucial, since these pose the greatest challenge to taxonomist and general biologist alike. Until they are understood, it seems beside the point to argue that general analysis of the variation shows that the discontinuities break down elsewhere. There are two aspects of diversity, global and local. If local diversity is underestimated because of the way in which we form taxonomic species, we are not going to get very far. Hence the apparent co-occurrence of two forms of *Calophyllum bicolor* on G. Mattang in Sarawak, and the different forms that have been included in the *C. blancoi* complex in Sabah, may pose biological problems of the greatest importance – assuming I have got my taxonomy straight. Similarly, the occurrence of *Kayea myrtifolia* and *K. hexapetala,* taxa with leaves very different in size, in the same general area of the First Division of Sarawak around Kuching, is confounded by apparent intergradation between the two as one proceeds northeast into Brunei; in the First Division, however, there are two discrete taxa. Ashton (1984) reports on local differences in the *Xerospermum noronhianum* complex (see Leenhouts 1983), while Leenhouts (1969) himself found that taxa that were distinct in India were to be included in the variable and widespread *Lepisanthes senegalensis.*

I am not denying that there are widespread and variable species; it is just that we need to be more careful when forming them. We need some detailed studies of variation to understand better the relationship between local discontinuity and more global apparent continuity. However, such studies are almost entirely lacking for plants in the Malesian region, as elsewhere in the tropics, Rogstad's recent study of the *Polyalthia glauca* group being a notable exception (Rogstad 1989). Vink (1970) provided some evidence to support his taxonomic decisions, as did Jessop (1978), but usually there is simply a brief description of the problem (e.g. Jacobs 1962; Leenhouts 1967; Whitmore 1976 and references; Stevens 1980). It is detailed studies that are needed, and yet in some cases we can no longer acquire the basic resources for the analysis. Indeed, it is simply no longer possible to analyse the variability throughout the ranges of some species; study of variation patterns in only parts of the range of a taxon, albeit detailed, can be extrapolated to the full range only with hesitation.

If there is evidence that there are discrete taxa locally, and that local discontinuities are caused by factors other than simple genetic polymorphism and the like, then formal taxonomic recognition is desirable. As Burtt (1970) remarked, without such recognition the variation is likely to be overlooked. Currently there is no evidence that there is intergradation between the variants of *Oreomyrrhis* in the northwest part of the Owen Stanleys; varietal rank is appropriate (but see below), even though the variation appears to break down elsewhere. If there had been no breakdown, species rank would be appropriate, although the differences between the taxa might seem slight (cf. Ashton 1988a; Ashton *et al.* 1984). Of course, 'biological' species are not characterisable by a particular level of divergence in morphology or in any other feature of the plant. In terms of understanding ecology and evolution, as well as in conservation, the local situation is of the greatest importance (Stevens 1988); what two taxa do elsewhere may be less immediately relevant.

Allopatric variation poses a problem, especially when the areas habitable by a taxon are necessarily fragmented, as with montane and oceanic taxa. However, I can only rarely recognise the mountain from which an unlocalised collection of the 'typical' form of *Oreomyrrhis linearis* was taken, and have only a little more success with some entities in *Drimys* and species of *Rhododendron* (e.g. the *R. vitis-idaea–R. vandeusenii* complex). Although the situation in *Garcinia pseudoguttifera* is not the norm, at least in my limited experience, it is not uncommon (see also van Steenis 1957; Burtt 1970). But what happens when we refine our methods of observation and analysis? What will be the ultimate recognisable unit then? We know practically nothing about population structure in general in tropical plants; molecular studies in the neotropics suggest that morphological and phylogenetic units may not be the same (see Sytsma & Schaal 1985a, 1985b; Sytsma 1988).

Nomenclature, data, and users of classifications

Clearly, the often stated assertion that neural classifications, at whatever level, are based on an evaluation of all the available evidence, needs to be qualified. That our minds do not evaluate all the evidence is perhaps paradoxically borne out by the relative stability of classifications when new characters bearing on relationships become available. The situation is accurately described by Hooker in a letter to Darwin in 1858 (Huxley 1918: 454):

> "We must never forget that Systematists have two very different ends to meet: 1. To provide a ready nomenclature without which the science cannot advance and which we change as little as possible – and further use every means to avoid even a necessary change – so important is it for all to get up the nomenclature [produce a synonymy], and so bulky and complicated is this literature. 2. To arrange the members of the Vegetable Kingdom scientifically, which is only done for the sake of scientific followers. Now we repeatedly find that to express our views scientifically we must break up the whole nomenclature, and rather than do this excessively, we confine ourselves to stating our views without acting on them."

It can become difficult to amass 'conclusive' evidence for changing the circumscription of taxa, because the nomenclatural *status quo* will always get the benefit of the doubt (Stevens 1986 for further references). The recent exchange between Pedley and Maslin over the proposal of the former to dismember *Acacia* contains elements of this probem (references in Maslin 1989; Pedley 1989). Stability then is not so much a feature of the way in which classifications incorporate data, but more of how they disregard it. This is clear from comments made by Gilmour (1958) to the effect that only information that does not destroy the general utility of the classification is to be included in it.

Along the same lines, it is difficult to evaluate the significance of comments such as "there is a very broad measure of agreement among practicing taxonomists about the limits of taxonomic species in well-studied areas, especially in North Temperate floras." (Davis & Heywood 1963 and references), since hundreds of new species have been recognised in Europe since the beginning of the Flora Europaea project. Walters (1988) saw little effect of systematic philosophy on the actual taxa recognised in the 'Flora Europaea', and comparable comments have been made about other supposed revolutions in plant taxonomy (Hagen 1983 for references). But consensus means little when we are so ignorant about what we are doing. Simple repeatability of results (Anderson 1957) is not good enough if the 'instinct' by which these results are produced is not understood, although Anderson saw that greater knowledge of how the mind functioned would benefit taxonomy.

It is interesting that ideas of how a 'natural' classification is constructed by the mind change with changing systematic philosophies; what a systematist does is *ex definitio* 'natural' (Stevens 1986). But even if the human mind does generally process information in a particular kind of way – and there seems to be no reason why there should be a single way – it is unclear why taxonomists should use that mental algorithm. The argument is further obscured when we are asked to do this because the man or woman in the street does this (Heywood 1988a) – everybody classifies in the same way; the same taxa are recognised; nomenclatural stability results. It would be considered strange if attempts were made to constrain any other discipline in a similar fashion. Even Linnaeus (e.g. 1751) was exercised to ensure that 'lubricious' distinctions such as 'tree' versus 'herb' (see Brown 1984 for their naming in folk taxonomies) did not creep into his systems, whether sexual or 'natural'.

The purpose of this paper is to make some people, at least, uneasy. Will there be a wholesale increase in names? My reply would be, quite possibly, but so what? And here we come back to the relationship between people and names. I have already emphasised the potential problem caused because the person in the street and the evolutionary biologist use the same set of names. Names, the groups that they circumscribe, and the relationships suggested by the system as a whole, are the common language of comparative biology, floras, the courtroom, and the legislative chamber. Yet the needs of these users may not be so different. Surely, when we conserve species, we do not want to conserve artefacts of our neural classification that misrepresent the biological universe to us.

Furthermore, that the users 'need' nomenclatural stability is possibly a red herring. A willingness or otherwise to accept name changes is as much an attitude that results from education as to what classifications represent, as anything else. Certainly, amateur ornithologists have accepted numerous changes to the checklist of names for North American birds over the years (Olson 1987), albeit with hearty grumbling (DeBenedictis 1983: it should, however, be noted that the American Birding Association has its own checklist – R.J. O'Hara, pers. comm.). We can usefully be on guard against the vagaries of our minds, realise that classifications and names are not facts graven in stone, and perhaps even indulge in some gentle introspection over how we form groups neurally, perhaps for a day or so every other year. This will surely result in some demystification of taxonomy, and may help to remove components of taxonomic decisions that are subjective, egotistic, and arbitrary (Ng 1988).

Nomenclatural stability can also be attained, even at the species level, so long as no particular significance is attached to taxonomic rank. We could accept a broad circumscription of *Drimys piperita* and treat the entities just as we treat species now. Biologists would no longer focus on rank, but on the appropriate pattern of variation at whatever rank that happened to be. Name changes would not result because somebody produced evidence suggesting that the entities were simply somewhat more or less discrete than had previously been believed, their limits remaining unchanged. This would be similar to the situation at higher taxonomic levels (van Steenis 1978), where it is generally conceded that rank is more arbitrary. Comparative biologists could develop ideas of relationship independently of so many changes in names (major misplacements should entail name changes). But we could keep the present names of *all* taxa, even those like *Agapetes, Vaccinium,* and *Mesua,* whatever is suggested about their phylogenetic relationships (*Agapetes* is wildly polyphyletic; *Mesua* more simply polyphyletic, *Vaccinium* is paraphyletic) if we could entirely divorce ideas of relationship from names.

But most of us think that taxa at the specific level, and even the generic level, represent a particular level of reality (e.g. van Steenis 1957; Burtt 1970; Ashton 1984; Heywood 1988b). Genera are particularly important units in biogeographical studies. The species is thought of as the fundamental unit of taxonomy, possessing properties that no other unit has, and is not simply the lowest obligate rank in the taxonomic hierarchy (for many tropical groups we have no evidence that a species is anything but the latter). In complex infraspecific hierarchies the lowest levels tend to be disregarded, given the prevailing attitude about the importance of species (Burtt 1970; Arbib 1973). Taxonomists deal with names in the context of hierarchies; it is through both that relationships are understood. It follows that without names, relationships are not perceived, and without changes in names, changes in relationships, of whatever sort, can be ignored. This is clear from Cronquist's comments about the lack of firm taxonomic (nomenclatural) conclusions in Donoghue's work on relationships in the Caprifoliaceae s.l. (Cronquist 1988; Donoghue 1983; see also Kellogg 1990).

In view of the preceeding comments, it is perhaps hardly surprising that discussion about priorities in taxonomic research, especially that in the tropics, is flawed. Conclusions that 'empirical and objective', 'intuitive' (neural) approaches are to be favoured over phylogenetic studies, which should not be funded (Prance & White 1988) leave something to be desired, both in general and in this particular case; we can hardly make sensible decisions about the relative usefulness of different approaches to classification without understanding clearly what these approaches do. Other authors also favour the production of monographs and floristic work over phylogenetic analyses (e.g. Davis 1978; Nooteboom 1988; Heywood 1988b; Bramwell 1989). But if a group is studied as if the organisms it contains have no history, the study may be of limited use in understanding phylogeny (Stevens 1990c; Chappill 1990). Data are not givens, but are shaped by the questions that are ultimately to be asked. However, at the level of taxonomic species and above, I see no necessary split between monographing and phylogeny reconstruction; even simply thinking about phylogeny heightens my perception of variation, and that is important for both occupations. The issue of whether there are time and resources to devote to phylogenies, or whether we should concentrate on field studies while they are still possible (Raven 1988), is a separate one. But we should not forget that a number of evolutionary questions are difficult or impossible to answer without phylogenies; we will need to answer these questions if our attempts to coexist with a major portion of diversity are to succeed. The detailed species-level studies that we need are likely to be carried out by a different set of workers than those who write monographs and floras, but take as their point of departure problems suggested by these people.

It is abundantly evident that the body of myth that has developed around neural classifications has to be removed if we are to make taxonomic practice a more accessible operation. We have to guard against using inappropriate methods, neural or otherwise, to analyse our data, and we have to be much more aware of the problem of local variation. We have to cut through the constraints of the Linnaean hierarchy. We have to understand the sad consequences for our discipline when we opt for nomenclatural stability, or 'multiple purpose' classifications. As students of the living world, we have to remember that we are studying organisms, and that organisms have histories.

Acknowledgements

I am grateful to P.S. Ashton, D. Boufford, J. Burley, J.S. Carpenter, J.A. Chappill, R.J. O'Hara, and E.A. Kellogg for helpful discussion and/or comments on the manuscript.

References

Anderson, E. 1957. An experimental investigation of judgements concerning genera and species. Evolution 11: 260–262.

Arbib, R. 1973. What the A.O.U. check-list committee has done to your life list. Am. Birds 27: 576–577.

Ashton, P.S. 1978. Flora Malesiana precursores: Dipterocarpaceae. Gdns' Bull., Singapore 31: 5–48.

Ashton, P.S. 1982. Dipterocarpaceae. In: C.G.G.J. van Steenis (ed.), Flora Malesiana I, 9^2: 237–552. Martinus Nijhoff, The Hague, Boston, London.

Ashton, P.S. 1984. Biosystematics of tropical forest plants: a problem of rare species. In: W.F. Grant (ed.), Plant Biosystematics: 497–518. Academic Press, Toronto.

Ashton, P.S. 1988a. Systematics and ecology of rain-forest trees. Taxon 37: 622–629.

Ashton, P.S. 1988b. Manual of the non-dipterocarp trees of Sarawak, volume 2. Forest Department, Sarawak.

Ashton, P.S., G. Yik-Yuen & F.W. Robertson. 1984. Electrophoretic and morphological comparisons in ten rain forest species of Shorea (Dipterocarpaceae). J. Linn. Soc. Bot. 89: 293–304.

Ball, G.E., & D.R. Maddison. 1987. Classification and evolutionary aspects of the species of the new world genus Amblygnathus Dejean, with descriptions of Platymetopsis, new genus, and notes about selected species of Selenophorus Dejean (Coleoptera: Carabidae: Harpalini). Trans. Am. Ent. Soc. 113: 189–307.

Bentham, G. 1827. Outline of a new system of logic, with a critical examination of Dr Whateley's "Elements of logic." Hunt & Clarke, London.

Bentham, G. 1832–1836. Labiatarum genera et species. James Ridgway and Sons, London.

Bisby, F.A. 1984. Information services in taxonomy. In: R. Allkin & F.A. Bisby (eds.), Databases in systematics: 17–34. Academic Press, London, New York.

Blackwelder, R.E. 1967. Taxonomy. John Wiley & Sons, New York.

Bramwell, D. 1989. Taxonomy: the sands of time. Taxon 38: 404–405.

Brown, C.H. 1984. Language and living things. Rutgers University Press, New Brunswick.

Burtt, B.L. 1970. Infraspecific categories in flowering plants. Biol. J. Linn. Soc. 2: 233–238.

Cain, A.J., & G.A. Harrison. 1958. An analysis of the taxonomist's judgement of affinity. Proc. Zool. Soc. Lond. 131: 85–98.

Cantino, P.D., & R.W. Sanders. 1986. Subfamilial classification of Labiatae. Syst. Bot. 11: 163–185.

Chappill, J.A. 1990. Cladistics and the Chrysobalanaceae. Taxon 39 (in press).

Crepet, W.L., & K.C. Nixon. 1989. Earliest megafossil evidence of Fagaceae: phylogenetic and biogeographic implications. Am. J. Bot. 76: 842–855.

Cronquist, A. 1987. A botanical critique of cladism. Bot. Rev. 53: 1–52.

Cronquist, A. 1988. The evolution and classification of flowering plants. Ed. 2. New York Botanical Garden, Bronx.

Davis, P.H. 1978. The moving staircase: an analysis of taxonomic rank and affinity. Notes R. Bot. Gdn. Edinb. 36: 325–340.

Davis, P.H., & V.H. Heywood. 1963. Principles of angiosperm taxonomy. Oliver & Boyd, Edinburgh, London.

Dawes, R.M., D. Faust & P.E. Meehl. 1989. Clinical versus actuarial judgment. Science 243: 1668–1674.

DeBenedictis, P.A. 1983. Coming! A new official checklist of North American birds – a revolution in avian nomenclature. Am. Birds 37: 3–8.

Donoghue, M.J. 1983. The phylogenetic relationships of Viburnum. In: N.I. Platnick & V.A. Funk (eds.), Advances in Cladistics, vol. 2: 143–166. Columbia University Press, New York.

Donoghue, M.J. 1989. Phylogenies and the analyses of evolutionary sequences, with examples from seed plants. Evolution 43: 1137–1156.

Estes, J.R., & R.J. Tyrl. 1987. Concepts of taxa in the Poaceae. In: T.R. Soderstrom *et al.* (eds.), Grass systematics and evolution: 325–333. Smithsonian Institution Press, Washington, D.C.

Farris, S. 1979a. On the naturalness of phylogenetic classification. Syst. Zool. 28: 200–214.

Farris, S. 1979b. The information content of the phylogenetic system. Syst. Zool. 28: 483–519.

Farris, S. 1983. The logical basis of phylogenetic analysis. In: N.I. Platnick & V.A. Funk (eds.), Advances in Cladistics, vol. 2: 1–36. Columbia University Press, New York.

Faust, D., P.E. Meehl & R.M. Dawes. 1990. Clinical and actuarial judgment – response. Science 247: 146–147.

Gilbert, M.G. 1986. Notes on East African Vernonieae (Compositae). A revision of the Vernonia galamensis complex. Kew Bull. 41: 19–35.

Gilmour, J.S.L. 1940. Taxonomy and philosophy. In: J. Huxley (ed.), The new systematics: 461–474. Oxford University Press, Oxford.
Gilmour, J.S.L. 1951. The development of taxonomic theory since 1851. Nature 168: 400–402.
Gilmour, J.S.L. 1958. The species: yesterday and tomorrow. Nature, Lond. 181: 379–380.
Gray, A. 1854. Introductory essay, in Dr. Hooker's Flora of New Zealand: vol. 1 [review]. Am. J. Sci. Arts 17: 241–252, 334–350.
Hagen, J.B. 1983. The development of experimental methods in plant taxonomy, 1920–1950. Taxon 32: 406–416.
Hawksworth, D.L., & F.A. Bisby. 1988. Systematics: the keystone of biology. In: D.L. Hawksworth (ed.), Prospects in systematics: 3–30. Clarendon Press, Oxford.
Heywood, V.H. 1984. Designing floras for the future. In: V.H. Heywood & D.M. Moore (eds.), Current concepts in plant taxonomy: 396–410. Academic Press, London.
Heywood, V.H. 1988a. Tropical taxonomy – who are the users? Symb. Bot. Upsal. 28 (3): 21–28.
Heywood, V.H. 1988b. The structure of systematics. In: D.L. Hawksworth (ed.), Prospects in systematics: 44–56. Clarendon Press, Oxford.
Hooker, J.D. 1853. Florae novae-zelandicae. Lovell Reeve, London.
Hooker, J.D., & T. Thomson. 1855. Flora indica. W. Pamplin, London.
Huxley, L. 1918. Life and letters of Sir Joseph Dalton Hooker. 2 vol. D. Appleton, New York.
Jacobs, M. 1962. Pometia (Sapindaceae), a study in variability. Reinwardtia 6: 109–144.
Jardine, N., & R. Sibson. 1971. Mathematical taxonomy. John Wiley, New York.
Jeffrey, C. 1987. The concept of the genus. Aust. Syst. Bot. Soc. Newsl. 53: 27–31.
Jessop, J.P. 1978. A revision of Peliosanthes (Liliaceae). Blumea 23: 141–159.
Kalkman, C. 1987. The two sides of a medal. In: P. Hovenkamp (ed.), Systematics and evolution: a matter of diversity: 13–21. Institute of Systematic Botany, Utrecht University.
Kay, Q.O.N. 1984. Variation, polymorphism and gene flow within species. In: V.H. Heywood & D.M. Moore (eds.), Current concepts in plant taxonomy: 181–199. Academic Press, London.
Kellogg, E.A. 1990. [Review: Genera graminum, W.D. Clayton & S.M. Renvoize; Grasses of the world, L. Watson & M.D. Dallwitz]. J. Arnold Arbor. 71: 271–273.
Kellogg, E.A., & A.L. Weitzman. 1986. A note on the oceanic species of Melicytus (Violaceae). J. Arnold Arbor. 66: 491–502.
Kleinmuntz, B. 1990. Clinical and actuarial judgment. Science 247: 146.
Leenhouts, P.W. 1967. A conspectus of the genus Allophylus (Sapindaceae). Blumea 15: 301–358.
Leenhouts, P.W. 1969. Florae Malesianae praecursores L. A revision of Lepisanthes (Sapindaceae). Blumea: 17: 33–91.
Leenhouts, P.W. 1983. A taxonomic revision of Xerospermum (Sapindaceae). Blumea 28: 389–401.
Levin, D.A. 1979. The nature of plant species. Science, N.Y. 204: 381–384.
Lewis, C.T., & C. Short. 1897. A new Latin dictionary. Oxford University Press, Oxford.
Linnaeus, C. 1751. Philosophia botanica. Godefr. Kiesewetter, Stockholm.
Maddison, W.P., M.J. Donoghue & D.R. Maddison. 1984. Outgroup analysis and parsimony. Syst. Zool. 33: 83–103.
Maslin, B. 1989. Wattle become of Acacia? Austr. Syst. Bot. Soc. Newsl. 58: 1–13.
McNeil, J. 1984. Numerical taxonomy and biosystematics. In: W.F. Grant (ed.), Plant biosystematics: 395–415. Academic Press, Toronto.

Melville, R. 1982. The biogeography of Nothofagus and Trigonobalanus and the genera of the Fagaceae. J. Linn. Soc. Bot. 85: 75–88.

Morin, N.R., R.D. Whetstone, D. Wilken & K.L. Tomlinson (eds.). 1989. Floristics for the 21st century. Monographs in Systematic Botany, vol. 28, Missouri Bot. Gard., Missouri.

Moss, W.W. 1971. Taxonomic repeatability: an experimental approach. Syst. Zool. 20: 309–330.

Nixon, K.C., & W.L. Crepet. 1989. Trigonobalanus (Fagaceae): taxonomic status and phylogenetic relationships. Am. J. Bot. 76: 828–841.

Ng, F.S.P. 1988. Problems in the organisation of plant taxonomic work. Fl. Males. Bull. 10: 39–44.

Nooteboom, H.P. 1988. What should botanists do with their time? Taxon 37: 134.

Olson, S.L. 1987. On the extent and source of instability in avian nomenclature, as exemplified by North American birds. Auk 104: 538–542.

Patterson, C., & A.B. Smith. 1989. Periodicity in extinction: the role of systematics. Ecology 70: 802–811.

Pedley, L. 1989. 'Racosperma' again. Austr. Syst. Bot. Soc. Newsl. 89: 1–2.

Prance, G.T., & F. White. 1988. The genera of Chrysobalanaceae: a study in practical and theoretical taxonomy and its relevance to evolutionary biology. Phil. Trans. Roy. Soc. B, 320: 1–184.

Raven, P.H. 1988. Tropical floristics tomorrow. Taxon 37: 549–560.

Riedl, R. 1978. Order in living organisms. John Wiley, New York.

Robson, N.K.B. 1974. Hypericaceae. In: C.G.G.J. van Steenis (ed.), Flora Malesiana I, 8^1: 1–29. Sijthoff & Noordhoff, Alphen a/d Rijn.

Rogstad, S.H. 1989. The biosystematics and evolution of the Polyalthia hypoleuca complex (Annonaceae) of Malesia, 1. Systematic treatment. J. Arnold Arbor. 70: 153–246.

Sattler, R.F. 1986. Biophilosophy. Springer-Verlag, Berlin.

Smith, A.C., & S.P. Darwin. 1974. Studies of Pacific Island plants, XXVIII. The Guttiferae of the Fijian region. J. Arnold Arbor. 55: 215–263.

Sneath, P.H.A., & R.R. Sokal. 1973. Numerical Taxonomy. W.H. Freeman, San Francisco.

Sokal, R.R., & F.J. Rohlf. 1970. The intelligent ignoramus, an experiment in numerical taxonomy. Taxon 19: 305–319.

Steenis, C.G.G.J. van. 1949. General considerations. In: C.G.G.J. van Steenis (ed.), Flora Malesiana I, 4^2: XIII–LXIX. P. Noordhoff, Groningen.

Steenis, C.G.G.J. van. 1957. Specific and infraspecific delimitation. In: C.G.G.J. van Steenis (ed.), Flora Malesiana I, 5^3: CLXVII–CCXXXIV. P. Noordhoff, Groningen.

Steenis, C.G.G.J. van. 1978. On the doubtful virtue of splitting families. Bothalia 12: 425–427.

Stevens, P.F. 1980. A revision of the Old World species of Calophyllum (Guttiferae). J. Arnold Arbor. 61: 1–699.

Stevens, P.F. 1984. Metaphors and typology in the development of botanical systematics 1690–1960, or the art of putting new wine in old bottles. Taxon 32: 169–211.

Stevens, P.F. 1986. Evolutionary classification in botany, 1960–1985. J. Arnold Arbor. 67: 313–339.

Stevens, P.F. 1987. Genera, what and why – some thoughts. Aust. Syst. Bot. Soc. Newsl. 53: 31–38.

Stevens, P.F. 1988. New Guinea. In: D.G. Campbell & H.D. Hammond (eds.), Floristic inventory of tropical countries: 120–132. New York Botanic Garden, New York.

Stevens, P.F. 1990a. Species. In: E.F. Keller & E. Lloyd (eds.), Keywords in evolutionary biology.

Stevens, P.F. 1990b. Some remarkable new species of Oreomyrrhis (Apiaceae) from New Guinea, with notes on species concepts.

Stevens, P.F. 1990c. Character states, continuous variation, and phylogenetic analysis, a review.

Sytsma, K.J. 1989. Taxonomic revision of the Central American Lisianthus skinneri species complex (Gentianaceae). Ann. Missouri Bot. Gard. 75: 1587–1602.

Sytsma, K.J., & B.A. Schaal. 1985a. Genetic variation, differentiation, and evolution in a species complex of shrubs based on isozyme data. Evolution 39: 582–593.

Sytsma, K.J., & B.A. Schaal. 1985b. Phylogenetics of the Lisianthus skinneri (Gentianaceae) species complex based on isozyme data. Evolution 39: 594–608.

Vink, W. 1970. The Winteraceae of the Old World. 1. Pseudowintera and Drimys – morphology and taxonomy. Blumea 18: 225–354.

Walters, S.M. 1988. The purposes of systematic botany. Symb. Bot. Upsal. 28 (3): 13–20.

Whitmore, T.C. 1973. Tree flora of Malaya. Vol. 2. Longman Group, London.

Whitmore, T.C. 1976. Natural variation and its taxonomic treatment within tropical tree species as seen in the Far East. In: J. Burley & B.T. Styles (eds.), Tropical trees: variation, breeding and conservation: 25–34. Academic Press, London.

Index to plant genera and species

MIX
Papier aus verantwortungsvollen Quellen
Paper from responsible sources
FSC® C105338

If you have any concerns about our products,
you can contact us on
ProductSafety@springernature.com

In case Publisher is established outside the EU,
the EU authorized representative is:
Springer Nature Customer Service Center GmbH
Europaplatz 3, 69115 Heidelberg, Germany

Printed by Libri Plureos GmbH
in Hamburg, Germany